Effects of Mineral Dusts on Cells

NATO ASI Series

Advanced Science Institutes Series

A series presenting the results of activities sponsored by the NATO Science Committee, which aims at the dissemination of advanced scientific and technological knowledge, with a view to strengthening links between scientific communities.

The Series is published by an international board of publishers in conjunction with the NATO Scientific Affairs Division

A Life Sciences **B Physics**	Plenum Publishing Corporation London and New York
C Mathematical and Physical Sciences **D Behavioural and Social Sciences** **E Applied Sciences**	Kluwer Academic Publishers Dordrecht, Boston and London
F Computer and Systems Sciences **G Ecological Sciences** **H Cell Biology**	Springer-Verlag Berlin Heidelberg New York London Paris Tokyo Hong Kong

Series H: Cell Biology Vol. 30

Effects of Mineral Dusts on Cells

Edited by

Brooke T. Mossman

University of Vermont
College of Medicine
Department of Pathology
Burlington, Vermont 05405, USA

Raymond O. Bégin

Service de Pneumologie
CHUS
Sherbrooke, Québec, Canada, JIH 5N4

Springer-Verlag Berlin Heidelberg New York
London Paris Tokyo Hong Kong
Published in cooperation with NATO Scientific Affairs Division

Proceedings of the NATO Advanced Research Workshop on In Vitro Effects of Mineral Dusts on Cells held at Auberge Estrimont, Orford, Québec, Canada, September 20–23, 1988

ISBN-13:978-3-642-74205-7 e-ISBN-13:978-3-642-74203-3
DOI: 10.1007/978-3-642-74203-3

Library of Congress Cataloging-in-Publication Data. NATO Advanced Research Workshop on In Vitro Effects of Mineral Dusts on Cells (4th : 1988 : Orford, Québec) Effects of mineral dusts on cells / edited by Brooke T. Mossman, Raymond O. Bégin. p. cm.—(NATO ASI series. Series H, Cell biology ; vol. 30) "Proceedings of the NATO Advanced Research Workshop on In Vitro Effects of Mineral Dusts on Cells held at Auberge Estrimont, Orford, Québec, Canada, September 20–23, 1988"—T.p. verso. "Published in cooperation with NATO Scientific Affairs Division."
ISBN-13:978-3-642-74205-7 (U.S.)
1. Mineral dusts—Toxicology—Congresses. 2. Mineral dusts—Physiological effect—Congresses. 3. Toxicity testing—In Vitro—Congresses. I. Mossman, Brooke T., 1947- . II. Bégin, Raymond O. III. North Atlantic Treaty Organization. Scientific Affairs Division. IV. Title. V. Series. RA1231.M55N38 1988
615.9'25—dc20 89-11499

Softcover reprint of the hardcover 1st edition 1989

2131/3140-543210 – Printed on acid-free paper

PREFACE

The Fourth International Workshop on "*In Vitro* Effects of Mineral Dusts on Cells" was held on September 20 - 23, 1988 in Auberge Estrimont, Orford, Québec, Canada. The emphasis of the N.A.T.O. Advanced Research Workshop was the use of cell and organ culture and lavage cell populations obtained from man and laboratory animals to elucidate cellular and molecular events occurring after their interaction with fibrous and nonfibrous particulates, including metal compounds. In seven sessions, an international representation of scientists from 17 countries (Austria, Belgium, Canada, France, Federal Republic of Germany, India, Japan, Netherlands, Norway, Poland, Portugal, Union of South Africa, Sweden, Switzerland, United Kingdom, United States of America, and Yugoslavia) presented recent research findings in the following areas:

- Epithelial cell injury and proliferation by minerals.
- Physico-chemical properties of minerals in relation to their biologic effects.
- Mechanisms of dust-induced pneumoconioses. Clinical and experimental studies.
- Mechanisms of cytotoxicity, mutagenicity and carcinogenicity.
- Oxidants, cytotoxicity and disease.
- Mechanisms of mineral-induced inflammation.
- Mechanisms of carcinogenesis.

A session on "Questions, risk and public policy" provided a lively discussion on the relevance of *in vitro* experiments to carcinogenicity studies in man and their implications in the formation of regulatory policies.

The organizing committee for this workshop was:

Co-Chairs: B.T. Mossman (USA) and R. Bégin (Canada)

E.G. Beck (FRG)

A. Brody (USA)

R.C. Brown (UK)

J. Bignon (France)

G. Fisher (FRG)

A. Lange (Poland)

P. Nettesheim (USA)

Q. Rahman (India)

K. Robock (FRG)

The workshop was sponsored by the Advance Study ARW Program of N.A.T.O. under the directorship of Dr. Craig Sinclair. A conference grant from the National Heart, Lung and Blood Institute with contributions from the National Institute of Environmental Health Sciences and National Cancer Institute also played a major role toward the success of this meeting. We thank Dr. Anthony Kalica, National Institutes of Health for his support of this proposal. A number of other organizations contributed financially to the workshop and deserve special mention. These include:

- Fonds de la Recherche en Santé du Québec
- Institut de Recherche en Santé et en Sécurité du Travail du Québec
- Thermal Insulation Manufacturers Association
- The Asbestos Institute
- Eurotalc
- American Petroleum Institute
- BP America, Inc.
- Chevron U.S.A., Inc.
- Mobil Oil Corporation
- The Council for Tobacco Research

The editors also wish to thank Ms. Claudette Laplume, the Conference Secretariat who also has assisted in preparation of these proceedings. Finally, we thank Springer-Verlag for their interest in publishing this state-of-the-art volume on cellular and molecular interactions between mineral dusts and cells.

The Fifth International Workshop on "*In Vitro* Effects of Mineral Dusts" will be held in France in 1992 and co-chaired by J.M.G. Davis (UK) and M.C. Jaurand (France).

TABLE OF CONTENTS

D. MECHANISMS OF CYTOTOXICITY, MUTAGENICITY AND CARCINOGENICITY

E. OXIDANTS, CYTOTOXICITY AND DISEASE

EFFECTS OF CIGARETTE SMOKE ON UPTAKE OF ASBESTOS AND IRON OXIDE BY RAT TRACHEAL EXPLANTS

Andrew Churg and Joanne L Wright
Department of Pathology and University Hospital
University of British Columbia
2211 Wesbrook Mall
Vancouver, BC, Canada V6T IW5

Introduction:

A variety of studies in man and animals have shown that cigarette smoke increases pulmonary mineral dust retention. This phenomenon has been demonstrated demonstrated both for inert dusts such as iron oxide (Cohen et al 1979; Gilks et al 1988) and toxic dusts such as asbestos (McFadden et al 1986a). Histopathologic studies in animals indicate that the effect of increased dust retention is increased penetration of dust into the walls of small airways. Again this is true for both inert and toxic dusts (McFadden et al 1986b; Gilks et al 1988).

Any process which increases the penetration of toxic or carcinogenic particles into the pulmonary tissues potentially may increase the fibrogenic and/or carcinogenic effects of such particles. This may be the reason why asbestos workers who smoke have considerably higher incidences of carcinoma of the lung (Hammond et al 1979) and probably higher incidences of asbestosis (Weiss, 1984) than asbestos workers who do not smoke.

The mechanism of smoke mediated increases in tissue particle penetration are unknown. Both smoke and dust particles evoke an inflammatory reaction in the lung (Bowden 1987; Plowman 1982), and mineral particles such as asbestos or silica also cause macrophages to produce fibroblast growth factors and a variety of other mediators of fibrosis (Kumnar et al 1988; Begin and Cantin 1988). Thus it is possible that the combined effects of smoke and dust arise from their effects on inflammatory cells. However, it is also possible that smoke can directly affect the way in which pulmonary epithelia handle dust particles. In the experiments described below we test this latter hypothesis.

Supported by grant MA8051 from the Medical Research Council of Canada

NATO ASI Series, Vol. H30
Effects of Mineral Dusts on Cells
Edited by B. T. Mossman and R. O. Bégin

Materials and Methods:

For these studies tracheal explants were prepared from 200g male Sprague-Dawley rats according to the method of Mossman and Craighead (1979). Briefly, the tracheas were excised, opened, and cut into segments of two tracheal rings each. The segments were then laid serosal side down on sterile Millipore filters and the filter/tracheas placed mucosal side up in Petri dishes containing just enough culture medium (Dulbecco's minimum essential medium supplemented with 10% fetal calf serum, 1% glutamine, and 1% fungizone-streptomycin-penecillin, GIBCO, Grand Island, NY) to cover the filter.

Tracheal segments were then exposed, mucosal side up, to either cigarette smoke or air (sham smoke). The smoke was the whole smoke from 1 commercial nonfilter cigarette delivered as ten 20cc puffs over 10 minutes into a 2 liter humidified chamber housing the segments.

Explants were subsequently exposed, mucosal side up, to the mineral dusts in serum free culture medium for 1 hour at 37°C. In the first experiment the dust was a suspension of 10mg/ml UICC Amosite asbestos; in the second experiment 40 mg/ml Pfizer R2999R iron oxide (a very finely divided iron oxide) was used. At the end of the exposure time the segments were transferred to fresh filters and maintained in an atmosphere of 95% O_2/5% CO_2 at 37°C. Again, just enough culture medium was added to the Petri dishes to cover the filters while leaving the mucosa exposed to air.

At 24hr, 72hr, and 1 week after exposure, 5 or 6 segments from each exposure group were removed from the incubators, fixed and embedded for histologic examination. Five micron thick sections were cut and stained with Nuclear Fast Red; this stain allows easy visualization of the dust and the underlying tissue structure. To evaluate dust penetration, asbestos fibers longer than 4 microns or optically visible iron particles were counted in or under the tracheal epithelium. The length of epithelium was measured using a computer assisted digitizer, and a value of fibers or particles/mm of epithelium generated. Separate evaluation was made of hyperplastic lesions, foci of proliferating epithelial cells containing very large amounts of dust. These foci were evaluated as the percent of total epithelial length occupied by hyperplastic lesions. Illustrations of the appearance of the tracheas and of the hyperplastic lesions can be found in Hobson et al (1988).

Data evaluation was carried out by log transformation of the data followed by analysis of variance with Bonferroni type correction for multiple comparisons.

Results

In general there was good preservation of the tracheal epithelium in our cultures with maintenance of a predominantly ciliated cell type. By itself, asbestos treatment resulted in the penetration of fibers into the epithelium as well as formation of hyperplastic lesions, responses which have been previously reported from several laboratories (Mossman et al 1983; Haugen et al 1982; Topping et al 1980). Iron oxide particles also penetrated the epithelium in nonsmokers. Although the statistical analysis was performed on log-transformed numbers, the means of the original values express the sense of the data, and were used to construct Tables 1-3. Table 1 shows the mean ± SE numbers of asbestos fibers/mm of epithelial length, and Table 2 the extents of hyperplastic lesions in the asbestos exposed segments. Table 3 shows the number of iron oxide particles/mm of epithelium; virtually no hyperplastic lesions were seen in the iron oxide exposed animals. Smoke exposure greatly increased the number of asbestos fibers penetrating the tracheal epithelium, as well as the extent of hyperplastic lesions, and these differences become more marked over time. By contrast, smoke had no effect on the penetration of iron oxide particles.

Table 1: Asbestos Fibers/MM Epithelium (Values as Mean ± SE)

Treatment	24 hrs	72 hrs	1 week
Asbestos	0.4 ± .1	1.1 ± .3	4.9 ± 2.4
Asbestos + Smoke	1.7 ± .3	8.5 ± .8	32 ± 4.6
P <	NS	.001	.01

Table 2: Length of Hyperplastic Lesions as % Total Epithelial Length (Values as Mean ± SE)

Treatment	24 hrs	72 hrs	1 week
Asbestos	0.6 ± 0.2	4.3 ± 1.2	6.0 ± 1.9
Asbestos + Smoke	4.0 ± 1.3	19 ± 5.5	21 ± 4.8
P <	.04	.04	.04

Table 3: Iron Particles/MM Epithelium (Values as Mean $\pm$ SE)

Treatment	24 hrs	72 hrs	1 week
Asbestos	46 $\pm$ 8	34 $\pm$ 5	36 $\pm$ 11
Asbestos + Smoke	44 $\pm$ 8	54 $\pm$ 11	43 $\pm$ 9
P <	NS	NS	NS

Discussion

In these studies we have examined the question of whether cigarette smoke can directly affect the uptake of mineral particles by tracheobronchial epithelium. As noted in the Introduction, smoke effects may be mediated by inflammatory cells or by direct effects on epithelium, or both. Data from our laboratory indicate that cigarette smoke does interfere with macrophage clearance of asbestos fibers (Churg et al 1987), and suggest that re-release of phagocytized fibers from macrophages into the alveoli ultimately leads to increased fiber incorporation into tissues. The possibility that exposure to both smoke and dust increases release of macrophage derived factors which might affect fiber uptake has also mentioned above.

The present system of tracheal explants is, for all practical purposes, free of inflammatory cells, and provides a simple system for examining smoke effects on the epithelium. Our studies show that, at least for asbestos, smoke directly increases the number of fibers penetrating the epithelium and that this process occurs in the absence of inflammatory cells. It is particularly interesting to note that not only is the smoke exposure in this system very brief, but smoke exposure and asbestos exposure need not be concurrent for increased uptake to occur. This observation implies that smoke produces some rapidly developing and persistent change in the epithelial cells.

By contrast, there was no increase in iron oxide uptake in this system. One might question whether this observation is an artifact of the size and dose of particle used, but in the intact animal smoke does increase uptake of this particular iron oxide (Gilks et al 1988). These findings suggest that the mechanisms of uptake of asbestos and iron oxide are different. Since iron oxide is considered a prototypical "inert" dust, while asbestos by itself is clearly toxic, our observations raise the possbility that smoke has different effects on the epithelial reaction to inert compared to pathogenic dusts.

References

Begin R, Cantin A (1988): Fibrogenic activity in early radiologic silicosis. Amer Rev Respir Dis 137 (Pt2): 352.

Bowden DH (1987): Macrophages, dust and pulmonary diseases. Exp Lung Res 12:89-107.

Churg A, Tron V, Wright JL (1987): Effects of cigarette smoke on retention of asbestos fibers in various morphologic compartments of the guinea pig lung. Amer J Pathol 129: 385-393.

Cohen D, Arai SF, Brain JD (1979): Smoking impairs long-term clearance from the lung. Science 204:514-516.

Gilks B, Wright J, Churg A (1988): Effects of Cigarette Smoke on Tissue Uptake and Retention of Iron Oxide. Amer Rev Respir Dis 137: 1382-1384.

Hammond EC, Selikoff IJ, Seidman H (1979): Asbestos exposure, smoking, and death rates. Ann NY Acad Sci 330:473-492.

Haugen A, Schafer PW, Lechner JF, Stoner GD, Trump BF, Harris CC (1982): Cellular ingestion, toxic effects, and lesions observed in human bronchial epithelial tissue and cells cultured with asbestos and glass fibers. Int J Cancer 30: 265-272.

Hobson J, Gilks B, Wright J, Churg A (1988): Cigarette Smoke Directly Enhances Asbestos Fiber Penetration and Asbestos-Induced Epithelial Proliferation in Rat Tracheal Explants. JNCI 80: 518-521.

Kumnar RK, Bennett RA, Brody AR (1988): A homologue of platelet-dervied growth factor produced by rat alveolar macrophages. FASEB J 2: 2272-2277.

McFadden D, Wright JL, Wiggs B, Churg, A (1986a): Smoking inhibits asbestos clearance. Am Rev Respir Dis 133:372-374.

McFadden D, Wright J, Wiggs B, Churg A (1986b): Cigarette smoke increases the penetration of asbestos fibers into airway walls. Am J Pathol 123:95-99.

Mossman BT, Craighead JE (1979): Use of hamster tracheal organ cultures for assessing the cocarcinogenic effects of inorganic particulates on the respiratory epithelium. Prog Exp Tumor Res 24: 37-47.

Mossman BT, Eastman A, Landesman JM, Bresnick E (1983): Effects of crocidolite and chrysotile asbestos on cellular uptake and metabolism of benzo(a)pyrene in hamster tracheal epithelial cells. Environ Hlth Perspect 51: 331-335.

Plowman PN (1982): The pulmonary macrophage population of human smokers. Ann Occup Hyg 25: 393-405.

Topping DC, Nettesheim P, Martin DH (1980): Toxic and tumorigenic effects of asbestos on tracheal mucosa. J Environ Pathol Toxicol 3: 261-275.

Weiss W (1984): Cigarette smoke, asbestos, and small irregular opacities. Am Rev Respir Dis 130:293-301.

CYTOTOXICITY OF ASBESTOS AND CALCIUMSILICATES TO VARIOUS LUNG EPITHELIAL CELL TYPES IN VITRO

V. Skaug, A. Haugen
Department of Toxicology
National Institute of Occupational Health
P.O.Box 8149 Dep
0033 Oslo 1
Norway.

Introduction

Prolonged exposure to mineral particles can lead to inflammation and fibrosis in the peripheral airways and lung parenchyma (Churg and Wright 1983). Carcinoma of the lung may also occur (Churg and Wiggs 1985). The naturally occurring calciumsilicate, wollastonite, may have a slight fibrogenic potential in humans (Huuskonen et al 1983), but carcinogenic effects have not been shown.

Eight epithelial cell types are described in the human lung. Differences in response of various epithelial cell types to the toxic effects of mineral particles may be an important factor for understanding the early events in tumor developement and fibrogenesis in the lung. The purpose of the following investigation was to examine the sensitivity of various lung epithelial cell types to asbestos and calciumsilicates. Both rabbit and human lung epithelial cells served as target cells.

Materials and Methods

Mineral dusts

UICC asbestos reference samples (crocidolite, chrysotile B) were kindly donated by Dr Wagner, MRC Pneumoconiosis Unit, Penarth, UK. Wollasonite was obtained from a Finnish quarry (Partec OY, Finland). Tobermorite, an industrial synthetic calciumsilicate, was obtained from Ytong AB, Sweden.

NATO ASI Series, Vol. H30
Effects of Mineral Dusts on Cells
Edited by B.T. Mossman and R.O. Bégin

Respirable size fractions of tobermorite and wollastonite were prepared and characterized as described previously (Skaug and Gylseth 1984). In brief, the two calciumsilicates contain mineral fibers along with non fibrous particles. The lengths of the wollastonite fibers were mainly < 4μm and 40 % of the fibers had a diameter between 0.2 - 0.4 μm. The lengths of tobermorite fibers were mainly < 3 μm and 50 % of them had a diameter < 0,2 μm.

Cell cultures

Alveolar type 2 cells and Clara cells were isolated from the lungs of German CHBB:CH rabbits as previously described (Devereux and Fouts 1980). The viability, examined by Trypan Blue exclusion test, was 90% for alveolar type 2 cells and 95% for Clara cells. The two cell types were cultured in α-MEM medium as described previously (Haugen and Aune 1986, Deilhaug et al. 1985). At the time of mineral dust exposure, the purity was 80-90% for Clara cells and 90-95% for alveolar type 2 cells.

Rabbit trachea tissue handling, disaggregation and culture of the epithelial cells has been described (Wu and Smith 1982). The cells were propagated in F-12 medium supplemented with epidermal growth factor (10 μg/ml), insulin (5μg/ml) transferrin (5μg/ml) and 0,01% fetal calf serum.

Human bronchial epithelial cells were obtained from pulmonectomy or lobectomy specimens immediately after surgical removal, immerced in ice-cold L-15 medium and transported to the laboratory. Primary cultures were propagated by the explant outgrowth culture procedure (Lechner et al. 1981). Medium and solutions required for this procedure has been described (Lechner et al 1982).

Determination of cytotoxicity

The rabbit tracheal cells and Clara cells were seeded at 0,5 x 10^4 and $1 x 10^5$ cells/ml medium, respectivly, into 24- multiwell plates (Flow Laboratories Inc., Hamden, CT), coated with FVB (fibronectin, 10μg/ml; vitrogen, 30μg/ml; bovine serum albumin, 10μg/ml) in complete medium (Lechner et al. 1982). Rabbit alveolar type 2 cells and human bronchial epithelial cells were seeded at $1 x 10^4$ cells/ml medium into 35 mm dishes (Costar, Cambridge, Mass). The cultures were incubated for 24 hours in 37^0C, 4% CO_2 and 95% humidity. The mineral dusts were then suspended in medium without serum, sonicated to obtain homogeneous dust suspensions, and added to the cell cultures in appropriate consentrations. The Clara cells were cultured for 5 days; the other cells for 8 days. The toxicity of mineral fibers was defined by determining the inhibition of growth rate.

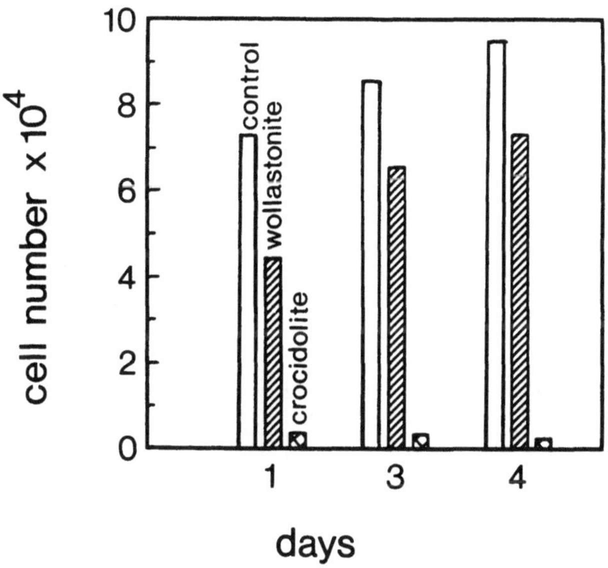

Fig. 1. Growth of Clara cells exposed to wollastonite and crocidolite (50μg/ml)

RESULTS AND DISCUSSION

After 5, 24 and 48 hours of exposure *in vitro*, mineral fibers were readily observed within the cytoplasm of the cells by scanning electron microscopy (back scattered electron image) and transmission electron microscopy. The Wollastonite tended to gather in the perinuclear area, wheras the longer crocidolite fibers were observed through the entire cytoplasm.

Studies with Clara cells demonstrated that crocidolite was far more cytotoxic than wollastonite. Figure 1 shows comparative responses as a function of time in a typical experiment.

As can be seen in Fig. 2, crocidolite and chrysotile asbestos inhibited proliferation of alveolar type 2 cells. There was no inhibition with wollastonite and a slight decrease in cellular growth with tobermorite.

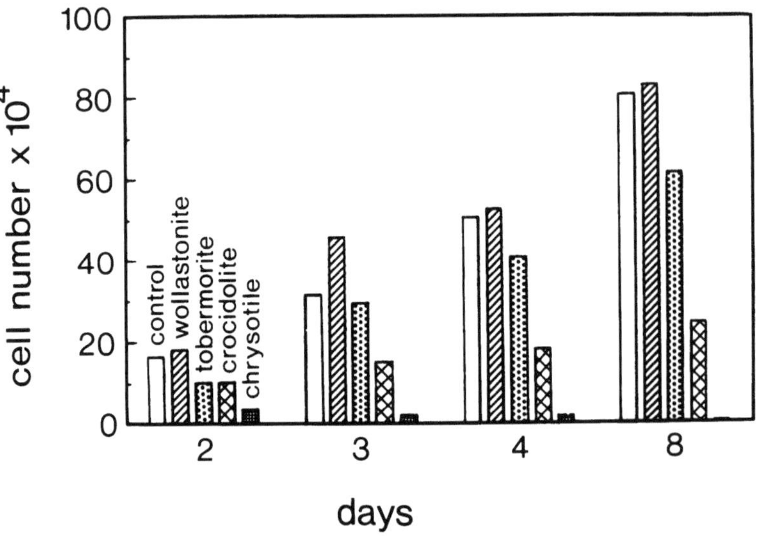

Fig. 2. Effect of mineral dust particles (50μg/ml) on growth of alveolar type 2 cells

The effects of calciumsilicates and crocidolite on Clara cells and on alveolar type 2 cells are compared in Fig. 3. There is no significant difference in response between the two cell types.

The inhibition of growth rate of rabbit tracheal epithelial cells by wollastonite and crocidolite (Fig. 4) was in the same order as for Clara cells, and consequently not significant different from the response of alveolar type 2 cells. Although the human bronchial epithelial cells have a higher growth rate, the relative toxicity of the dusts were the same as for the other cell types examined (Fig. 5).

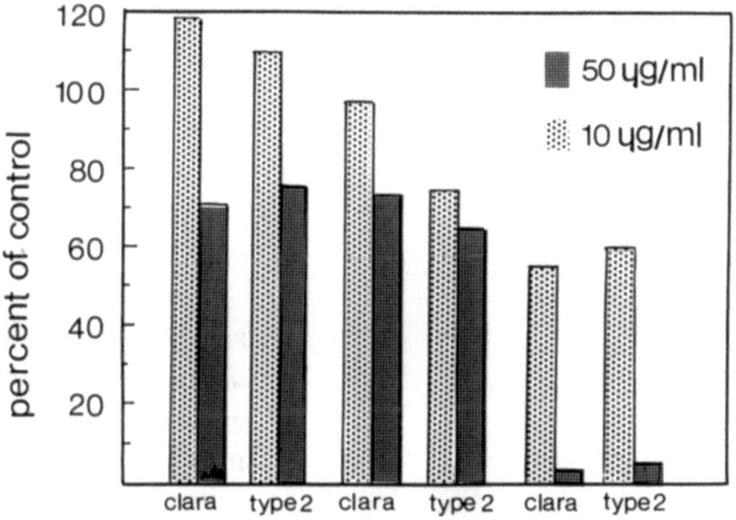

Fig. 3. Growth of Clara cells and alveolar type 2 cells exposed to calciumsilicates and crocidolite

There was no evident difference in sensitivity to the minerals for the various lung cell types _in vitro_. The observed asbestos cytotoxicity may be due to both acute cytotoxic effects during the initial contact between minerals and cells, and to an inhibition of cell division and proliferation due to the phagocytosed material.

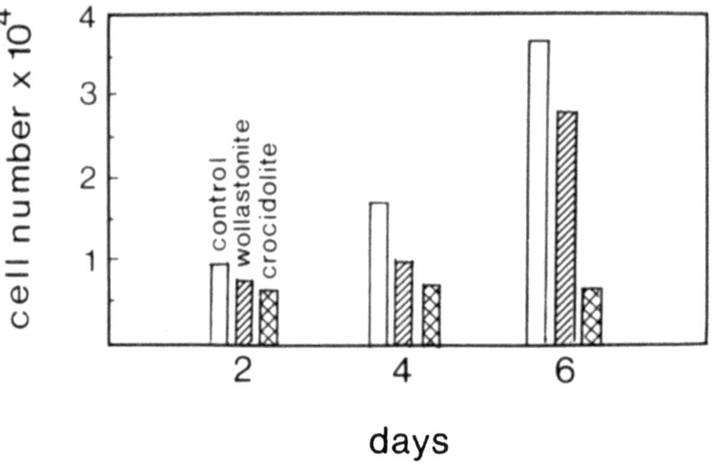

Fig. 4. Growth of rabbit tracheal epithelial cells exposed to wollastonite and crocidolite (50μg/ml)

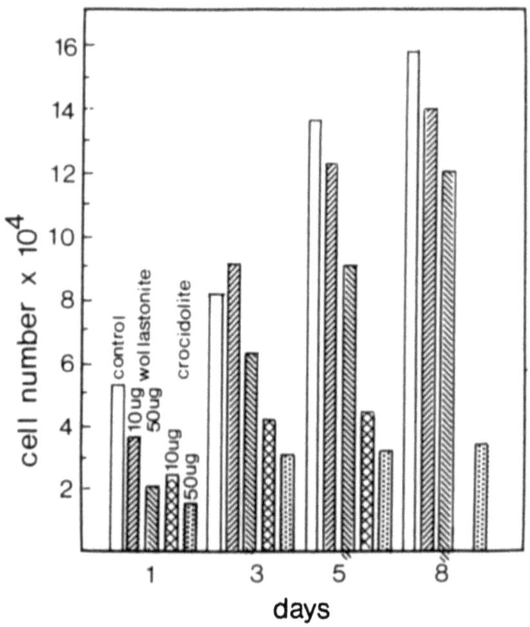

Fig. 5. Effect of wollastonite and crocidolite on proliferation of human bronchial epithelial cells. Dust concentrations are 10 and 50 μg/ml

Data presented in the study indicate that calciumsilicates have little toxic effect on these epithelial cells. The acute cytotoxic effect of calciumsilicates have been evaluated previously (Skaug and Gylseth 1983). These studies has shown that toberminite is slightly more toxic than wollastonite in an *in vitro* hemolytic assay.

In conclusion, calciumsilicates are less toxic than asbestos fibers in the lung epithelial cell types studied. No differences in cell specific cytotoxicity is observed.

References

Churg A, Wiggs B (1985) Mineral particles, mineral fibres, and lung cancer. Environ Res 37:364-372

Churg A, Wright JL (1983) Small-airway lesions in patients exposed to nonasbestos mineral dusts. Human Pathology 14:688-693

Deilhaug T, Myrnes T, Aune T, Krokan H, Haugen A (1985) Differential capacities for DNA repair in Clara cells, alveolar type 2 cells and macrophages of the lung. Carcinogenesis 6:661-663

Devereux TR, Fouts JR (1980) Isolation and identification of Clara cells from rabbit lung. In Vitro 16:958-969

Haugen A, Aune T (1986) Culture of rabbit pulmonary Clara cells. Proc Soc Exp Biol Med 182:277-281

Huuskonen MS, Tossavainen A, Koskinen H, Zitting A, Korhonen A, Nickels J, Korhonen K, Vaaranen V (1983) Wollastonite exposure and lung fibrosis. Environ Res 30:291-304

Lechner JF, Haugen A, Autrup H, McClendon IA, Trump BF, Harris CC (1981) Clonal growth of epithelial cells from normal adult human bronchus. Cancer Res 41:2294-2304

Lechner JF, Haugen A, McClendon IA, Pettis E (1982) Clonal growth of normal adult human bronchial epithelial cells in a serum-free medium. In Vitro 18:633-642

Skaug V, Gylseth B (1983) Hemolytic activity of five different calcium silicates. Env Health Perspectives 51:195-203

Skaug V, Gylseth B (1984) In vitro macrophage cytotoxicity of five calcium silicates. Br J Ind Med 41:116-121

Wu R, Smith D (1982) Continious multiplication of rabbit tracheal epithelial cells in a defined, hormonsupplemented medium. In Vitro 18:800-812

Injurious Effects of Mineral Dust-elicited Bronchoalveolar Leukocytes on Epithelial Cells *in vitro*: The Role of Extracellular Matrix Components

K. Donaldson, Joan Slight, Geraldine M. Brown.
The Institute of Occupational Medicine
8 Roxburgh Place, Edinburgh, Scotland.

Introduction

Deposition of mineral dust associated with the development of pneumoconiosis causes a range of changes in the lung parenchyma including alveolar inflammation and, in the longer term, septal fibrosis and epithelial abnormalities (Begin *et al.* 1986; Gibbs *et al.* 1984). We have been studying the role of inflammatory leukocytes in the development of pathology in dust-exposed lung and have previously reported on the ability of inflammatory bronchoalveolar leukocytes from rat lung, exposed by intratracheal installation to quartz, to cause proteolytic injury to cells of an alveolar epithelial cell line *in vitro* (Donaldson *et al.* 1988[c]). In the present paper we extend these studies by examining inflammatory bronchoalveolar leukocytes lavaged from chrysotile asbestos-exposed lung, in terms of ability to cause epithelial injury *in vitro*. We further report on the ability of bronchoalveolar leukocytes from lung exposed by the more appropriate inhalation route to coalmine dust, with the regard to causing epithelial injury and proteolytic degradation of fibronectin. In order to study the role of the extracellular matrix in epithelial injury *in vitro*, we have examined the ability of exogenous protease and mineral dust-elicited inflammatory bronchoalveolar leukocytes, to injure epithelial cells cultured on surfaces coated with purified extracellular matrix components.

Materials and Methods

Models of mineral dust exposure

Mineral dusts used in the study were quartz (DQ_{12} standard), chrysotile asbestos (UICC sample A) and titanium dioxide (TiO_2: rutile form; Tioxide Ltd., Stockton on Tees). For intratracheal instillation, dusts were suspended at 2mg/ml in sterile saline and 0.5ml (1mg) instilled into the lungs of anaesthetised rats. For inhalation exposure using coalmine dust, the dusts were collected from the air of a colliery mining medium rank

NATO ASI Series, Vol. H30
Effects of Mineral Dusts on Cells
Edited by B. T. Mossman and R. O. Bégin

coal (26% ash) containing 0.7% quartz. Dust was generated in purpose-built exposure chambers as a monodispersed airborne cloud at 50mg/m^3 - full details of the generation and monitoring procedures are given elsewhere (Donaldson *et al.* (1988[a]). At selected time points after instillation of dust or 45 days after airborne exposure to coalmine dust, rats were killed by barbital overdose and their lungs lavaged with 4 x 8ml volumes of warm saline with massaging to increase the yield of leukocytes (Donaldson *et al.* (1988[a]).

Assay of bronchoalveolar leukocyte-mediated injury to alveolar epithelial cells

This assay has been described in detail elsewhere (Donaldson *et al.* (1988[c]). Briefly cells of the A549 alveolar epithelial cell line were prelabelled with ^{51}Cr overnight and washed. The labelled cells were co-cultured with inflammatory bronchoalveolar leukocytes at varying effector:target ratios and leukocyte-mediated detachment injury and lytic injury was assessed.

Assay of bronchoalveolar leukocyte-mediated proteolysis of ^{125}I fibronectin

This assay is described in detail elsewhere (Brown and Donaldson, 1988[b]). Briefly ^{125}I-labelled fibronectin were prepared and adhered, as a matrix, to the bottom of microtitre plate wells. The bronchoalveolar leukocytes under test were added to the microtitre plate wells and their ability to degrade ^{125}I fibronectin was assessed.

Results

Detachment injury caused to alveolar epithelial cells by co-culture with dust-elicited bronchaolveolar leukocytes

Figure 1 shows the epithelial-injuring activity of bronchoalveolar leukocytes from the lungs of rats exposed, by intratracheal instillation, to chrysotile asbestos or titanium dioxide. It shows clear detachment injury with both chrysotile asbestos and TiO_2-elicited leukocytes at day 1 with chrysotile leukocytes producing more detachment activity at the higher effector:target ratios. The ability of the TiO_2-elicited bronchoalveolar leukocytes to injure epithelial cells was virtually over by 3 days whilst the chrysotile asbestos bronchoalveolar leukocytes were still able to cause epithelial injury at 3 days although this had returned almost to normal control levels by 7 days. The percentage of neutrophils in each of the populations tested was:- TiO_2 - 1 day 48%, 3 days 15%, 7 days 0%; chrysotile - 1 day 40%, 3 days 12%, 7 days 4%; the remainder of the cells were macrophages with <5% lymphocytes.

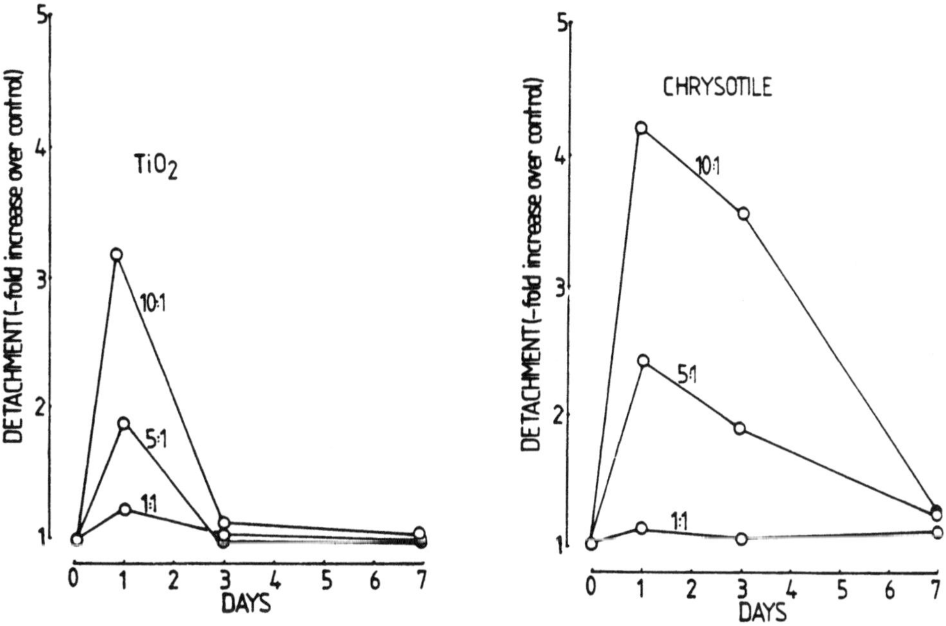

Fig. 1. Left. Detachment injury caused to alveolar epithelial cells by TiO_2- (left hand panel) or chrysotile asbestos (right hand panel) elicited bronchoalveolar leukocytes at varying effector:target ratios; rats received 1mg of dust intratracheally on day 0

Exposure to coalmine dust (50mg/m^3) by inhalation elicited a bronchoalveolar leukocyte population with ability to cause detachment injury to alveolar epithelial cells *in vitro* (Figure 2); concomitant with the increased detachment caused by these inflammatory leukocytes there was increased proteolysis of ^{125}I fibronectin. The coalmine dust-elicited population contained 56% neutrophils.

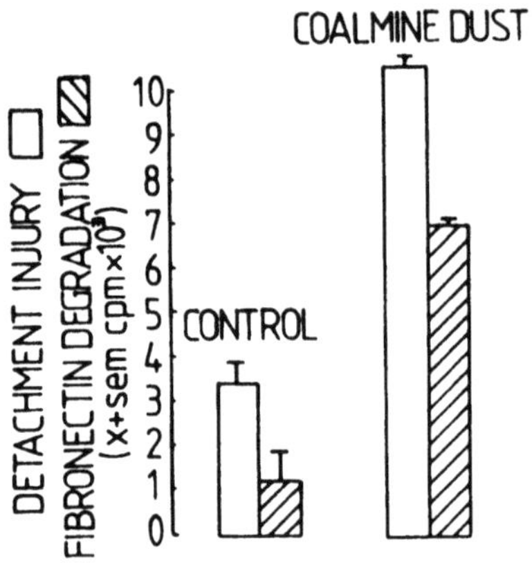

Fig. 2. Epithelial cell detachment injury and breakdown of fibronectin caused by bronchoalveolar leukocytes from control rats or rats inhaling coalmine dust for 44 days.

Role of extracellular matrix in detachment injury

Culture of the target epithelial cells on microtitre plates coated with purified extracellular matrix components profoundly altered their ability to be detached on culture with exogenous elastase (Figure 3). It is clear that this microbial elastase preparation causes rapid detachment of cells grown on plastic alone even at very low doses but growing cells on fibronectin or Type IV collagen matrix renders the alveolar epithelial cells resistant to detachment injury even at very high levels of this exogenous enzyme. Figure 4 shows that, in the presence of quartz-elicited bronchoalveolar leukocytes, culture on extracellular matrices also causes protection against detachment injury. Clear differences exist in the ability of different components to produce protection, with Type IV collagen promoting greatest attachment and resistance to inflammatory cell damage and laminin and fibronectin being less protective.

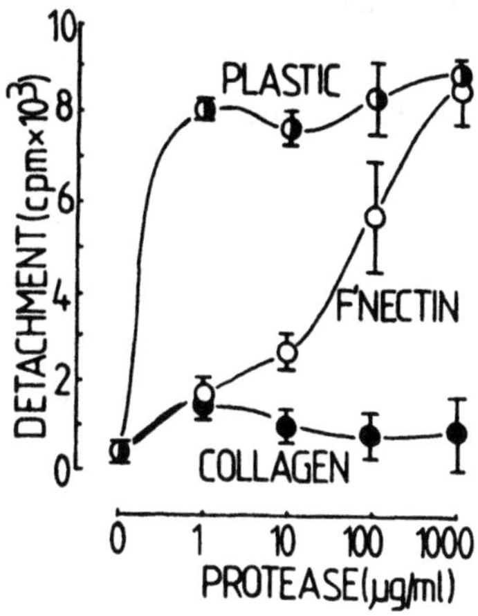

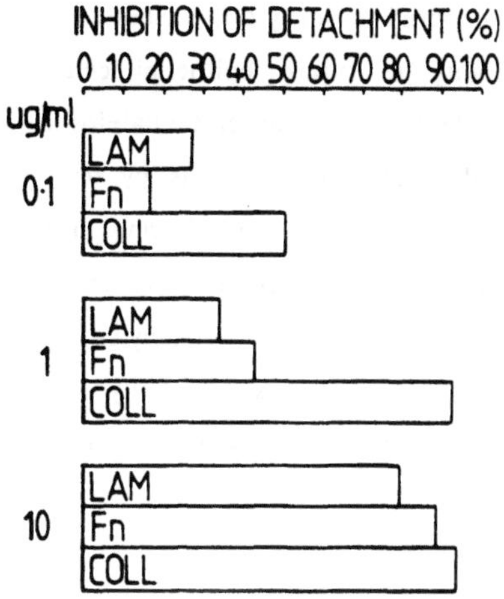

Fig. 3. Epithelial cell detachment injury caused by exogenous elastase to alveolar epithelial cells cultured on plastic, fibronectin or collagen-coated plates.

Fig. 4. Epithelial cell detachment injury caused by quartz-elicited bronchoalveolar leukocytes to alveolar epithelial cells cultured on laminin, fibronectin or collagen-coated plates. Results shown as inhibition of detachment compared to cells cultured on plastic.

Discussion

We have been examining the relationship between inflammatory leukocyte recruitment into the lungs following the deposition of pathogenic mineral dust, and subsequent development of pathological change. This paper examines, in particular, the ability of inflammatory leukocytes to injure alveolar epithelial cells *in vitro* and the role of the extracellular matrix has been studied in this phenomenon. Injury to epithelial cells could lead to alveolar epithelial hyperplasia and, possibly through failure to attain re-epithelialisation, also contribute to alveolar fibrosis (Witschi et al. 1980). Our previous results have shown that bronchoalveolar leukocytes from quartz-exposed lung have the ability to injure alveolar epithelial cells by causing detachment through a largely protease-mediated mechanism (Donaldson et al. 1988[c]) and that these leukocytes can also cause degradation of extracellular matrix components (Brown and Donaldson, 1988[b]). In the present paper we have extended these findings and have demonstrated that inflammatory leukocytes from lungs exposed to chrysotile asbestos or TiO_2 by instillation both have, in the short-term, the ability to injure epithelial cells. The single intratracheal instillation of asbestos elicited leukocytes with epithelial-injuring activity which was present at both 1 and 3 days and still above control levels by day 7. With intratracheal TiO_2, a dust not associated with pneumoconiosis and having little inflammatory potential (Donaldson et al. 1988[b]), epithelial-detaching activity was present at 1 day but had returned to the same level as control bronchoalveolar leukocytes by 3 days. Delivered by intratracheal instillation, chrysotile asbestos is substantially less active than quartz in causing leukocyte recruitment to the rat lung and a similar pattern is found with inhalation exposure (Donaldson et al. 1988[b]).

We report here also that bronchoalveolar cells from rats exposed to coalmine dust by inhalation show increased epithelial-injuring activity with concomitant increased fibronectin degrading activity; details of the time relatedness of this response are described elsewhere (Brown and Donaldson, 1988[a]). It therefore seems likely that bronchoalveolar leukocytes from rats exposed to any mineral dust which cause inflammation on inhalation exposure will have the potential to cause epithelial injury and extracellular matrix damage. Our experience has shown that such dusts include the classical pneumoconiosis-generating dusts quartz, asbestos (Donaldson et al. 1988[b]) and coalmine dust (Donaldson et al. 1988[a]). TiO_2, which does not cause pneumoconiosis, does not cause inflammation when inhaled except after very long exposure at very high dose (Donaldson *et al.* 1988[a]). We suggest therefore that some of the epithelial abnormalities and hyperplasia commonly present in association with pneumoconiotic fibrosis may be due to injurious effects on the epithelial cells caused by inflammatory bronchoalveolar leukocytes.

The present study has also demonstrated an important modulating role of extracellular matrix elements in detachment injury caused to the alveolar epithelial cells by inflammatory bronchoalveolar leukocytes. Epithelial cells appear to have affinity for extracellular matrix which may be receptor-mediated particularly in the case of Type IV collagen which was particularly protective against proteolytic injury. Leukocyte proteases may be excluded from interacting with the areas of contact between the cell and the matrix which could be important targets for the proteases during detachment *in vitro.* Epithelial cells lay down their own basement membrane containing extracellular matrix elements but in the time scale of this assay it is unlikely that they have had time to do so. These results suggest that, if a proper basement membrane could be produced by the target cells, their susceptibility to detachment injury could be much less. However, studies on whole basement membrane and mixtures of purified extracellular matrix components are necessary to clarify this response. The results suggest that epithelial detachment injury *in vivo* might not be so severe as suggested by detachment experiments carried out with the target cells grown on plastic and highlights the importance of the culture conditions in assessing epithelial cell response to leukocyte-mediated injury. During inflammation *in vivo*, however, activated leukocytes are trafficking through the alveolar epithelium during which the basement membrane and other extracellular matrix associated with the epithelial barrier may be more susceptible to injury. An assay system which would allow migration of the effector leukocytes through a barrier of epithelial cells would provide a better model.

Acknowledgement. Research funded in part by the Colt Foundation and in part by the Commission of the European Communities.

References

Begin R, Bisson G, Boileau R, Masse S (1986) Assessment of disease activity by Gallium ^{67} scan and lung lavage in the pneumoconioses. Sem Resp Med 7:275-280

Brown GM, Donaldson K (1988[a]) Inflammatory response in lungs of rats inhaling coalmine dust: enhanced proteolysis of fibronectin by bronchoalveolar leukocytes. Env Res (submitted)

Brown GM, Donaldson K (1988[b]) The degradation of connective tissue components by lung-derived leukocytes *in vitro*: the role of proteases and oxidants. Thorax 43:132-139

Donaldson K, Bolton RE, Brown DM, Brown GM, Cowie H, Jones AD, Robertson MD, Slight J, Davis JMG (1988[a]) Studies on the cellular response in lung tissue to the inhalation of mineral dust. Institute of Occupational Medicine Report No. TM/88/01

Donaldson K, Bolton RE, Jones AD, Brown GM, Robertson MD, Slight J, Cowie H, Davis JMG (1988[b]) Kinetics of the bronchoalveolar leukocyte response in rats following exposure to equal airborne mass concentration of quartz, chrysotile asbestos and titanium dioxide. Thorax 43:525-533

Donaldson K, Slight J, Brown GM, Bolton RE (1988[c]) The ability of inflammatory bronchoalveolar cell populations elicited with microbes or mineral dust to injure alveolar epithelial cells and degrade extracellular matrix *in vitro*. Br J Exp Path 69:327-338

Gibbs AR, Seal RME, Wagner JC (1984) Pathological reactions of the lung to dust. In: Morgan WKC, Seaton A (Eds) Occupational Lung Diseases. Saunders, Philadelphia 129-162

Witschi H, Haschek WM, Meyer KR, Ollrich RL, Dalbey WE (1980) A pathogenetic mechanism in lung fibrosis. Chest 78:395-399

TOXICITY OF CHRYSOTILE ASBESTOS THROUGH SURFACE MODIFICATION WITH CHELATING AGENTS

M. Martin, M. Pagé, F. Habashi et F.T. Awadalla
Department de Biochimie, Faculté de Médecine
Université Laval, Ville de Québec
Québec C1K 7P4, Canada

Introduction

This work is the fruit of a collaboration between the department of Mines and Metallurgy and the department of biochemistry of Université Laval. The modified asbestos fibers were prepared in the laboratory of Dr. Fathi Habashi and the cytotoxicity assays were made in the labretori of Dr. Michel Page.

One of the principal uses of asbestos fibers is in the manufacture of asbestos cement which is used to make water pipes, panels, tiles and other materials of construction. The process involves preparing a thick water slurry of asbestos and cement of certain proportions, forming it in a desired shape, rapidly filtering off the water, then letting the mixture set to harden. Two problems face this technology: 1) the speed, with which the water must be removed from the system and 2) the potential toxicity of the product due to the presence of asbestos fibers.

To solve this problem, we wanted to develop a new chrysotile product using the present industrial technology while eliminating dust and to reduce the toxicity of chrysotile while giving an added value to the new product.

In a similar approach, Flowers showed that when chrysotile asbestos was thoroughly mixed with a solution of ferric ammonium sulfate and the slurry was filtered and dried, the product showed a much decreased cytotoxicity. Reports and literature indicate that the toxicity of asbestos is decreased

NATO ASI Series, Vol. H30
Effects of Mineral Dusts on Cells
Edited by B. T. Mossman and R. O. Bégin

when magnesium is completely leached by acid but of course, the fiber loses it valuable properties since it is transformed into a silica skeleton.

Since chrysotile asbestos is a hydrated magnesium silicate with a lot of hydroxyl groups on the surface, we selected a certain number of chelating agents that react with the magnesium hydroxide to form lakes and tested the cytotoxicity of these products in vitro.

Materials and Methods.

Several chelating agents were used for reaction with asbestos. Most of the agents used were cotton dyes or fabric dyes which are used currently in industry. These new compounds are vividly colored and stable. The process of fixing the organic compound on the surface of asbestos is very similar to the old method for dyeing cotton fabrics. In this method, the fabric is soaked in a solution of aluminum or magnesium acetate, squeezed to drain excess solution, then heated with steam in a special chamber to decompose the acetate and precipitate the metal hydroxide on the surface. Fabrics are then immersed in dye solution whereby a colored chelate is precipitated on the surface. In this case, asbestos already bears on its surface the magnesium hydroxide and it is simply soaked in these chelating dyes. Various conditions were tested to determine the amount of dye that could be absorbed and chelated on the fiber and the toxicity. The deciding factor for choosing any of the compounds were the toxicity of the product, the cost of the reagent and also the color it produces on asbestos.

The toxicity tests that we used are the already classic hemolysis test and the cytotoxicity assay in vitro using P388 cells. Red blood cells are incubated with various concentration of fibers and free hemoglobin is then measured in the supernatant by spectrophotometry. The results that we obtain by the hemolysis test using some typical new compounds are the following: 1) using 4T30 asbestos as a standard as 100% in these conditions, we found that thiazole yellow covered asbestos at a concentration of about 2% of dye is no longer toxic; 2) with methyl blue, at a concentration of 2.4% of chelating agent, the product is barely toxic; 3) hydroxyquinoline at 4% or 10% is as toxic as asbestos. This is not surprising since hydroxyquinoline by itself is toxic.

The in vitro cytotoxicity assay was carried out with mouse leukemia P388D1 cells. Various concentrations of fibers were used and the percentage of non viable cells was evaluated by the trypan blue dye exclusion method. Again, we noticed that thiazole yellow modified fibers were almost not toxic when covered with 2% dye as compared to 100% toxicity for free chrysotile. The results are in agreement with the hemolysis test where also hydroxyquinoline covered asbestos is even more toxic than raw asbestos.

Discussion

It is then possible to greatly decrease the toxicity of chrysotile asbestos by dyeing it with suitable dyestuffs that form colored lakes or chelates with magnesium. Dyeing is accomplished in aqueous medium at ambient temperature. This method would be suitable not only for preparing colored nontoxic asbestos cement but also for treating asbestos fiber during its separation from the bearing rock by the wet method as practiced now at Woodreef Mines in Australia.

These compounds are obtained in conditions that could be applied for industrial use.

A COMPARISON OF THE *IN VIVO* AND *IN VITRO* TOXICITY OF FIBROUS AND NON-FIBROUS SILICON NITRIDE

G. L. Fisher, K. L. McNeill, A. W. Singer, and J. T. Smith*

Battelle Columbus Division
505 King Avenue
Columbus, Ohio 43201-2693

ABSTRACT

In vitro and *in vivo* evaluations were conducted to assess the pulmonary cytotoxicity and fibrogenicity of fibrous and non-fibrous (milled) silicon nitride (Si_3N_4). The milled powder was derived from the fibrous material by standard ceramic communition techniques. For the *in vitro* evaluation, bovine pulmonary alveolar macrophage cells were exposed to 0.3, 0.1, 0.03, and 0.01 mg/ml of fibrous or non-fibrous Si_3N_4, glass beads (an inert control), or alpha-quartz (an active control). Functional, biochemical, and morphological analyses revealed that fibrous Si_3N_4 is as cytotoxic as quartz, and that the non-fibrous Si_3N_4 particles are, like glass beads, relatively inert. For the *in vivo* evaluation, a 25 mg dose of test or control article was intratracheally instilled in male rats, which were serially sacrificed at 3, 7, 14, and 28 days. Functional, biochemical and histopathological analyses supported the conclusion of the *in vitro* evaluation that the non-fibrous Si_3N_4 was relatively inert. Alpha-quartz induced significant pulmonary toxicity including fibrosis, edema, and coalescing granulomas. The fibrous Si_3N_4 produced minimal pulmonary changes because physical limitations associated with instilling the fibrous Si_3N_4 resulted in only a fraction of the test dose being delivered to the rat lung. However, the pulmonary response to the fibrous Si_3N_4 was greater than the response to the non-fibrous material and, shortly after initial dosing, similar to that of α-quartz. Since both test articles were the same material, differing only in particle morphology, these studies further support the critical importance of physical form as a determinant of biological activity.

---

*GTE Laboratories, Inc., Waltham, Massachusetts 02254

NATO ASI Series, Vol. H30
Effects of Mineral Dusts on Cells
Edited by B. T. Mossman and R. O. Bégin

INTRODUCTION

Exposure to silica may cause severe exudative or chronic inflammatory pulmonary disease, depending upon exposure intensity and particle size (Lee, 1985). Potentially fatal pulmonary fibrosis may result from chronic exposure to silica dusts. Occupational exposure to naturally occurring pathogenic crystalline forms of silica, such as quartz, cristobalite, and tridymite, represents a possible hazard in a variety of occupations, including mining operations, stone cutting and polishing, glass manufacturing, foundry work and sand blasting. Amorphous siliceous dusts such as glass beads or diatomaceous earth, however, appear to be relatively inert.

A variety of *in vitro* and *in vivo* techniques have been employed as models of silica-induced pulmonary toxicity (Lee, 1985; Chamberlain *et al.*, 1982; Fisher and Placke, 1987; and Reiser and Last, 1979). Results of *in vitro* tests, which include functional and biochemical assays of isolated alveolar macrophages incubated with toxic compounds, have been shown to provide a good correlation with *in vivo* fibrogenicity (Chamberlain *et al.*, 1982; and Fisher and Placke, 1987). This paper summarizes the effects of *in vitro* exposure of bovine pulmonary alveolar macrophages to fibrous and non-fibrous silicon nitride (Fisher *et al.*, 1988a), and compares these results with changes induced in rats exposed to these materials by intratracheal instillation (Fisher *et al.*, 1988b).

MATERIALS AND METHODS

Test Articles

Commercially prepared silicon nitride (Si_3N_4) compounds served as the test articles. Fibrous Si_3N_4 (surface area 4.4 m^2/mg, equivalent diameter 10 μm) was milled to produce an equiaxial non-fibrous Si_3N_4 (surface area 13.5 m^2/mg, equivalent diameter of 3.5 μm). The fibrous Si_3N_4 (type SN-502) was manufactured by GTE Chemical and Metallurgical Division, Towanda, Pennsylvania; manufacture was discontinued in early 1986. A commercially manufactured (Denko, Japan), equiaxial (non-fibrous) Si_3N_4 (surface area, 10.1 m^2/g; equivalent diameter, 5 μm) was also evaluated. Alpha-quartz (SiO_2, Whittaker, Clark and Daniels, Inc., South Plainfield, New Jersey) with a mass median diameter (MMD) of 2.0 $\pm$ 1.8 μm was employed as a positive control. Glass beads (MMD of 2.1 $\pm$ 72.0 μm) were used as a negative (inert) control (Particle Information Services, Bremerton, Washington). When appropriate, saline was used as a non-particulate, or vehicle, control.

In Vitro Assays

Pulmonary alveolar macrophages (PAM) were harvested from the cardiac lung lobe of a healthy adult steer after slaughter. Using methods described by Fisher, *et al.* (1983), the lobe was lavaged with aliquots of cold phosphate buffered saline (PBS). Lavage fluid was centrifuged and cells resuspended in incomplete media (MEM). A sample of the lavaged cells was evaluated for cell concentration, viability and differential composition.

Macrophage function evaluations were conducted by exposing the PAM to 0.3, 0.1, 0.03, or 0.01 mg/ml of glass beads, quartz, fibrous Si_3N_4, or non-fibrous Si_3N_4 using methods previously described (Fisher *et al.*, 1978).

In vitro biochemical assays of lactate dehydrogenase (LDH), acid phosphatase (AP) and total protein levels were also conducted. LDH was assayed using a diagnostic kit (Sigma Co., Grand Island, New York) and AP was assayed using a colorimetric kit (Boehringer Mannhein, Houston, Texas), along with a Gilford 2400-S spectrophotometer. Total protein was evaluated using the method of Lowry, *et* al. (1951).

In Vivo Evaluations

Six- to eight-week old male Fischer 344 rats were obtained from Charles River Breeding Laboratories, Inc., Kingston, New York. Rats were randomized by body weight into 5 treatment groups of 40 rats each, corresponding to fibrous, non-fibrous and milled Si_3N_4, SiO_2, glass beads, and saline-control groups.

Five rats from each group were sacrificed with sodium pentobarbital at 3, 7, 14, and 28 days post instillation with 25 mg of the test articles, for lavage, lung weight and biochemical evaluations. The lungs were removed and the right apical lobe was closed by ligation and removed for dry weight determination. Free cells were lavaged from the remaining lobes with a total of 25 ml of cold PBS. Viability was determined by staining with trypan blue and examining for nuclear exclusion. A differential count was performed on smears prepared with Wright's stain.

Remaining lobes were frozen and subjected to biochemical analysis for hydroxyproline. Five rats per group were also sacrificed at 3, 7, 14, and 28 days after dosing for histopathologic examination.

RESULTS

In Vitro Assays

Quartz and fibrous Si_3N_4 caused marked dose-related decreases in cellular viability. The effect of glass beads or non-fibrous Si_3N_4 on PAM viability was minimal (Figure 1).

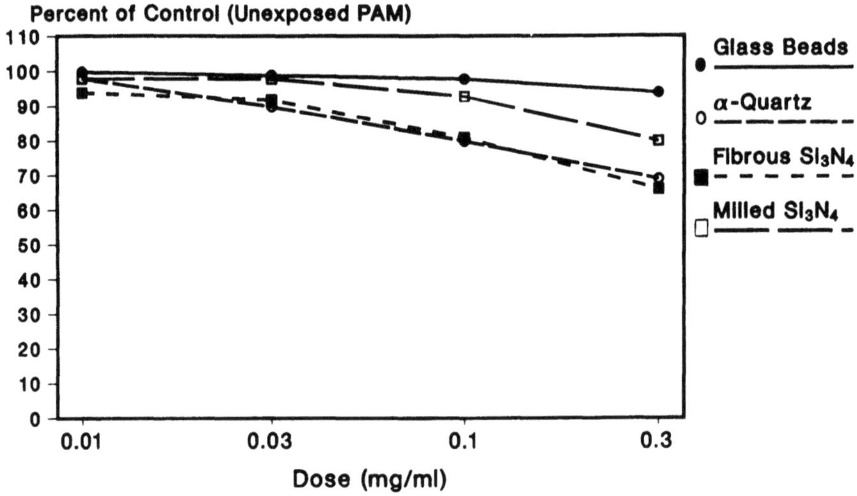

FIGURE 1. EFFECT OF *IN VITRO* EXPOSURE TO TEST ARTICLE ON MACROPHAGE VIABILITY

Loss of adherence by the cells after a 20-hour exposure to the test and control materials was variable but similar for all groups. For all test particles, an increase in the number of cells lost was apparent at the two highest particle concentrations.

Exposure of PAM to any of the test particles, including the glass beads, caused a dose-related decrease in phagocytosis (Figure 2). This effect was markedly greater for PAM exposed to either quartz or fibrous Si_3N_4 than for cells exposed to glass beads or non-fibrous Si_3N_4.

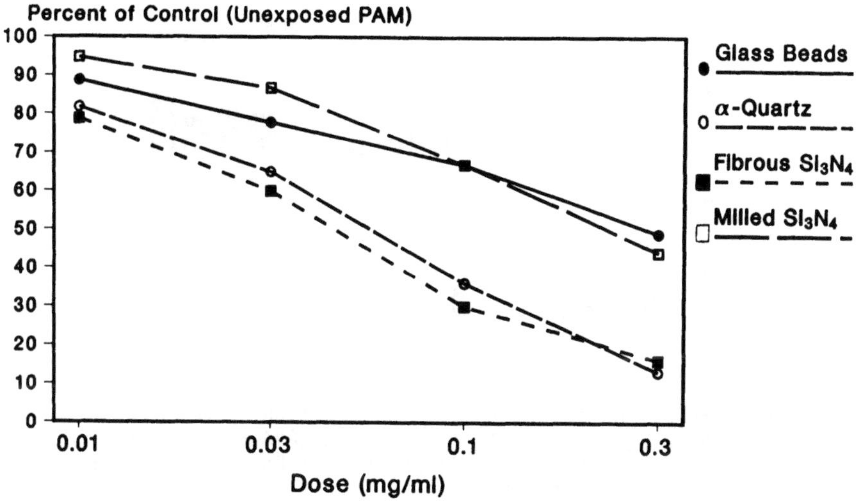

FIGURE 2. EFFECT OF *IN VITRO* EXPOSURE TO TEST ARTICLE ON MACROPHAGE PHAGOCYTOSIS OF INERT BEADS

Exposure to quartz caused a significant ($p < 0.05$) decrease in total protein of the cell pellet, an increase in LDH, and an increase in AP levels. In comparison, fibrous Si_3N_4 induced a significant ($p < 0.05$) increase in LDH, but no difference in total protein or AP levels. The non-fibrous Si_3N_4, like the glass beads, showed no effect on LDH, AP or total protein content when compared to control values.

In Vivo Evaluations

Total macrophage yield (Figure 3) and viability were similar for all treatment groups at all periods examined except for the group exposed to quartz. Lungs from rats exposed to the quartz averaged higher overall macrophage counts and lower overall cell viability than any of the other test or control groups. Quartz-, Si_3N_4-, and glass-bead-exposed rats demonstrated increases in polymorphonuclear cells recovered by lavage from their lungs, when compared to the saline-instilled rats. The numbers of polymorphonuclear cells recovered returned to normal (saline-group levels) by the 28-day sacrifice period for all groups of rats except quartz. In quartz-exposed rats, the number of polymorphonuclear cells recovered by lavage was greatest from the 28-day necropsy period (Figure 4).

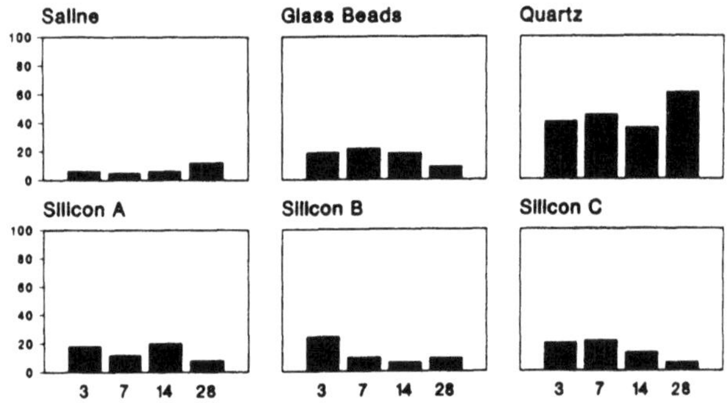

FIGURE 3. EFFECT OF INTRATRACHEAL INSTILLATION ON MACROPHAGE NUMBERS ($x10^6$) OBTAINED BY LUNG LAVAGE 3, 7, 14 AND 28 DAYS AFTER TEST ARTICLE EXPOSURE. (SILICON A, B, AND C EQUALS FIBROUS, MILLED, AND NON-FIBROUS Si_3N_4, RESPECTIVELY.)

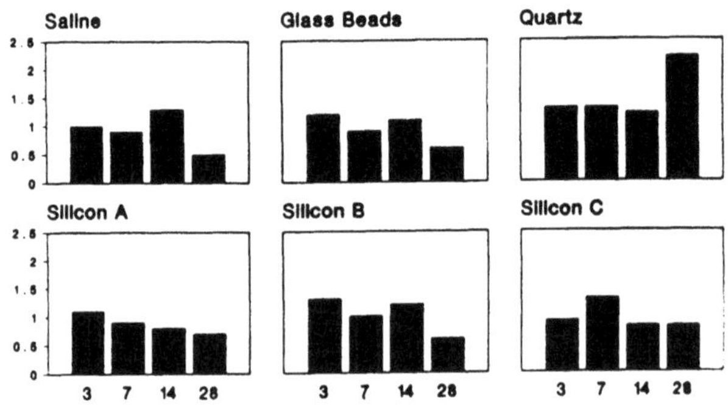

FIGURE 4. EFFECT OF INTRATRACHEAL INSTILLATION ON NEUTROPHIL DIFFERENTIAL COUNT (%) OBTAINED BY LUNG LAVAGE 3, 7, 11 AND 28 DAYS AFTER TEST ARTICLE EXPOSURE. (SILICON A, B, AND C EQUALS FIBROUS, MILLED, AND NON-FIBROUS Si_3N_4, RESPECTIVELY.)

Rats instilled with quartz had increased ($p < 0.05$) water content of the right apical lobe at all necropsy periods. Due to a concurrent increase in the dry lung weight for quartz-exposed rats, however, the wet:dry lung weight ratios were similar across all groups. Significant ($p < 0.05$) differences in mean hydroxyproline content were found only in the quartz-instilled group and only at Day 14 and 28.

Few microscopic lesions of significance were found in the saline control rats. Principally, these consisted of rare foci of acute inflammation found shortly after dosing. The lungs of rats instilled with glass beads had minimal to moderate accumulation of neutrophils and macrophages in alveolar spaces and an admixture of mononuclear cells within thickened alveolar septae on post-treatment Day 3. With each successive sacrifice, the lungs displayed lesion progression and resolution.

Rats treated with quartz had mild to marked acute and chronic inflammation at Day 3, much more severe than that seen in rats treated with the glass beads. In addition, all rats had mild to moderate granuloma formation. By post-treatment Day 28, all rats had severe coalescing granulomas and mild fibrosis. All rats also had large granulomas present in bronchus-associated lymphophoid tissue. Irregular birefringent crystals were common in phagocytes.

On Day 3, the fibrous Si_3N_4 produced minimal to mild chronic alveolar inflammation and minimal granuloma formation. Lung lesions became less severe and affected fewer rats at later sacrifice periods.

Rats instilled with either the milled or the non-fibrous Si_3N_4 had minimal or moderate chronic inflammation of the lungs at Day 3, with a moderate amount of the instilled material both in alveolar macrophages and in the alveolar spaces. By Day 28, all five had minimal or mild chronic inflammation and no evidence of fibrosis was found.

DISCUSSION

Results of the *in vitro* assays indicate that fibrous Si_3N_4 is of approximately equal toxicity to PAM as α-quartz, a known fibrogenic material. Both quartz and fibrous Si_3N_4-exposed cells displayed cytotoxicity, evidenced by decreased viability, marked reduction in phagocytic uptake, and alterations in cellular enzymes. Quartz and fibrous Si_3N_4 toxicity differed principally in the specific results of the biochemical assays. For quartz, the enzyme alterations (increased AP and LDH) are compatible with a mechanism of action involving phagolysosomal lysis. Liberation of lysosomal enzymes within the cytosol might contribute to cell death, resulting in elevations of both LDH and AP. Decreased protein measurements for the quartz-exposed cells are consistent with decreased protein synthesis, as might be expected of toxic cells. Cells incubated with the fibrous Si_3N_4 released increased levels of LDH but not AP, and showed no effect on total protein. These changes suggest fibrous Si_3N_4 has little effect on the phagolysosome, and acts through some other pathway to damage the cell membrane.

The overall effects of milled Si_3N_4 on PAM *in vitro* were similar to these of glass beads, considered to be an inert or "nuisance" dust.

The observed results of the *in vivo* evaluations generally correlated well with those of the *in vitro* assays. Alpha-quartz, cytotoxic *in vitro*, produced evidence of pulmonary edema, acute and chronic inflammation, granuloma formation, fibrosis, decreased macrophage viability and increased hydroproline content *in vivo*. Glass beads and milled Si_3N_4, which were of low toxicity *in vitro*, produced corresponding minimal pulmonary responses in rats.

The *in vivo* response to fibrous Si_3N_4 was minimal, however, despite significant cytotoxicity in the *in vitro* assays. This is due to physical difficulties in delivering the dosage of the fibrous Si_3N_4 to the rat lung. During the intratracheal instillation procedures, the fibrous Si_3N_4 particles tended to clump and fall out of solution, necessitating a larger diameter needle for the administration. All of the animals receiving fibrous Si_3N_4 developed tracheal obstructions after dosing and required resuscitation. Some portion of the dosage was cleared during resuscitation and never reached the lung tissues. Similar problems with immediate clearance have been noted by other investigators using the intratracheal technique (Finch *et al.*, 1987). Even though the effective dose of fibrous Si_3N_4 was significantly less than the other test articles, at Day 3 the lung lesions associated with the fibrous material were equivalent to α-quartz and greater than the equiaxial Si_3N_4.

The results of these experiments add further evidence of the importance of physical form as a determinant of biological activity. Further, *in vitro* assays can provide useful information on the relative toxicities of different substances, and can be used to provide additional insight into potential health-related issues and pathogenic mechanisms.

ACKNOWLEDGEMENT

This study was sponsored by the GTE Precision Materials Group.

REFERENCES

Chamberlain, M., Davies, R., Brown, R. C., and Griffiths, D. M. (1982). *In vitro* tests for the pathogenicity of mineral dusts in inhaled particles. (W. H. Walton, Ed.). Pergamon Press, Oxford.

Finch, G. L., Fisher, G. L., and Hayes, T. L. (1987). The pulmonary effects and clearance of intratracheally instilled Ni_3S_2 and TiO_2 in mice. *Environ. Res.* **42**, 83-93.

Fisher, G. L., McNeill, K. L., and Democko, C. J. (1983). Application of bovine macrophage bioassays in the analysis of toxic agents in complex environmental mixtures. *In* "Short-Term Bioassays in the Analysis of Complex Environmental Mixtures III," (M. D. Waters, S. S. Sandhi, J. Lewtas, L. Claxton, N. Chernoff and S. Nesnow, Eds.). pp. 257-268. Plenum Press, New York.

Fisher, G. L., McNeill, K. L., Whaley, C. B., and Fong, J. (1978). Attachment and phagocytosis studies with murine pulmonary alveolar macrophages. *J. Reticuloendothel. Soc.* **24**, 243-252.

Fisher, G. L., and Placke, M. E. (1987). *In vitro* models of lung toxicity. *Toxicology* **47**, 71-93.

Fisher, G. L., McNeill, K. L., and Smith, J. T. (1988a). *In vitro* effects of fibrous and non-fibrous silicon nitride on bovine pulmonary macrophages. *Environ. Res.*, Submitted.

Fisher, G. L., McNeill, K. L., Singer, A. W., and Smith, J. T. (1988b). Pulmonary response to intratracheally instilled silicon nitride in rats. *Inhal. Toxicol.*, Submitted.

Lee, K. P. (1985). Lung response to particulates with emphasis on asbestos and other fibrous dusts. *CRC Crit. Rev. Toxicol.* **14**, 33-86.

Lowry, O. H., Rosebrough, N. J., Farr, A. L., and Randall R. J. (1951). Protein measurement with folin phenol reagent. *J. Biol. Chem.* **193**, 265-275.

Reiser, K. M., and Last, J. A. (1979). Silicosis and fibrogenesis: fact and artifact. *Toxicology* **13**, 51-72.

MEMBRANOLYTIC ACTIVITY AND MORPHOLOGICAL CHARACTERIZATION OF PALYGORSKITE AND SEPIOLITE

R.P. Nolan, Ph.D.
A.M. Langer, Ph.D.
G.B. Herson*
Environmental Sciences Laboratory
Applied Sciences Institute of
The City University of New York
At Brooklyn College
Brooklyn, NY 11210

INTRODUCTION

The hormitic clays, palygorskite (attapulgite) and sepiolite, are being commercially used for many different applications. In 1983, approximately 1.1 million tons of palygorskite was used in industrial countries of the western world, as compared to 400,000 tons of sepiolite in 1984 (see IARC, 1987, for production, consumption, and application data). Concern about the potential health hazard of these minerals has arisen due to their fibrous nature and extensive application as asbestos substitutes (Bignon *et al*, 1980).

The use of both *in vivo* and *in vitro* methods for the evaluation of these two mineral fibers have shown that the mineral name alone is insufficient for defining its carcinogenic and cytotoxic properties. For example, *in vivo* testing of several palygorskite specimens using different routes of administration have been negative, although palygorskite specimens from Torrenjon, Spain, have induced tumors in rats by both intrapleural and intraperitoneal inoculation (Wagner *et al*, 1987; Pott *et al*, 1987). The *in vivo* experimental evidence is insufficient to evaluate the carcinogenic activity of sepiolite. *In vitro* assays used to aid in health hazard evaluation found the cytotoxic properties of palygorskite and sepiolite vary depending on the origin of the

*Andre Meyer Department of Physics and Nuclear Medicine, Mount Sinai School of Medicine, 1 Gustave L. Levy Place, New York, NY, 10029.

NATO ASI Series, Vol. H30
Effects of Mineral Dusts on Cells
Edited by B.T. Mossman and R.O. Bégin

specimen. Although very little is known about the surface properties of palygorskite, it was found to have binding properties quantitatively comparable to chrysotile for three organic carcinogens (Harvey et al, 1984). Nine specimens of palygorskite and one sepiolite were selected for this study. The palygorskite specimens were selected from different geological locales in an effort to be representative of the natural diversity of the minerals' physico-chemical states. The single sepiolite specimen was included for comparison. The preliminary characterization of the specimens involved the determination of the size distribution and membranolytic activity.

MATERIALS AND METHODS

Mineral Specimens

The following specimens were obtained from the Department of Mineral Sciences, American Museum of Natural History, New York, NY, 10024:

Palygorskite, AMNH #24478, Dartmoor, Western Australia;

Palygorskite, AMNH #C57076, Zermatt, Switzerland;*

Palygorskite, AMNH #27270, Yalu River, Korea;*

Palygorskite, AMNH #C57079, Brazil;*

Palygorskite, AMNH #17887, Nizhni, Novgorod, USSR;

Palygorskite, Area #1, New Melones Lake, Calaveras, CA;*

Palygorskite A and B were obtained from industrial specimens obtained from the National Institute of Occupational Safety and Health, Cincinnati, OH;

Attapulgite, API #43, Attapulgite, GA;

Sepiolite, AMNH #30121, Vallesas, Madrid, Spain;

The UICC chrysotile A (Zimbabwe) and B (Canada) were obtained from the Pneumoconiosis Research Unit, Penarth, Wales, UK;

---

*These specimens were opened in a high-speed blade mill.

The Calidria RG-144, a naturally occurring, predominantly short fiber chrysotile from the New Idria, CA, was obtained in a commercially available form from Union Carbide.

Inhibitors

The inhibitors of the membranolytic activity of the different mineral specimens studied were obtained from the following sources:

Albumin from Bovine Serum, #A-4378, Sigma Chemical Company, St. Louis, MO;

2(Poly-vinylpyridine-N-oxide) (2-PVPNO), #1564, Polyscience, Inc., Warrington, PA;

Chondroitin Sulfate, R.J. Schnitzer, Mount Sinai School of Medicine, New York, NY.

Size Distribution Measurements

The size distribution of the palygorskite, sepiolite, and chrysotile specimens used experimentally were determined by transmission electron microscope (see Table 1). To prepare the specimens, each was dispersed in a mixture of amylacetate and nitrocellulose to form a thin film, transferred to a copper or nickel locator grid, and carbon-coated (see Langer et al, 1978, for details of method). The size distributions were determined from photographs taken at 2,000x direct magnification and photographically enlarged to 5,000x. Overlapping photographs were taken so that photomontages could be made to permit sizing of long fibers.

Human Hemolytic Model

The ability of each of the mineral specimens to alter the permeability of human erythrocytes was determined quantitatively. The HC50 is the concentration of particulate (given in mg/ml) to lyse 50% of the erythrocytes in a suspen-

Table 1: Size distribution of palygorskite, sepiolite, and chrysotile specimens determined by transmission electron microscope.

Specimen	Population	Fiber Length ≤1.0µm	Fiber Length 1.1–5.0µm	Fiber Length 5.1–10.0µm	Fiber Length >10.0µm	Aggregates Undetermined Size	Fiber Diameters <0.15µm N	Fiber Diameters <0.15µm %
Palygorskite Australia	Total* N=2554	28.2%	2.9%	0.0%	0.0%	68.8%	-	-
	Fiber N=797	90.2%	9.3%	0.3%	0.3%	-	29	3.6
Palygorskite Switzerland	Total* N=4306	64.7%	19.3%	1.7%	0.5%	13.8%	-	-
	Fiber N=3710	75.1%	22.4%	2.0%	0.6%	-	49	1.3
Palygorskite Korea	Total* N=2409	39.4%	3.0%	0.1%	0.0%	57.5%	-	-
	Fiber N=1023	92.7%	7.1%	0.0%	0.0%	-	11	1.1
Palygorskite Brazil	Total* N=3518	34.3%	12.6%	0.8%	0.2%	52.0%	-	-
	Fiber N=1687	71.5%	26.3%	1.7%	0.5%	-	92	5.5

*Total is defined as the size of the greatest dimension of all particles (both fibrous and non-fibrous).

Table 1: Continued

Specimen	Population	Fiber Length ≤1.0μm	Fiber Length 1.1–5.0μm	Fiber Length 5.1–10.0μm	Fiber Length >10.0μm	Aggregates Undetermined Size	Fiber Diameters <0.15μm N	Fiber Diameters <0.15μm %
Palygorskite USSR	Total* N=2406	60.7%	16.6%	0.6%	0.2%	22.1%	-	-
	Fiber N=1874	78.0%	21.3%	0.7%	0.2%	-	23	1.2
Palygorskite California	Total* N=2360	50.2%	31.7%	2.2%	0.5%	15.5%	-	-
	Fiber N=1995	59.4%	37.5%	2.6%	0.6%	-	49	2.5
Attapulgite Georgia	Total* N=5070	44.9%	4.3%	0.0%	0.1%	50.7%	-	-
	Fiber N=2500	91.1%	8.7%	0.1%	0.1%	-	12	0.5
Palygorsk. A NIOSH	Total* N=1960	56.0%	11.1%	-	-	32.9%	-	-
	Fiber N=1315	83.4%	16.6%	-	-	-	44	3.4
Palygorsk. B NIOSH	Total* N=3211	64.7%	13.1%	-	-	22.1%	-	-
	Fiber N=2500	83.1%	16.8%	-	-	-	13	0.5

Table 1: Continued

Specimen	Population	Fiber Length ≤1.0μm	Fiber Length 1.1–5.0μm	Fiber Length 5.1–10.0μm	Fiber Length >10.0μm	Aggregates Undetermined Size	Fiber Diameters <0.15μm N	Fiber Diameters <0.15μm %
Sepiolite Spain	Total* N=689	42.7%	38.0%	0.7%	0.1%	17.8%	-	-
	Fiber N=569	52.4%	46.6%	0.9%	0.2%	-	12	2.1
Chrysotile California	Total* N=3259	69.6%	18.5%	1.6%	0.5%	9.8%	-	-
	Fiber N=2939	77.2%	20.5%	1.8%	0.5%	-	27	0.9
Chrysotile Canada	Total* N=3386	73.1%	11.8%	0.5%	0.3%	14.3%	-	-
	Fiber N=2903	83.9%	13.5%	0.6%	0.4%	-	49	1.7
Chrysotile Zimbabwe	Total* N=3599	86.9%	10.4%	0.4%	0.2%	2.1%	-	-
	Fiber N=3523	88.8%	10.6%	0.4%	0.2%	-	17	0.5

sion containing $1.8x10^8$ cells/ml (see Nolan et al, 1981, for details).

EXPERIMENTAL

The membranolytic activity of nine palygorskite specimens and one sepiolite specimen was determined. For comparison, both of the UICC chrysotile specimens were included, as well as a naturally occurring short chrysotile from California. The stock suspension of each mineral was sonicated to uniform dispersion prior to determination of the HC50. The necessity for this is illustrated by the experiment shown in Table 2. Attapulgite without sonication was virtually inactive, while a few minutes of sonication increased the membranolytic activity significantly. The effect of sonication is to break open the aggregates, thereby exposing the cells to a much greater surface area per unit mass (Figure 1). The HC50 of each mineral specimen was determined

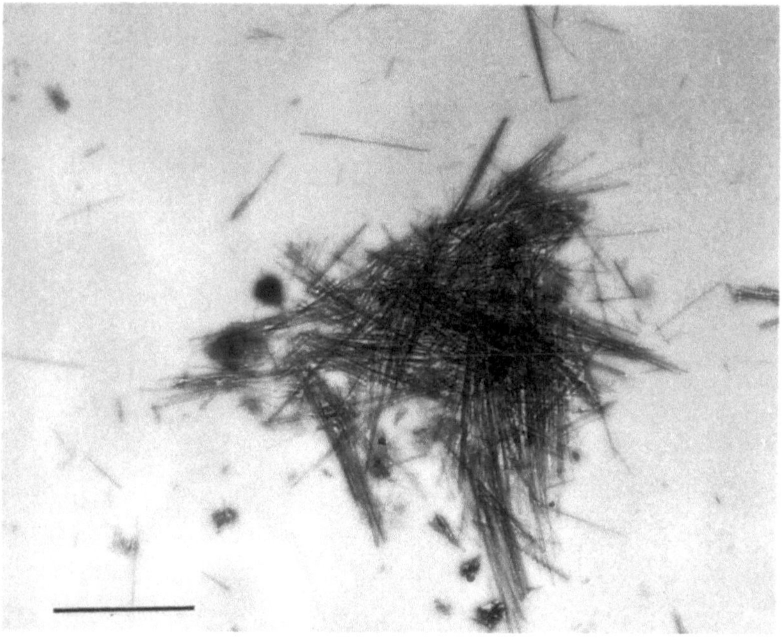

Figure 1. A transmission electron micrograph showing the naturally occurring fiber aggregation in a palygorskite specimen from Attapulgus, GA. Bar represents 1μm.

Table 2: Variation in the membranolytic activity of attapulgite Georgia with time of sonication.

Attapulgite Georgia µg/ml	Without Sonication %H	Time of Sonication* 1 Min %H	2 Min %H	4 Min %H	8 Min %H
-	0.7±0.0	0.7±0.0	0.7±0.0	0.7±0.0	0.7±0.0
25	0.7±0.1	2.6±0.8	12.9±0.2	8.2±0.5	22.7±0.5
50	1.8±1.4	4.6±1.1	30.4±0.1	21.5±1.2	41.3±0.9
75	1.1±0.2	6.6±1.2	44.4±1.8	38.1±0.2	50.8±0.2
100	1.0±0.2	15.5±1.3	55.8±1.3	55.3±3.9	59.4±1.4

*Sonication power field constant at 50 Watts.

using at least three different erythrocyte donors. The HC50 was determined for each donor, averaged, and the results are shown with their standard deviation in Table 3. With the exception of the Korean specimen, all of the palygorskites and the sepiolite were found to be active.

The membranolytic activity was further characterized to determine if any of the commonly known antagonist-to-mineral lysis would inhibit the effect. Of the three antagonists studied, chondroitin sulfate, bovine serum albumin, and 2-PVPNO, only the hydrogen-bonding polymer, 2-PVPNO, had an inhibitory effect. Further study showed all the active palygorskites and sepiolite to be inhibited by 2-PVPNO (data not shown).

DISCUSSION AND CONCLUSIONS

The membranolytic activity of the palygorskite specimens ranged from relatively inert to active, comparable with specimens of the most hemolytic chrysotile asbestos mineral (see Table 3). Among the active palygorskite specimens, the range of the HC50s were from 51 ± 10 to $746 \pm 159\mu g/ml$ while the chrysotile specimens ranged from 41 ± 3 to $82 \pm 26\mu g/ml$. A factor of almost 14 separates the lowest from the highest HC50 value for the palygorskite specimens from different geological locales.

Assuming the specimens to be pure palygorskite, the variability in the HC50s is related to either differences in 1) the surface area per unit mass, or 2) the chemical nature of the mineral's surface which imparts the membranolytic activity per unit mass. Inhibition of the membranolytic activity of palygorskite with the hydrogen bonding polymer, 2-PVPNO, implies a quartz-like surface and the possibility that quartz and palygorskite lyse erythrocytes by the same mechanism. The single specimen of Spanish sepiolite was only slightly less active than the chrysotile specimens.

Experimental pathologists have tried to correlate certain palygorskite morphologies with carcinogenic potential (Wagner *et al*, 1987; Pott *et al*, 1987). For all the palygor-

Table 3: The HC50s of the various palygorskite, sepiolite, and chrysotile specimens were determined. Three different erythrocyte donors were used for the HC50 determinations, which were then averaged and are shown with their standard deviations.

Specimen	HC50 µg/ml
Palygorskite, Australia	746±159
Palygorskite, Switzerland	369±75
Palygorskite, Korea	Inactive
Palygorskite, Brazil	400±101
Palygorskite, USSR	211±168
Palygorskite, California	51±10
Attapulgite, Georgia	76±35
Palygorskite A, NIOSH	83±26
Palygorskite B, NIOSH	109±12
Sepiolite, Spain	150±17
Chrysotile, California	41±3
Chrysotile, UICC A, Zimbabwe	82±19
Chrysotile, UICC B, Canada	59±3

skite specimens examined and averaged, 80.5% of the fibers were <1μm and 99.0% of the fibers were <5μm. The ≥5μm fibers represent a very small fraction of the fibers present. For example, of the 17,401 palygorskite fibers sized, only 51 were ≥10μm. Very few of the total fibers and the total surface area are represented by the ≥10μm. The palygorskite specimen from Korea, found to be membranolytically inactive, contained the greatest number of short fibers and had an amphibole-like morphology (Figure 2).

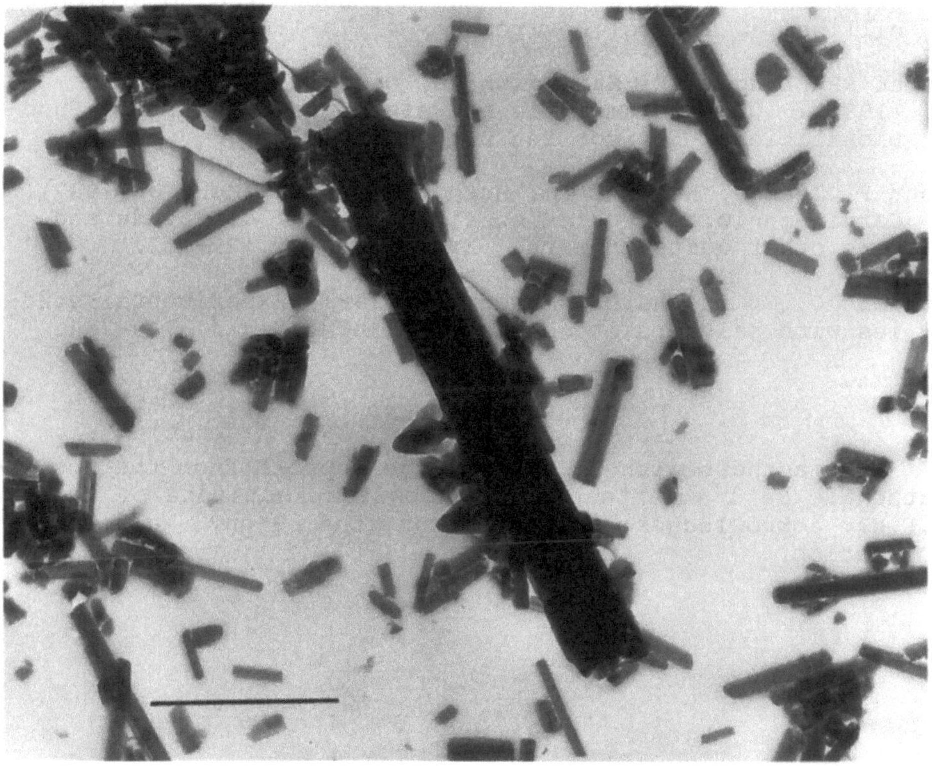

Figure 2. A transmission electron micrograph showing the distinctly different morphology of the palygorskite specimen from Yalu River, Korea. Bar represents 0.5μm.

REFERENCES

Bignon J, Sebastien P, Gaudichet A, Jaurand MC (1980) Biological Effects of Attapulgite. In: Wagner JC (ed) Biological Effects of Mineral Fiber, Vol. 1. IARC Scientific Publication No. 30, Lyon, France, p. 163.

Chamberlain M, Davis R, Brown RC, Griffiths DM (1982) In vitro tests for the pathogenicity of mineral dusts. Ann Occup Hyg 26:583-592.

Harvey G, Page M, Dumas L (1984) Binding of environmental carcinogens to asbestos and mineral fibers. Br J Ind Med 41:396-400.

IARC (1987) Silica and silicates. Monographs on the evaluation of carcinogenic risk of chemicals to health. IARC Scientific Publication No. 40, Lyon, France, pp. 159-173, pp. 175-183.

Langer AM, Wolff MS, Rohl AN, Selikoff IJ (1978) Variation of properties of chrysotile asbestos, subjected to prolonged milling. J Toxicol Environ Sci 4:173-188.

Nolan RP, Langer AM, Harington JS, Oster G, Selikoff IJ (1981). Quartz hemolysis as related to its surface functionalities. Environ Res 26:503-520.

Pott F, Ziem U, Reiffer FJ, Huth F, Ernst H, Mohr U (1987). Carcinogenicity studies on fibers, metal compounds and some other dusts in rats. Exper Path 32:129-152.

Wagner JC, Griffiths DM, Munday DE (1987) Experimental studies with palygorskite dusts. Br J Ind Med 44:749-763.

ACKNOWLEDGEMENTS

The authors wish to acknowledge support from the Societé Nationale de l'Amiante and the Asbestos Institute of Canada. R.P.N. acknowledges support from the Stony Wold-Herbert Foundation.

MINERAL SURFACE-SPECIFIC DIFFERENCES IN THE ADSORPTION AND ENZYMATIC REMOVAL OF SURFACTANT AND THEIR CORRELATION WITH CYTOTOXICITY

W.E.Wallace, M.J.Keane, P.S.Mike, C.A.Hill, V.Vallyathan
National Institute for Occupational Safety and Health,
Division of Respiratory Disease Studies; and
West Virginia University, Morgantown, West Virginia, 26505, USA

INTRODUCTION

Respirable quartz dust and a kaolin clay dust have been found to be of comparable cytotoxic potential *in vitro* on a specific surface area basis (Wallace, et al,1985), despite the distinctly different potentials of quartz and clay for causing pneumoconiosis or pulmonary fibrosis (Sheers, 1984). A respired particle depositing in a respiratory bronchiole or alveolus will contact a pulmonary surfactant hypophase, of which a primary component is diacyl glycerophosphorylcholine, or diacyl lecithin (DPL) (Clements, et al, 1970). Several studies have found that adsorption of such surfactant suppresses *in vitro* cytotoxic effects of quartz dust (Marks, 1957). Subsequent phagocytosis and digestion by a macrophage may remove the coating, restoring cytotoxic potential and initiating disease processes (Iler, 1979). Adsorption of dipalmitoyl lecithin (DPL) from emulsion in physiologic saline by respirable sized quartz dust and kaolin dusts, and consequent diminution of the dusts' hemolytic and macrophage cytotoxic potential have been reported (Wallace, et al, 1985). This report presents additional data on the cell-free *in vitro* phospholipase A2 enzymatic digestion of dipalmitoyl lecithin adsorbed on a quartz and a kaolin respirable sized dust (Wallace, et al, 1988) to investigate mineral specific differences in the rates of digestion and of consequent restoration of dust membranolytic potential.

This work was sponsored in part by the U.S. Department of Interior's Mineral Institute Program administered by the U.S. Bureau of Mines through the Generic Mineral Technology Center for the Study of Respirable Dusts, on Grant #G1135142 to West Virginia University.

NATO ASI Series, Vol. H30
Effects of Mineral Dusts on Cells
Edited by B.T. Mossman and R.O. Bégin

MATERIALS AND METHODS

Respirable crystalline alpha quartz used was eighty percent smaller than 5 micorometer aerodynamic diameter with a nitrogen adsorption specific surface area of 3.97 m^2/g. Respirable kaolin dust was ninety percent smaller than 5 micrometer diameter and had a specific surface area of 13.25 m^2/g.

L-alpha-dipalmitoyl glycerophosphorylcholine (lecithin) (DPL) obtained commercially was ultrasonically emulsified into 0.165 M NaCl. Phospholipase A_2 (PLA_2) enzyme obtained commercially from porcine pancreas was dissolved in 0.165 M saline at a pH between 6.5 and 7.0, with 2.0 mM calcium chloride. Quartz and kaolin dusts were incubated with DPL in physiologic saline emulsion for 1 hr at 37°C at a ratio of 100 mg DPL/g quartz and 200 mg DPL/g kaolin. After incubation, the dusts were washed twice in phosphate buffered saline (PBS), and the DPL-coated dusts finally suspended at a concentration of 2 mg dust/ml PBS. Parallel sets of dusts were used for hemolysis and for phosphate assay of residual adsorbed lecithin and lysolecithin. Saline dilution of lecithin stock emulsion was used for tests which measured the suppression of dust hemolytic potential as a function of adsorbed DPL.

Incubations of DPL-coated dusts with enzymes, carried out for periods of 2, 20, 44, and 72 hours, were started sequentially so that all assays for an experiment were performed at the same time. Following incubation with enzyme, samples were resuspended twice in PBS containing 2.0 mM ethylene diamine triacetic acid (EDTA) to stop the enzymatic digestion. Samples for assay of hemolytic potential were resuspended in PBS.

Hemolytic potential of native dusts, dust plus the maximum value of PLA_2 applied, dusts incubated with DPL/saline emulsion, and DPL-coated dust subsequently incubated with PLA_2, were measured by the method of Harington, et al. (1971). Sheep erythrocytes were washed six times in PBS and prepared as a 4% by volume suspension in PBS. These suspensions were mixed with equal volumes of dust in PBS suspension to result in 2% by volume erythrocyte suspensions with dust concentraion of 1 mg/ml. The mixtures were incubated at 37°C for 1 hour. After incubation the optical density of the supernatant at 540 nm wavelength was determined to measure hemoglobin release. Samples for phosphate deterinations were vacuum-dried at room temperature, and then eluted with chloroform/methanol (2/1 by volume). Amounts of phospholipid adsorbed to dust samples were quantitated as lecithin or lysolecithin (L-alpha-lecithin, beta palmitoyl). Aliquots were separated by thin layer chromotography, and the bands for lecithin and lysolecithin quantitated for phosphate by the method of Bartlett (1959).

RESULTS

Incubation of these quartz and kaolin dusts with increasing amounts of DPL results in neutralization of hemolytic potential between 15 and 20 mg DPL adsorbed per gram quartz, and between 75 to 85 mg DPL/g for kaolin (Keane et al, 1988). This quartz adsorbs about 60 mg DPL/g dust and this kaolin adsorbs about 150 mg/g dust at high lecithin dispersion concentrations. But saline washing removes all but a constant specific adsorption amount retained for each dust. For a series of 24 such measurements the residual adsorbed value for this quartz was 16.5 mg DPL/g quartz, and a series of 12 tests gave the value of 83.0 mg DPL/g kaolin. Using the nitrogen adsorption specific surface areas for these dusts, these lecithin retention values are 5.7 micromoles DPL/square meter of quartz surface, and 8.5 micromoles DPL/m^2 of kaolin surface.

The digestive removal of DPL from quartz and kaolin is shown in Figure 1. for applied specific activities of 0.82 units PLA2 per micromole DPL on quartz, and 0.96 units PLA2 per micromole DPL on kaolin. Data were taken in three separate experiments with triplicate points for each; standard deviations shown are for the nine values at each time point. Parallel measurements show that most of the produced lysolecithin remains adsorbed on both dusts at the 2 hour time point, with a significant amount still retained on the quartz at 20 hours. It is gone from both dusts at 44 hours. Figure 2 shows the resultant hemolytic potential of the treated dusts.

Digestive removal of DPL when the specific activity applied to quartz is decreased to 0.082 units PLA2 per micromole adsorbed DPL, and when the specific activity applied to kaolin is incresed to 2.9 units PLA2 per micromole kaolin-adsorbed DPL is shown in Figure 3. Parallel measurements show the hemolytic potentials for the two dusts track each other from approximately 30% native dust level at 2 hrs digestion to about 50% native dust level at 72 hrs digestion.

In our thin layer chromatographic separation of DPL and lysolecithin we note a band running at the solvent front which is visible after heating but not after staining for phosphate. Two sets of experiments using C-14 radiolabel on the palmitic acid moiety of DPL have found about 50% of the digestion-produced fatty acid stays associated with quartz at the 2 hour time point, and about 70% with the kaolin.

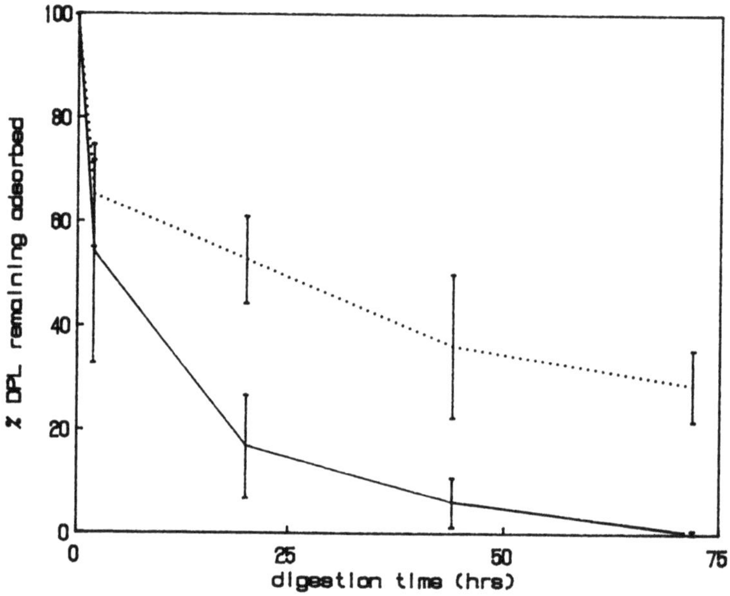

Figure 1. Digestive removal of dipalmitoyl lecithin from quartz (solid line) and from kaolin (dashed line) for applied phospholipase A2 specific activities of 0.82 and 0.96 units per micromole adsorbed lecithin on quartz and kaolin, respectively.

Limited infrared spectroscopic examinations were made of kaolin incubated with critical amounts of DPL in saline, then dried under mild conditions, and analyzed in potassium bromide pellets by grating IR spectroscopy. Contrasting the spectra of kaolin, of DPL, and of DPL adsorbed to kaolin indicates suppression of the broad DPL band in the region of the 3400 cm^{-1} for DPL adsorbed on kaolin. This band is assigned to the hydrated acidic phosphate group of DPL. Subsequent FTIR examination has indicated this band suppression occurs for DPL on kaolin but not significantly for DPL on quartz, while a weak band attributed to the trimethylamine of DPL is diminished for both dusts (Keane, et al, 1988).

DISCUSSION

The quartz and kaolin used in this work are comparably hemolytic on a surface area basis. The amounts of DPL needed to fully suppress the hemolytic potentials of the dusts is about equal to the amounts of DPL retained on the dusts after mild saline rinsing. The ratio of these critical amounts of DPL for kaolin to quartz is about equal to 1.5 times the ratio of their nitrogen

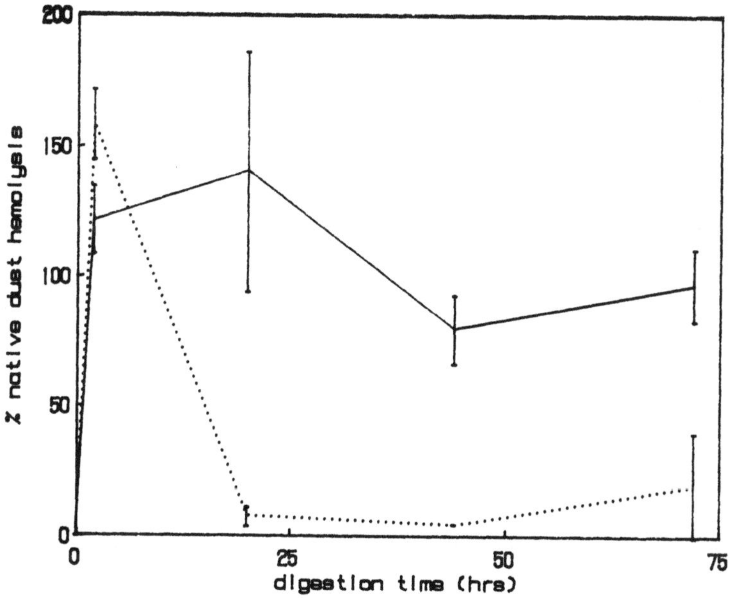

Figure 2. Hemolytic potential of lecithin coated quartz (solid line) and kaolin (dashed line), expressed as percent of native dust potential, following the digestion by PLA2 shown in Figure 1.

adsorption specific surface areas. The critical DPL values are about half the amounts of DPL adsorbed before rinsing. They are a factor of two or more greater than estimates of amounts needed for monolayer coverage using the nitrogen adsorption surface areas and information on molecular size or on water-air interfacial surface spreading of DPL. This is consistent with a picture of the stable DPL coating being a monolayer coverage of the dust surface by DPL molecules oriented with their zwitterionic head groups toward the dust surface, and backed by a second layer, with palmitate tails of the two layers in association and with ionic phosphorylcholine heads of the second layer pointed outward into the surrounding aqueous media. Such an arrangement would permit ionic interaction of acidic or basic dust surface sites with the zwitterionic DPL phosphorylcholine head group, while minimizing contact of hydrophobic acyl groups with the surrounding aqueous medium.

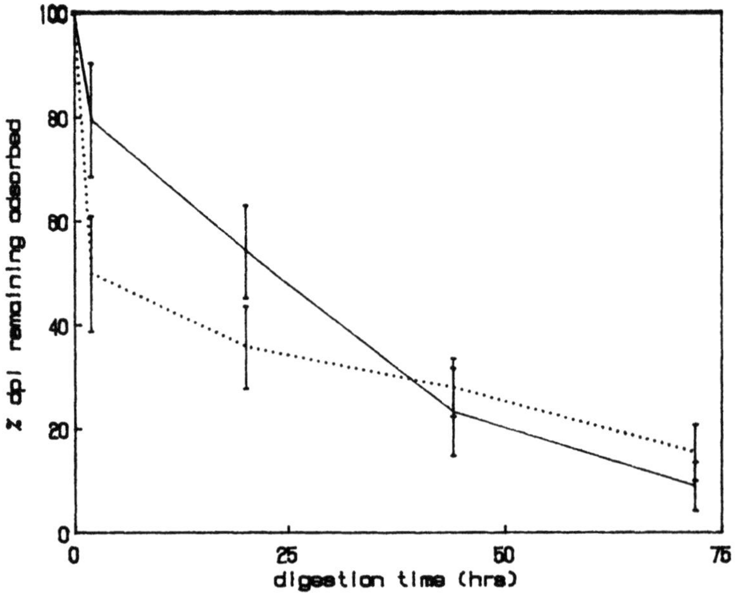

Figure 3. Digestive removal of dipalmitoyl lecithin from quartz (solid line) and from kaolin (dashed line) for applied phospholipase A2 specific activities of 0.082 and 2.9 units per micromole adsorbed lecithin on quartz and kaolin, respectively.

Application of phospholipase A2 to DPL coated dusts results in digestion and eventual removal of DPL and its phosphate-bearing products of digestion. The produced and initially adherent lysolecithin is itself a membranolytic agent. This correlates with exaggerated levels of hemolytic potential seen for all samples at the two hour time point. As the DPL is digested from the dust surface the lysolecithin may be lost by dissolution into the surrounding water, or it may be subject to further enzymatic hydrolytic attack. Non-phosphate fatty acids are also present as measured at the two hour time point. Thus, the hemolytic potential at any time is due to a mixture of effects: direct potential of mineral surface partially bared of lecithin, modified by further suppression of that surface potential by adsorbed palmitic acid product, and modified by additional lytic activity of adsorbed lysolecithin product.

For the applied specific PLA2 activities of about 1 unit per micromole of adsorbed DPL, the quartz is fully stripped in three days, while kaolin retains about 30% to 40% of the DPL. This specific enzyme activity level is an order of magnitude estimate of the total lysosomal phospholipase activity which a single cell contains (Hostetler, et al, 1982) in ratio to the critical DPL coating for a 1 micrometer diameter quartz particle. After initial rapid decrease in adsorbed DPL in the first two hours of digestion, there is a slower rate of digestion over the three day period. This longer time behavior may represent digestive removal of the mineral- surface bound layer of DPL, and therefore be most significant for the restoration of dust surface membranolytic potential. This is indicated by Figures 1. and 2., in which quartz hemolytic potential is fully restored with no contribution from lysolecithin, while kaolin is still significantly suppressed at two and three days digestion. Figure 3. shows that a 30- fold increase in applied enzymatic activity is needed to restore the surface potential of kaolin at the same rate as quartz. This suggests a greater hindrance to enzymatic hydrolysis of DPL bound to kaolin. This is consistent with the spectroscopic indication of association of the DPL-phosphate moiety with kaolin. PLA2 induced hydrolysis of DPL occurs at the glycerol carbonyl ester linkage of the palmitic acid residue adjacent to the phosphate of DPL. This phosphate is acidic, and therefore more likely to associate with basic aluminol groups on the kaolin surface rather than acidic silanol groups on the kaolin or quartz surface. The positively charged trimethylamine head group of DPL may associate with the silanol groups of either dust. A model which suggests itself is that outer adsorbed bilayer DPL molecules are readily enzymatically digested from both dusts; while for the inner, dust surface associated DPL, a clay surface basic aluminol group interaction with the DPL acidic phosphate group results in greater hindrance for PLA2 digestion of kaolin- in contrast to quartz-adsorbed lecithin. This is a much simplified model of cellular phagolysosomal digestion which additionally involves phospholipase C, acidic pH media, and membrane bound enzymes. To determine if such mineral specific differences occur in phagolysosomal digestion within a pulmonary macrophage, and to determine if such dust-specific differences fall on either side of characteristic clearance times for respired particles from the acinar region of the lung, requires data from a more representative digestion system. We are attempting measurements of radiolabelled surfactant digestion from phagocytized dusts and measurement of associated cytotoxic response.

REFERENCES

Bartlett GR (1959) "Phosphorous Assay in Column Chromatography". J. Biolog. Chem. 234:466.

Clements JA, Nellenbogen J, Trahan HJ (1970) "Pulmonary Surfactant and Evolution of the Lungs". Science 169:603.

Harington J, Miller K, Macnab G (1971) "Hemolysis by Asbestos". Environ. Res. 4:95-117.

Hostetler KY, Yazaki PJ, van den Bosch H "Purification of Lysosomal Phosphplipase A. Evidence for Multiple Isoenzymes in Rat Liver". J. Biol. Chem. 257:13367-73.

Iler RK (1979) "The Chemistry of Silica". John Wiley & Sons. ISBM 0-471-02404-X.

Keane MJ, Wallace WE, Seehra M, et al (1988) "Respirable Particulate Surface Interactions with the Lecithin Component of Pulmonary Surfactant". In: Proceedings - VII International Pneumoconiosis Conference; International Labor Organization. In Press.

Marks J (1957) "The Neutralization of Silica Toxicity in vitro". Br. J. Ind. Med. 14:81.

Sheers G (1964) "Prevalence of Pneumoconiosis in Cornish Kaolin Workers". Br. J. Ind. Med. 21:218.

Wallace WE, Vallyathan V, Keane MJ, Robinson V (1985) "In vitro Biologic Toxicity of Native and Surface-Modified Silica and Kaolin". J. Toxicol. Environ. Health 16:415-424.

Wallace WE, Keane MJ, Hill CA, et al (1988) "The Effect of Lecithin Surfactant and Phospholipase Enzyme Treatment on Some Cytotoxic Properties of Respirable Quartz and Kaolin Dusts". In: Respirable Dusts in the Mineral Industries (Frantz RL, and Ramini RV, eds.) pp.154-166. American Congress of Governmental Industrial Hygienists Monograph. ISBN 0-936712-76-7.

BRONCHOALVEOLAR LAVAGE CELLS IN OCCUPATIONAL EXPOSURE TO MINERAL FIBERS

A. Forni, G. Rivolta and G. Chiappino
Research Center on the Biological Effects
of Inhaled Dusts
Institute of Occupational Health
University of Milan
Via S. Barnaba 8
20122 Milan
Italy

Introduction

In the last decade, the possibility of obtaining informations on the cellular, biochemical and mineralogical content of the deep lung by studying bronchoalveolar lavage (BAL) fluid has largely increased our knowledge on the pulmonary reactions to inhalation of dusts. The lung effects of asbestos exposure are among the most extensively studied, but several problems remain open. In our institution, workers occupationally exposed to mineral fibers represent approximately one third of over 300 subjects investigated by BAL since 1983.

In our experience, in agreement with the literature data, asbestos exposure may induce an alveolitis characterized by a moderate increase of total cells, an increase of polymorphs and/or of lymphocytes, presence of iron-laden macrophages (indicators of deep lung microhemorrhages) and of asbestos bodies, a marker of exposure (Bignon et al, 1978; Bégin et al, 1985; Forni et al, 1985, 1988; Gellert et al, 1985 a,b; Robinson et al, 1986).

The aim of this paper is to evaluate the prevalence of cytological indicators of alveolitis in a large, highly selected group of workers occupationally exposed to mineral fibers, in relation to exposure history and clinical status.

NATO ASI Series, Vol. H30
Effects of Mineral Dusts on Cells
Edited by B.T. Mossman and R.O. Bégin

Materials and Methods

Subjects. From a group of 133 workers occupationally exposed to mineral fibers studied by BAL for a total of 147 adequate samples, 93 workers (70 males, 23 females) engaged in jobs involving significant direct or indirect exposure to asbestos (29 asbestos-cement, 18 insulation, 15 asbestos textile, 4 miners, 4 friction products, 23 various jobs), and 12 males exposed only to man-made mineral fibers (MMMF) were selected. Criteria for exclusion were concomitant occupational exposure to iron dusts and fumes (which increases the total number of cells and of iron-laden macrophages in BAL) and, for MMMF workers, possible even minimal concomitant or past occupational exposure to asbestos. Twelve workers exposed to asbestos and one to MMMF were sampled twice 1 to 3 years apart, for a total of 105 and 13 BAL samples respectively.

Out of the 93 asbestos workers, 28 (25 males, 3 females) had clinical and radiological signs of asbestosis, with or without pleural plaques, while the remainders had either pleural plaques or thickenings with no X-ray signs of parenchymal involvement (39 cases), or were just exposed with no evidence of asbestos-related disease. The median age was 52 (range 27-65) years for the whole group, and did not differ for the asbestosis group (median 53; range 39-62). Fifty four subjects (14 with asbestosis) were non-smokers or had stopped smoking since at least 5 years. Duration of exposure to asbestos was <2 to 37 years (median 15 years) and did not differ for the asbestosis group. Fifty seven were no longer exposed since 2 to 37 years, and 38 were not exposed since more than 10 years. The 12 MMMF workers were 39-63 years old (median 53), 6 were non-smokers. Duration of exposure was 7-28 years (median 18); 6 were no longer exposed since 2 to 20 years. Four subjects presented at X-rays pleural plaques or thickenings.

Techniques. BAL was performed according to the method described elsewhere (Forni et al, 1985), by introducing 50 ml x 3 sterile physiological saline at 37° C into a segmentary or subsegmentary bronchus of the middle lobe or, less frequently, of the lingula, and by immediately recovering the fluid under negative pressure. The median recovery was 47% (range 27-78) for the 105 samples obtained from the asbestos workers and did not differ for the 31 samples of the asbestosis cases (median 44%; range 27-77). The recovery for the 12 MMMF workers was 25-78% (median 52%).

BAL fluid was immediately brought to the laboratory, filtered through sterile gauze, cooled in ice bath, and processed according to the methods described elsewhere (Forni et al, 1985). Total cells were counted in Bürker's chamber. Differential counts were performed on Papanicolaou- and Giemsa-stained cytocentrifuge preparations. Iron-laden macrophages were routinely counted on cytocentrifuge preparations stained by the Perls' method for iron. T-lymphocyte phenotyping, when appropriate, was performed on cytocentrifuge preparations by monoclonal antibodies and a peroxidase-antiperoxidase method (Forni et al, 1988). Asbestos bodies were counted on cytocentrifuge preparations and on Millipore filter preparations of 10% antiformin-digested BAL fluid, and the results were adjusted to 1 ml BAL fluid.

Indicators of alveolitis. On the basis of our overall experience on 390 BAL samples, including "controls", the following cut-off values have been established as indicators of alveolitis: total cells $>200 \times 10^3$/ml for non-smokers or $>700 \times 10^3$/ml for smokers; iron-laden macrophages $\geq 3\%$ (indicators of microtraumatic alveolitis if other causes, e.g., exposure to iron dusts, mitral valve disease, recent hemoptysis, are ruled out); lymphocytes $>15\%$ or $>30 \times 10^3$/ml; neutrophils + eosinophils $\geq 3\%$ (excluding from evaluation samples with possible contamination of material from large bronchi). The prevalence of indicators in the groups or subgroups has been calculated, and the results have been statistically evaluated by the chi square test.

Results and Discussion

The results obtained in the 105 samples of the 93 asbestos workers are summarized in Table 1. Table 1A shows that approximate-

Table 1. Prevalence of indicators of alveolitis in 105 BAL samples from 93 asbestos workers

	No. of cases/samples (and percentage) with				
	All parameters normal	Increased total cells	Iron-laden macrophages ≥ 3%	Increased lymphocytes	Neutrophils+ eosinophils ≥ 3% *
A					
93 cases (1st BAL)	10 (11%)	30 (32%)	67 (72%)	33 (35%)	35 (42%)
105 samples	11 (10%)	35 (33%)	75 (71%)	34 (32%)	37 (39%)
B					
Current exposure (42 samples)	7 (17%)	14 (33%)	27 (64%)	11 (26%)	12 (30%)
Exposure ceased (63 samples)	4 (6%)	21 (33%)	48 (76%)	23 (37%)	25 (45%)
C					
Asbestosis (31 samples)	2 (7%)	15 § (49%)	20 § (64%)	12 (39%)	22 **(73%)
No asbestosis (74 samples)	9 (12%)	20 § (27%)	55 § (74%)	22 (30%)	15 (23%)

Percentages approximated to the nearest unit

* Calculated in 93 cases and 96 samples respectively (excluding samples with suspect pollution with material from large bronchi) ** p <.001

§ One case "No asbestosis" at 1st BAL, "Asbestosis" at 2nd BAL

ly 10% of the case (1st BAL sampling) showed all cytological parameters normal, according to smoking habits, while the remainders presented one or more positive indicators of alveolitis. There were no differences when considering samples instead of cases, since there was often consistency between results of 1st and 2nd BAL samples obtained 1 to 3 years apart.

The prevalence of the single indicators of alveolitis did not differ for samples obtained from workers still exposed at time of BAL and those from workers no longer exposed since ≥2 years (Table 1B). It seems of interest to pinpoint the fact that the results did not change when considering the data of the 39 samples from the 38 cases whose exposure had ceased 10 or more years before BAL (data not shown).

Subdiving the group on the basis of clinical status (Asbestosis/ No asbestosis), the subgroup with X-ray evidence of parenchymal disease had a significantly higher prevalence of increased polimorphs in BAL (chi square = 20.21; $p <.001$), while the frequency of the other indicators of alveolitis did not differ significantly (Table 1C).

Taking into consideration the number of positive indicators per sample, there was absolutely no difference between the currently exposed and those with past exposure, while the prevalence of cases with 3 or 4 positive indicators was significantly higher in the asbestosis group than in the group with no X-ray evidence of asbestosis (chi square = 4.72; $p < .05$) (Table 2).

Table 2. Number of positive indicators of alveolitis in 105 BAL samples of 93 asbestos workers in relation to: (A) exposure status and (B) clinical status.

(A)	% samples with positive indicators	
	0-2	3-4
Current exposure 42 samples	76	24
Exposure ceased 63 samples	76	24

(B)	% samples with positive indicators	
	0-2	3-4
Asbestosis 31 samples	58	42
		$p<.05$
No asbestosis 74 samples	84	16

BAL T-lymphocytes have been phenotyped in 25 cases (22 with relative or absolute increase of lymphocytes). The results showed normal CD4/CD8 ratios in 9 cases, decreased in 10 cases, definitely increased in 6 cases, and seemed independent from smoking habits. Low and high values were present both in the group with asbestosis and in those with no parenchymal asbestos disease,

Table 3. CD4/CD8 lymphocyte ratio in BAL samples of 25 asbestos workers in relation to asbestos pathology and smoking habits

Cases	No.		No. of cases with CD4/CD8 ratio		
			0.1-1	1.5-2.1	3.7-4.2
Asbestosis	8		5	2	1
	5	NS	3	1	1
	3	S	2	1	
Benign pleural disease	10		4	2	4
	7	NS	3	2	2
	3	S	1		2
Exposed, no disease	7	NS	1	5	1

NS, non-smoker S, smoker

but normal values seemed to prevail in the group of exposed subjects with no evidence of asbestos-related disease (Table 3). In the literature there are some conflicting reports on decreased (Gellert et al, 1985 c) or increased (Costabel et al, 1987) helper/suppressor ratios in asbestos workers, but the problem seems to be more complicated. Different life habits, different types of exposure and different phase of disease might play a role. Therefore the significance of increased and imbalanced T-lymphocytes in deep lung for the pathogenesis of asbestosis is still to be clarified.

The prognostic significance of BAL cell changes (and particularly of increased polymorphs) in asbestos workers is still poorly understood. Among the 12 cases followed-up for 1-3 years, 5 out of 8 subjects with increased polymorphs at 1st BAL showed no progression and 3 showed progression of parenchymal disease at chest X-rays. One case progressing from benign pleural disease to asbestosis s/t 1/0 had increased total cells at 1st BAL and increased total cells + increased polymorphs at 2nd BAL two years later. Another case progressing in two years from benign pleural disease to asbestosis t/s 1/1 had increased iron-laden macrophages, lymphocytes and polymorphs at 1st BAL and only increased iron-laden macrophages at 2nd BAL.

Asbestos bodies, indicators of exposure, were present in 98 samples from 87 cases in concentrations from <1 to 15,900/ml BAL (median 50/ml for the overall cases, 180/ml for the asbestosis cases). These findings agree with the data by De Vuyst et al (1987) and Sébastien et al (1988).

As far as concerns MMMF, long term exposure does not seem to be completely devoid of cellular effects on deep lung. In fact, out of 12 workers occupationally exposed only to MMMF, 3 (2 non-smokers) had in their BAL's increased total cells, and 9 had definitely increased iron-laden macrophages (4-19%; median 10%), a sign of deep lung microhemorrhages. On the contrary, inflammatory cells were not increased. Short, relatively thin, often iron-protein-coated glass fibers, engulfed in macrophages, were present in 11 samples from 10 cases, 4 of whom were no longer exposed since two or more years.

Conclusions

In our series of asbestos workers, indicators of alveolitis in BAL seem to persist for years and decades from cessation of exposure. BAL polymorphs seem to represent the only indicator of alveolitis more frequently found in subjects with asbestosis, compared to asbestos-exposed with no radiological evidence of fibrosis. Further follow-up in a larger number of cases is warranted in order to try to understand the role of inflammatory and immune competent cells of deep lung in the development of pneumoconiosis, and the prognostic significance of indicators of alveolitis at the individual level.

Exposure to MMMF seems to induce deep lung microhemorrhages, but not to induce an increase of inflammatory cells.

Acknowledgment. The authors wish to acknowledge Mrs E Ortisi and Mr M Fornoni for skilled technical help.

References

Bégin R, Massé S., Rola-Pleszczynski M, Drapeau G (1985) Bronchoalveolar and lung tissue analyses in asbestos-exposed humans and sheep. In: Beck EG, Bignon J (eds) In vitro effects of mineral dusts. Third Int Workshop, NATO ASI Series, Series G: Ecological Sciences, vol. 3. Springer, Berlin Heidelberg New York, p 359

Bignon J, Atassi K, Jaurand MC, Yamine J, Kaplan H, Geslin P, Solle R, Bientz M (1978) Etude cytologique et biochimique du liquide du lavage bronchoalvéolaire dans la fibrose pulmonaire idiopathique et l'asbestose. Rev Fr Mal Respir 6: 353-358

Costabel U, Bross KJ, Huck E, Guzman J, Matthys H (1987) Lung and blood lymphocyte subsets in asbestosis and in mixed dust pneumoconiosis. Chest 91: 110-112

De Vuyst P, Dumortier P, Moulin E, Yourassowsky N, Yernault JC (1987) Diagnostic value of asbestos bodies in bronchoalveolar lavage fluid. Am Rev Respir Dis 136: 1219-1224

Forni A, Guerreri MC, Chiappino G (1985) Nuovi metodi di indagine nelle pneumopatie professionali: il lavaggio broncoalveolare. Med Lav 76: 11-16.

Forni A, Ortisi E, Rivolta G, Chiappino G (1988) Bronchoalveolar lavage T-lymphocyte subpopulations in occupational and non-occupational lung diseases. In: Goerttler K, Feichter G, Witte S (eds) New frontiers in cytology. Modern aspects of research and practice. Springer, Heidelberg New York Tokyo, p 440

Gellert AR, Langford JA, Uthayakumar S, Rudd RM (1985a) Bronchoalveolar lavage and clearance of 99m-Tc-DTPA in asbestos workers without evidence of asbestosis. Br J Dis Chest 79: 251-257

Gellert AR, Langford JA, Winter RJD, Uthayakumar S, Sinha G, Rudd RM (1985b) Asbestosis: assessment by bronchoalveolar lavage and measurement of pulmonary epithelial permeability. Thorax 40: 508-514

Gellert AR, Macey MG, Uthayakumar S, Newland AC, Rudd RM (1985c) Lymphocyte subpopulations in bronchoalveolar lavage fluid of asbestos workers. Am Rev Respir Dis 132: 824-828

Robinson BWS, Rose AH, James A, Whitaker D, Musk AW (1986) Alveolitis of pulmonary asbestosis. Bronchoalveolar lavage studies in crocidolite- and chrysotile-exposed individuals. Chest 90: 396-402

Sébastien P, Armstrong B, Monchaux G, Bignon J (1988) Asbestos bodies in bronchoalveolar lavage fluid and in lung parenchyma. Am Rev Respir Dis 137: 75-78

ALVEOLAR MACROPHAGES SECRETORY DYSFUNCTIONS IN COAL WORKERS'S PNEUMOCONIOSIS. COMPARISON BETWEEN SIMPLE PNEUMOCONIOSIS AND PROGRESSIVE MASSIVE FIBROSIS.

P. Lassalle, P. Gosset, C. Aerts, M. Benhamou*, F. Fortin, B. Wallaert, A.B. Tonnel, C. Voisin
Institut Pasteur de Lille, 1, rue Calmette, 59019 Lille FRANCE
* INSERM U 200, 92140 Clamart FRANCE.

INTRODUCTION

Coal worker's pneumoconiosis (CWP) comprises a variety of pulmonary radiological and pathological changes resulting from the inhalation of coal dust. Coal worker's pneumoconiosis is usually divided in two groups : simple pneumoconiosis (SP) and progressive massive fibrosis (PMF). This distinction is of importance in clinical practice since PMF is definitively disabling. However, why some workers should develop PMF when exposed to dust concentrations similar to those experimencied by workers who develop only SP is unknown. Current concepts emphazise the role of alveolar macrophages in the pathogenesis of chronic interstitial lung diseases (Voisin, 1983; Davis, 1986). In this way, the cells recovered by bronchoalveolar lavage were extensively used to evaluate the distribution of cell type and the secretory activities of alveolar macrophages that could be relevant to the pathogenesis of CWP, such as superoxide anion, neutrophil chemotactic factors, fibronectin (Wallaert, 1986). Here we have extended our study on macrophage secretory dysfunctions by the measurement of interleukin-1 and tumor necrosis factor alpha spontaneously released by alveolar macrophages.

PATIENTS AND METHODS

The study population consisted of pneumoconiotic coal workers either with simple pneumoconiosis (SP) or with progressive massive fibrosis (PMF). Our results were compared with those obtained from healthy controls. Bronchoalveolar lavage was performed under local anesthesia with lignocaine using a wedged fiberoptic bronchoscope and 250 ml of sterile saline solution was applied in five 50 ml aliquots with immediate gentle aspiration after each aliquot, as previously described (Voisin, 1982).

Production of superoxide anion by alveolar macrophages (AM) has been appreciated by chemiluminescence using lucigenin as light amplyfier (Williams, 1981; Aerts, 1986). The amounts of fibronectin contained in supernatants of cultured AM were measured by a radioimmunoassay kindly provided from A. OUAISSI (Ouassi, 1986). Neutrophil chemoattractant activity of AM supernatants was studied with a 48-well microchemotaxis chamber (NeuroProbe, Cabin John, MD, USA) as previously described (Gosset,1984).

NATO ASI Series, Vol. H30
Effects of Mineral Dusts on Cells
Edited by B. T. Mossman and R. O. Bégin

The spontaneous release of interleukin-1 (IL-1) and Tumor necrosis factor (TNF) contained in supernatants of 3-hours-cultured AM were also investigated. Interleukin-1 was detected by a proliferation assay using thymocytes from C3H-HeJ mice strain as previously described (Gosset). Tumor necrosis factor was detected by both radioimmunoassay (IRE- MEDGENIX, Fleurus, Belgium) and cytotoxicity assay using the L929 mouse fibroblasts (Gosset).

RESULTS

The results of cytological analysis of three hundred and seventeen B.A.L. revealed a significant increase of the total number of cells, when compared to healthy controls (Fig. 1A). The percentage of alveolar macrophages, of lymphocytes and of neutrophils in patients with SP and in patients with PMF were in normal range whereas the percentage of neutrophils was slightly increased in non smoking patients (Fig. 1B).

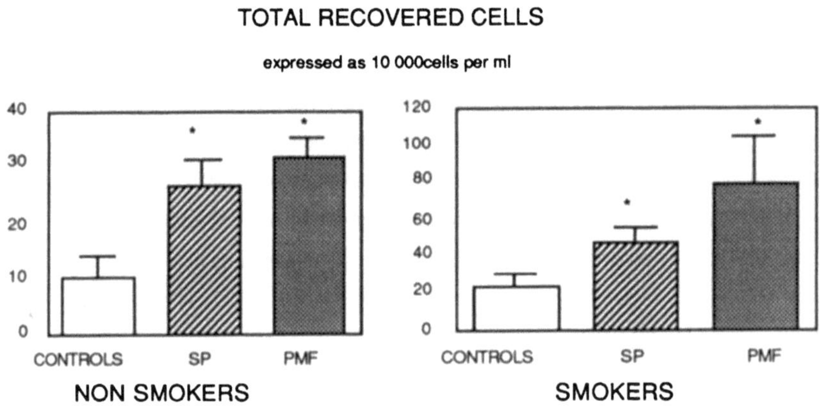

Fig.1A. Total recovered cell count in bronchoalveolar lavage from 317 patients with coal worker's pneumoconiosis (*:p<0,05).

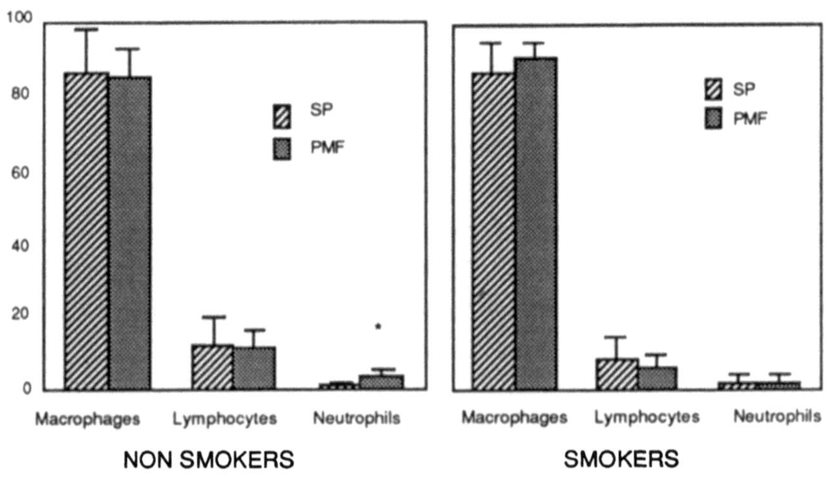

Fig.1B. Cytology of bronchoalveolar lavage in patients with CWP.

The cellular viability of these cells at the time of recovery did not differ between controls and pneumoconiotic coal workers demonstrating that inhalation of coal dust has no direct cytotoxic effects in humans. The alveolitis associated with chronic inhalation of coal dust comprised an increased number of cells, predominantly alveolar macrophages, suggesting the presence of a macrophage alveolitis in coal worker pneumoconiosis.

Evaluation of spontaneous release of superoxide anion by alveolar macrophages by means of chemiluminescence using lucigenin as light amplifier has indicated that macrophages from patients with PMF produced significant high levels of superoxide anion when compared to macrophages from patients with SP and controls (Fig. 2A). The same pattern of results was shown after stimulation of alveolar macrophages by PMA as well in non smokers than in smokers (Fig. 2B).

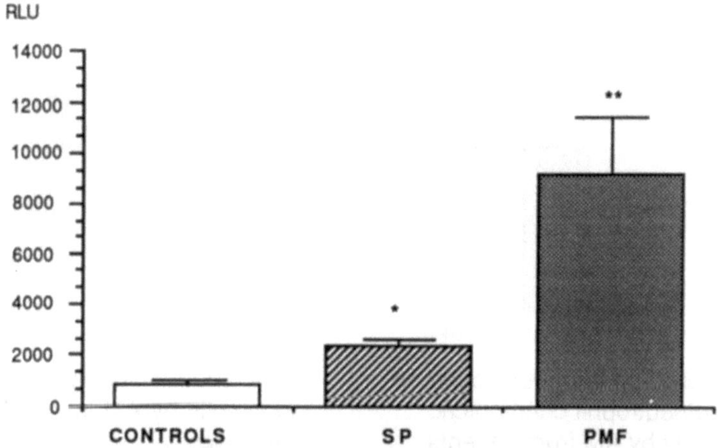

Fig.2A . Spontaneous production of superoxide anion by AM. Comparison between controls and patients with SP and PMF.

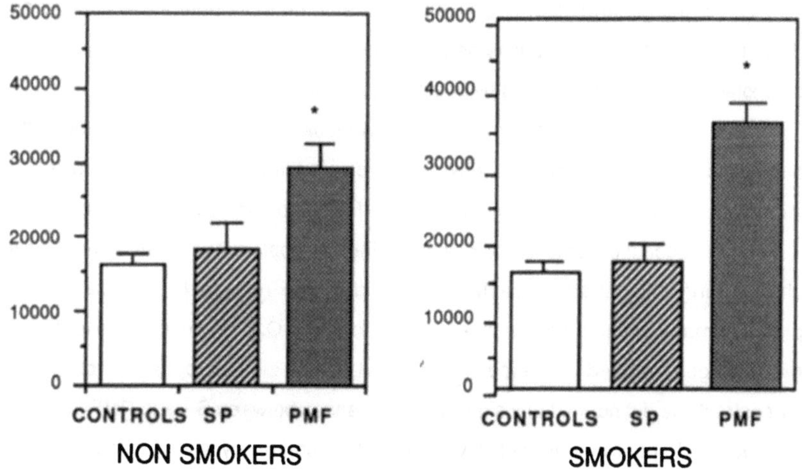

Fig.2B . PMA - induced production of superoxide anion by AM. Comparison between controls and patients with SP and PMF.

These results clearly demonstrate that alveolar macrophages from patients with PMF release increased amounts of free radicals. This was not due to a decrease in the antioxidant content of the cells, such as superoxide dismutase, which was not significantively different between the 3 groups (Voisin, 1985). This could be related to an increased metabolic activity of these cells indicating that these cells are activated.

In another hand, it appeared that alveolar macrophages from patients with PMF released more chemotactic activity than did macrophages from patients with SP (Fig. 3).

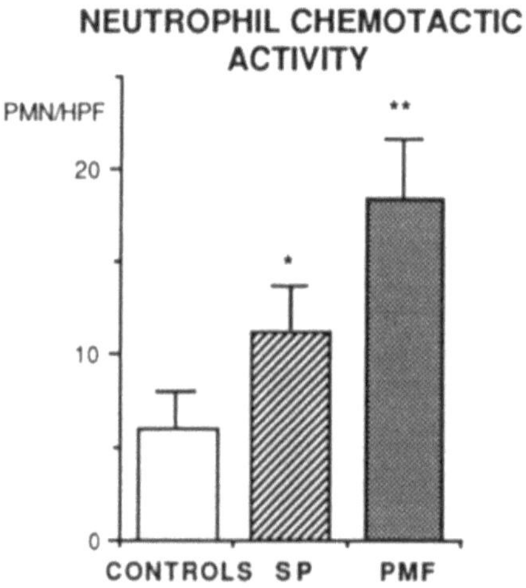

Fig 3 . Spontaneous secretion of a neutrophil chemotactic activity by AM from patients with PMF .

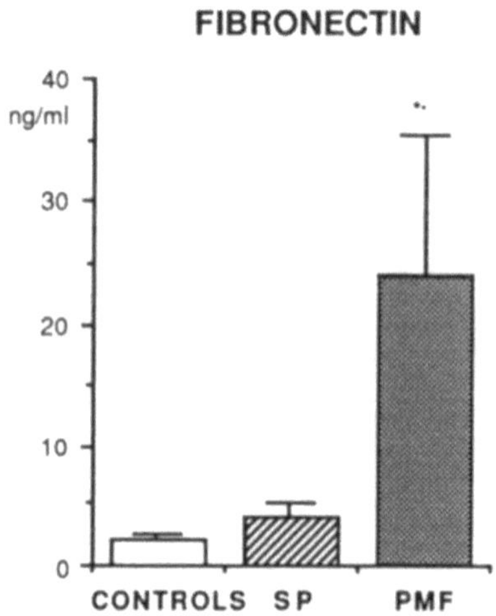

Fig.4. spontaneous secretion of fibronectin by AM from patients with PMF

We look at the generation of LTB4 which is one of the more potent chemotactic agent, but we did not find significant differences between the 3 groups. These results may suggest that other chemotactic agents are released by or activated by alveolar macrophages.

In another group of experiments, we look at the secretion by alveolar macrophages of mediators, such as fibronectin, which can promote the proliferation of fibroblast. We observed a dramatic increase in fibronectin release in PMF when compared to SP and controls (Fig. 4).

Because some pathological findings in coal worker's pneumoconiosis suggest a chronic lung inflammatory reaction, we studied the release by alveolar macrophages of proinflammatory mediators such as IL-1 and TNF. Our preliminary results indicated that macrophages released increased amounts of IL-1 than did macrophages from healthy controls. By contrast, we did not find any significant differences between SP and PMF (fig. 5A). In parallel, we observed spontaneous high levels of TNF secretion by alveolar macrophages than did macrophages from healthy controls. Here again, we did not find any significant differences between SP and PMF (fig. 5B).

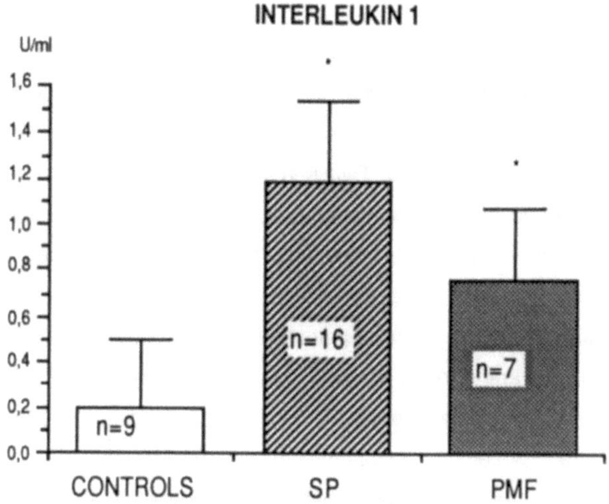

Fig. 5A. Spontaneous release of IL-1 by alveolar macrophages.
(*: p<0,05)

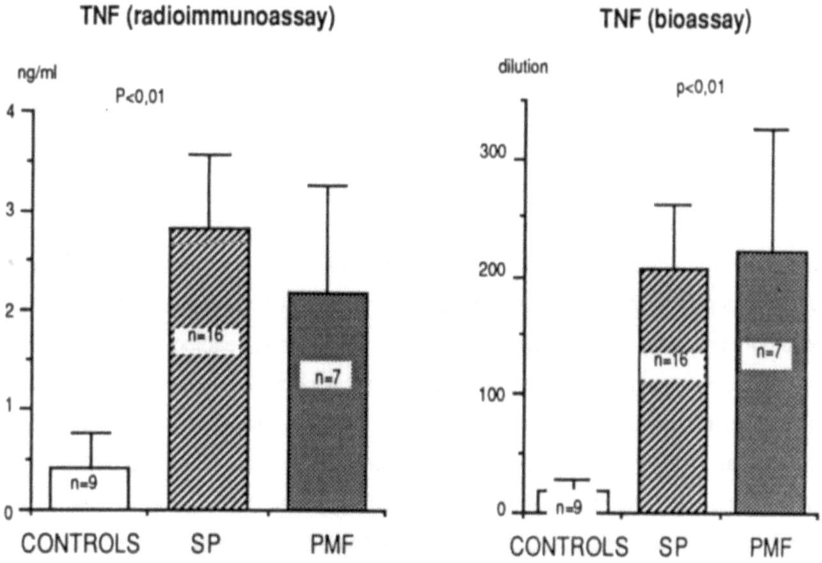

Fig. 5B. Spontaneous release of TNF by alveolar macrophages.

DISCUSSION

There is now evidence that alveolar macrophages activated by environmental agents can release a battery of immune, proinflammatory and profibrotic mediators within the alveolar structures. Here, this work has attempted to compare the secretory dysfunctions of alveolar macrophages between SP and PMF versus healthy controls as reference. These results clearly demonstrate that alveolar macrophages are activated in the two groups of patients. The secretory dysfunctions of alveolar macrophages from patients with SP appear to be limited to the release of monokines such as IL-1 and TNF which would form the support of the inflammatory reactions observed. This is in agreement with previous reports describing an IL-1 secretion by monocytes and macrophages in response to silica exposure (Schmidt, 1984).

Beside the release of IL-1 and TNF, the alveolar macrophages from patients with PMF show the spontaneous secretion of high levels of superoxide anion, fibronectin and neutrophil chemotactic activity. A similar pattern of secretory dysfunctions of human and animal alveolar macrophages are found by several investigators (Lugano, 1981; Lugano, 1984; Rom, 1987). So, these results clearly support the idea that alvelar macrophages from patients with PMF become more activated than those from patients with SP.

Trying to integrate these results in an hypothelical scheme congruent with our knowledge of the pathology of coal worker's pneumoconiosis, we would like to distinguish three steps : 1) the inhalation of coal dusts is associated with a macrophage alveolitis as the result of the influx of monocytes from the blood into the lung ; 2) the alveolar macrophages become activated, thus releasing proinflammatory mediators such as IL-1 and TNF ; 3) under favour of unknown stimuli, the activation degree of alveolar macrophages dramatically increases, leading to the secretion of various mediators involved in lung injury, inflammation, fibrosis and emphysema.

This view support the concept that alveolar macrophages may play key role in the development of pneumoconiosis. Here, we have artificially separated two activation states of alveolar macrophages which would account for the distinction between simple pneumoconiosis and progressive massive fibrosis. However, the machanisms leading the passage from SP to PMF remain to be establish. The other alveolar cell populations, which were hidden here, may be involved in part in pathogenesis of pneumoconiosis towards either direct activation by inhaled coal dusts or indirectly by several mediators, in particular those produced by alveolar macrophages. But, the role of these cell populations needs to be precised.Thus it is of both theoretical and clinical importance to determine whether the complex responses of coal worker's pneumoconiosis and of pneumoconiosis in general can be modulated or aborted despite the presence of dust in the lung. There is evidence that better understanding of the mechanisms that govern the development of pneumoconiosis will provide better treatment for those who suffer from it.

REFERENCES

AERTS C, WALLAERT B, GROSBOIS JM, VOISIN C.(1986) Superoxide anion release by alveolar macrophages in pulmonary sarcoidosis. ANN. NY. ACAD. SCI. 465:193-200.

DAVIS GS.(1986) Pathogenesis of silicosis: current concepts and hypotheses. LUNG 164:139-154.

GOSSET P, LASSALLE P, GRAU G, PESTEL J, VASSALLI P, TONNEL AB, CAPRON A. Secretion and mRNA expression of Tumor necrosis factor alpha and interleukin-1 by human mononuclear phagocytes after IgG-dependent stimulation: obtention of a different pattern between alveolar macrophages and blood monocytes (in press).
GOSSET P,TONNEL AB, JOSEPH M, PRIN L, MALLART A, CHARON J, CAPRON A.(1984) Secretion of a chemotactic factor for neutrophils and eosinophils by alveolar macrophages fromasthmatic patients. J. ALL. CLIN. IMMUNOL. 74:827-834.
LUGANO EM, DAUBER JH, DANIELE RP.(1981) Silica stimulation of chemotactic factor release by guinea pig alveolar macrophages. J. RETICULOENDOTHEL. SOC. 30(5):381-390.
LUGANO EM, DAUBER JH, ELIAS JA, BASHEY RI, JIMENEZ SA, DANIELE RP.(1984) The regulation of lung fibroblast proliferation by alveolar macrophages in experimental silicosis. AM. REV. RESPIR. DIS. 129:767-771.
OUAISSI A, NEYRINCK JL, CAPRON A.(1986) Development of a competitive radioimmunoassay for human plasma fibronectin. INT. ARCH. ALL. APPL. IMMUNOL. 81:75-80.
ROM WN, BITTERMAN PB, RENNARD SI, CANTIN A, CRYSTAL RG.(1987) Characterization of the lower respiratory tract inflammation of nonsmoking individuals with interstitial lung disease associated with chronic inhalation dusts. AM. REV. RESPIR. DIS. 136:1429-1434.
SCHMIDT JA, OLIVER CN, LEPE-ZUNIGA JL, GREEN I, GREY I.(1984) Silica-stimulated monocytes release fibroblast proliferation factors identical to interleukin-1. Potential role for interleukin-1 in the pathogenesis of silicosis. J. CLIN. INVEST. 73:1462-1472.
VOISIN C, GOSSELIN B, RAMON PH, WALLAERT B, AERTS C, LENOIR L.(1983) Le lavage-bronchoalvéolaire dans la pneumoconiose des mineurs de charbon. Aspects cytologiques. REV. FRANC. MAL. RESPIR. 11:455-466.
VOISIN C, WALLAERT B, AERTS C, GROSBOIS JM.(1985). Broncho-alveolar lavage in coal worker's pneumoconiosis. Oxidant and anti-oxidant activities of alveolar macrophages. NATO A.S.I. Series, Vol. G3, In vitro effects of Mineral dusts Edited by EG BECK and J BIGNON. Springer-Verlag Berlin Herdelberg, pp. 93-100.
WALLAERT B, BART F, AERTS C, OUAISSI MA, NEYRINCK JL, VOISIN C.(1986). Spontaneous superoxide anion, neutrophil chemotactic activity and fibronectin secretion by alveolar macrophages in pneumoconiotic patients. SILICOSIS AND MIXED-DUSTS PNEUMOCONIOSIS, Journées d'études GERP, ed. LE BOUFFANT (INSERM), (155):115-121.
WILLIAMS AJ, COLE PJ.(1981) Investigation of alveolar macrophage function using lucigenin dependent chemiluminescence. THORAX. 36:866-869.

Supported by GERP,by Université de Lille II , by Commission of European Communities and by INSERM (reseau de Recherche Clinique 850027).

CELLULARITY IN BRONCHOALVEOLAR LAVAGE IN CEMENT DUST AGGRESSION

AJA Robalo-Cordeiro, MF Baganha, M Melo, JRG Almeida, L Chieira, I Mesquita, ACP Leite, A Pego, JP Sousa, A Saraiva, MA Lima, E Gaspar

Center of Pneumology
Coimbra University
3049 Coimbra Codex
Portugal

INTRODUCTION

The high hygroscopic characteristic of cement dust favours, when inhaled, its floculation in upper airways, so preventing the reach of the lower respiratory tract.
Therefore, the pulmonary involvement in cement workers has not been accepted by most researchers (Sayers 1937; Gardner 1939; Vaccareza 1950; Sander 1958; Jenny 1960; Bernardou 1971; Parkes 1982), although several studies (Doerr 1952; Prosperi 1957; Hublet 1968; Kalacic 1973) have pointed out this type of pneumoconiosis.
And studies carried out in Portugal since 1973 have agreed, under pathological as well as under epidemiological perspectives, with this occupational lung disorder (Pimentel 1973,1977; Villar 1975; Ávila 1975; Santos-Rosa 1975; Mendes 1977; Baganha 1979, 1981,1986; Couto 1981; Agarez 1981; Robalo-Cordeiro 1982-1983).
This study deals with two objectives: (1) to assess the capability of experimental acute lung aggressions by the cement dust; (2) to ascertain the deep lung involvement in cement workers with pulmonary disorders.

EXPERIMENTAL STUDY

In a rat model, we have described high rates of lung neutrophilic afflux, as assessed by lung lavages, following the intratracheal introduction of silica, crystalline as amorphous (Leite 1984; Robalo-Cordeiro 1985a, 1985b, 1987). So, we have profit, as comparative model, of the percentages of neutrophils reached 4 hours after the intratracheal introduction of crystalline silica ($79 \pm 24\%$), those of control animals being of $5 \pm 3\%$.
Wen a suspension of 5o mgs of cement dust in one milliliter saline was introduced in the trachea of rats, the mean percentages of neutrophils were normal, either at the 4th (9 animals), or at the 8th hours (8 animals) (Table 1). Only in 1 rat an afflux of 60% neutrophil was observed at the 4th hour.

NATO ASI Series, Vol. H30
Effects of Mineral Dusts on Cells
Edited by B.T. Mossman and R.O. Bégin

Table 1. Percentages of neutrophils, evaluated by lung washings, in rats submitted to intratracheal cement

	Control		Cement [b]	
	Normal	Silica [a]	4 hours	8 hours
% of neutrophils	5 ± 3	79 ± 24	4 ± 2	5 ± 3

[a] 4 hours after intratracheal instillation of 5o mgs of silica in 1 milliliter saline.

[b] 5o mgs of cement dust in 1 milliliter saline.

Nevertheless, when the total cellularities of bronchoalveolar lavage fluids (BALF) were plotted against their own percentages of neutrophils, they correlated significantly, either at 4 or at 8 hours after the intratracheal instillation of the cement dust ($p < 0.01$) (Figure 1).

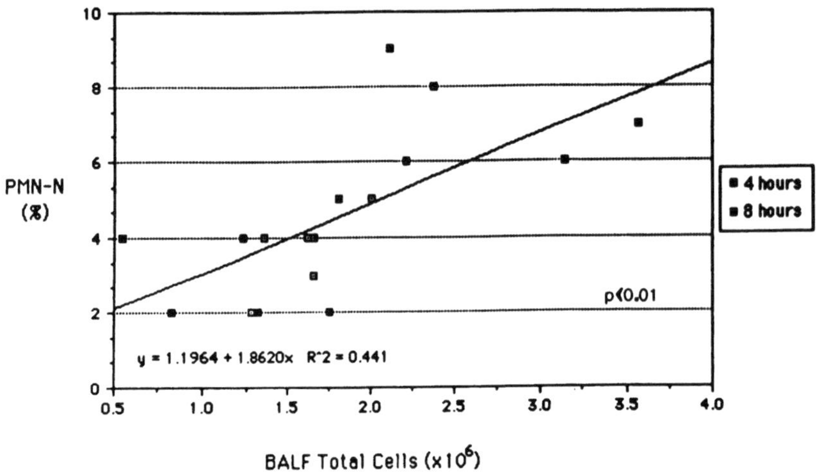

Fig.1 - BALF total cells and % PMN-N in intra-tracheal cement dust aggression in rats.

Therefore, it seems possible, according to these data, that the cement dust has reached, in some cases and with yarying intensity, the deep lung of rats. But its hygroscopic characteristics will generally prevent the high intensity cellular response of other experimental aggressions of the deep lung.

OCCUPATIONAL STUDY

Alyte, belyte and tobermoryte crystals, from the crystalline and gel phases of cement production, have been found in the deep lung of exposed people, as well as granulomatous and fibrosing lesions (Pimentel 1973,1977; Ávila 1975; Mendes 1977; Baganha 1979).
Also epidemiological studies in exposed workers have suggested their lung involvement (Kalacic 1971; Ávila 1975; Mendes 1977; Baganha 1981; Agarez 1981; Couto 1981).
In this study, 318 workers of a Portland cement plant (mean age 37 ± 8 years; mean duration of exposure 8 ± 6 years) were observed, presenting 7.2% of chest X-Rays alterations (Z - 5.3%; p - 0.3%; n = 1.6%) and 16% of lung function impairments (obstructive - 9.1%; restrictive - 5%; mixed - 1.9%). Among these workers and in order to a better understanding of mechanisms underlying to the respiratory involvement in early phases of the cement dust aggression, 10 volunteers were submitted to bronchoalveolar lavage: 3 smokers, 7 non smokers; 3 obstructive, 6 restrictive and 1 mixed ventilatory syndromes; 5 with normal and 5 with abnormal chest X-Rays; duration of exposure ranging in 9 workers between 6 and 12 years, the 10th referring a cummulative work of 50 years. As control groups, 17 normal volunteers were used for BALF's cellularity and immunoglobulins and 5 others for the lymphocytic subsets.

The cellularity of the BALFs showed high percentages of lymphocytes (and less of neutrophils) in several cement workers, the worker with the greatest duration of exposure presenting 26% of lymphocytes and 14% of neutrophils.
As concerns the mean values, the workers had significantly higher percentages of lymphocytes and neutrophils than the normal controls; and non smoking workers also presented augmented rates of lymphocytes (Table 2).

Table 2. Cellularity of BALF in cement workers

	Cells/ml	%		
	($\times 10^4$)	MØ	L	PMN-N
Controls				
Total	33 ±34	87 ± 7.3	8.3 ± 4.6	4.4 ± 2.7
Non Smokers	9 ± 3.9	81.8 ± 8.3	11.7 ± 4.7	6 ± 3.5
Smokers	46 ±36.4	89 ± 5	7 ± 4	4 ± 2
Cement workers				
Total	25.3±26.5	77.9±10.6[a]	14.4 ± 9.2[b]	7.7 ± 3.8[a]
Non Smokers	10.8 ± 7.5	74.1 ± 9.9	18.6 ± 7.6	7.3 ± 3.4
Smokers	59.1±23.5	87 ± 7	5 ± 2	9 ± 6[c]

Significant differences between homologous groups:
[a] $p < 0.02$; [b] $p < 0.03$; [c] $p < 0.05$

Cement workers with interstitial involvement, as expressed by an alteration of their chest X-Rays, presented higher percentages of lymphocytes and neutrophils (respectively, 18.6 ± 8.9) and 9.2 ± 4.9) than the workers with normal chest X-Rays (10.2 ± 8.1 and 6.2 ± 1.9).
So, it seems that chronic inhalatory aggression by cement dust may provoke an alveolointerstitial reaction predominantly of lymphocit type, such as we have found in the occupational lung aggression by crystalline and amorphous silicas. Besides, lymphocytosis above 15% were reached by 60% of our cement workers, against 15% of our silicotic patients.
As concerns T-lymphocytic subsets, an wide variation in the individual values of the T4 and T8-cells and of the *rationes* T4/T8 has been found in the BALFs as in the sera, but there was a general tendency to small figures in T4-cells and to reduced *rationes* of T4/T8, namely in the BALFs, with a significant decrease in the percentages of T-helper cells, T-supressor lymphocytes being almost inaltered. So, the ratio T4/T8 was lower in cement workers than in normal controls. These tendencies were maintained in non smokers alone (Table 3).

Table 3. T-lymphocytic subsets in cement workers

	Blood			BALF		
	T4	T8	T4/T8	T4	T8	T4/T8
Cement workers						
Total	29 ± 5.3[a]	22.9 ± 6.7	1.3 ± 0.4	21.9 ± 10[c]	28.8 ± 13	0.9 ± 0.5
Non smokers	29.6 ± 6[b]	21.2 ± 4.2	1.4 ± 0.4	24.6±9.8[d]	28.8 ± 12.6	1 ± 0.5
Controls (all non smokers)	44.4 ± 5	26.9 ± 6.5	1.7 ± 0.5	37.9±9.2	29.9 ± 5.1	1.3 ± 5.1

[a] $p < 0.001$; [b] $p < 0.002$; [c] $p < 0.02$; [d] $p < 0.05$

Concerning immunoglobulins synthesis, no alterations were present in its serum levels, although some deviations (besides, no significantly) have been found in its BALF values (Table 4). Also no differences have been seen between the blood B lymphocytes of cement workers and those of the controls (respectively, 19 ± ±6.4% and 18.7 ± 1.6%).
The levels of the Angiotensin Converting Enzyme were not so high, in the serum and in the BALF, as those of our silicotic workers (Table 5).

Table 4. Immunoglobulins in cement workers

	Serum (mg%)			BALF (mg/mg Alb %)		
	IgG	IgA	IgM	IgG	IgA	IgM
Controls	1.2 ± ± 0.2	0.28 ± ± 0.05	0.135 ± ± 0.062	0.4 ± ± 0.2	0.19 ± ± 0.11	0.119 ± ± 0.106
Cement workers	1.25 ± ± 0.25	0.27 ± ± 0.12	0.161 ± ± 0.159	0.24 ± ± 0.08	0.0207 ± ± 0.096	0.104 ± ± 0.057

Table 5. Angiotensin Converting Enzyme, in the Serum (SACE and in the BALF (LACE) of cement workers (nm/mn/ml)

	Controls	Workers	
		Silicotic	Cement
SACE	40.8 ± 9.5	60.3 ± 19[a]	45.1 ± 10
LACE	0.52 ± 0.2	0.82 ± 0.38	0.64 ± 0.25

[a] $p < 0.04$

COMMENT

It seems suggestive that the cement dust, besides its hygroscopic characteristics, may reach the deep lung of exposed people, and consequently be responsible for cellular alterations detectable in the BALFs of cement workers, although with lower expressiveness than other inhaled inorganic occupational dust do.

The involvement of neutrophils, whenever the deep lung be aggressed, is suggestive in early phases and probably in over exposed cases.

Concerning the detected lymphocytic patterns of our workers, high values have been frequently present, as such as in other pneumoconiosis. In any way, the frequent decrease of T-helper cells, namely in the lung compartment, may contribute to a non augmented immunological response; and possibly the normal values of T-suppressor lymphocytes will express a relative unnecessariness of counteracting immunological reactions indeed not expressively augmented in our cases.

In any way, many problems are yet unresolved, namely the induced depression of the T-helper lymphocytes.

Obviously, further studies are necessary for the understanding of the mechanisms underlying the lung response in the inhaled cement dust aggression.

REFERENCES

Agarez M, Maulide I, Vieira ML, Ascenção M, Cruz M (1981) Influên cia da poeira de cimento na etiopatogenia das doenças respirató rias. Análise de um inquérito epidemiológico numa fábrica de cimento. Arq.Soc.Port.Pat.Resp. 2:39-71.

Ávila R (1975) Granulomatoses pulmonares de causa inalatória. Tese, Lisboa, 217-238.

Baganha MF, Oliveira LC, Morais JCT, Mendes JA, Bernarda R, Robalo Cordeiro AJA (1979) Cancro do pulmão e actividade profissional. J.do Médico 99:461-467.

Baganha MF, Oliveira LC, Morais JCT, Mendes JA, Bernarda R, Oliveira M, Robalo Cordeiro AJA (1981) Alguns aspectos da patologia respiratória consecutiva à inalação do pó de cimento. O Médico 99: 787-820.

Bernardou M, Fréour P (1971). Méd. et Hyg.988:1947.

Couto A, Bugalho de Almeida A, Pires JP (1981) A função ventilatória em trabalhadores da indústria do cimento. Arq.Soc.Port. Pat.Resp. 2:153-164.

Doerr W (1952) Pneumokoniose durch Zementataub. Virchows Arch. Path.Anat.Physiol. 322:397-427.

Gardner LU, Durkan TM, Brumfield DM, Sampson Hl (1939) Survey in seventeen cement plants of atmosphere dusts and their effects upon the lungs of twenty-two hundred employees. J.Ind.Hyg. Toxicol. 21:279-318.

Hublet P (1968) Enquête relative au risque de pneumoconiose dans la fabrication des ciments de constrution. Arch.Belg.Med. Soc. 26:417-430.

Jenny M, Battig K, Horisberger B, Havas L, Grangdjen E (1960) Arbeitsmedizinische Intersuchung in Zementfabriken. Schweitz. Med.Wschr. 90:705-709.

Kalacic V (1973) Chronic nonspecific lung disease in cement workers. Arch.Environ Health 26:78-83.

Leite ACP, Gaspar E, Bernarda RA, Pego AM, Robalo Cordeiro AJA (1984) Pulmonary fibrosing site effects of drugs evaluated in an experimental model. Respiration 46,Suppl.157.

Mendes JJA (1977) Estudo dos efeitos das poeiras das Fábricas dos Cimentos da Região de Leiria. III - Estudo Médico Epidemiológico, Coimbra.

Parkes WR (1982) Occupational lung disorders. Butterworths, London, 2nd Ed.,11-22.

Pimentel JC, Gomes FB (1973) Pneumopatia por inalação da poeira do cimento. J.do Médico 81:11-19.

Pimentel JC, Menezes AP (1977) Doença granulomatosa hepática e pulmonar provocada pela inalação de poeiras de cimento e mica. J.do Médico 93:184-94.

Prosperi G, Barsi C (1957) Sulle pneumoconiosi dei lavatori del cemanto. Rass.Med.Ind. 26:16-24.

Robalo Cordeiro AJA, Baganha MF, Mendes JA, Morais T, Oliveira LC, Araujo AT (1982-83) Patologia respiratória por exposição a poeiras do cimento. Rev.Port.Med.Trabalho 2:47-57, 37-50; 3:53-78.

Robalo Cordeiro AJA, Baganha MF, Azevedo-Bernarda R, Leite ACP, Almeida JRG, Bairos VF, Gaspar E, Garção MF, Lima MAM, Rosa MAS, Pega AF, Bastos JMP (1985a) Biological effects of fume silica (amorphous type). In "In vitro effects of mineral dusts", Beck EG.Bignon J Eds. Nato ASI series G3, Springer-Verlag, Berlin Heidelberg:489-496.

Robalo Cordeiro AJA, Bernarda RA, Leite ACP, Gaspar E, Bairos VF, Pego AM (1985b) Bronchial, alveolar and interstitial reactions against acute pulmonary attacks. Italian J.of Chest Diseases.6: Suppl.402-408.

Robalo Cordeiro AJA, Bernarda RA, Leite ACP, Gaspar E, Pego AM (1987) Quimiotactismo pulmonar para os neutrófilos. Coimbra Médica 8 Supl.1:13-21.

Sander OA (1985) Roentgen re-survey of cement workers. Archs Ind.Hlth 17:96-103.

Santos Rosa MA, Biscaia J, Chieira C, Tavares C, Robalo Cordeiro AJA (1975) Aspectos pulmonares (epidemiologia). In "Problemas da Medicina do Trabalho nos trabalhadores de cimento". Hoesch Port. Ed:31-51.

Sayers RR, Dallavalle JM, Bloomfield SG (1973) Occupational and environmental analysis of the cement. Clay and Pottery Industries Public Hlth Rep.U.S. No. 238: 1-50.

Vaccareza RA (1950) Higiene y salubridad en la Industria del Cemento Portland. Su investigacion en las fabricas argentinas. Buenos Aires, Guillermo.

Villar TG, Ávila R, Araújo AT (1975) La pneumopathie des ouvriers exposés au ciment. Comptes Rend. XVII Congr.Nat.Tub.Mal. Resp., Clermont-Ferrand, Masson Ed, Paris; 369.

CARBORUNDUM PNEUMOCONIOSIS.

R BÉGIN, M.D. A Dufresne, Ph.D., A Cantin, M.D., S Massé, M.D., P Sébastien, Ph.D., P Durand, Ph.D., and G Perrault, Ph.D.
Université de Sherbrooke, Sherbrooke (QC), Dust Disease Unit, McGill School of Occupational Health, Montréal (QC) and Institut de Recherche en Santé et Sécurité au Travail, Montréal (QC).

INTRODUCTION

Silicon carbide (SiC) also called carborundum is a universally used abrasive produced by the fusion of high-grade silica and finely ground carbon in electric furnace at 2400°C. A recent review of the subject in Parkes textbook (6) lead to the conclusion that there was no evidence that exposure to silicon carbide dust gave rise to a pneumoconiosis.

Nonetheless, pneumoconiosis has been reported in long term workers engaged in the manufacturing of SiC and it was suspected that sick workers had been previously exposed to other dust hazards or to quartz dust in the raw materials of the manufactures. One pathological report raised the possibility of a carborundum pneumoconiosis (4).

Recent investigations in the Quebec carborundum industry have documented an excess of radiographic abnormalities compatible with pneumoconiosis, particularly in older workers of the industry (7). In view of these findings, collaboration with the industry was established and a thorough multidisciplinary investigation was initiated.

Pathological studies of available lung tissue from long term workers of the SiC industry were analyzed in detail (5). Recent chest radiographs of 1984 and 1979 of some 128 workers were reviewed (3). Evaluation of occupational exposures in the industry was conducted and reported in detail (1, 2). A review of the occupational hygiene lead to the preparation and characterization of reference samples of respirable particles collected in the SiC production industry for use in this experimental research.

The objectives of this study were to evaluate the in vivo lung biologic activity of the mineral dusts found in the carborundum manufacturing industry, to identify the offending dust in order to establish appropriate control of dust level in the workplace.

MATERIALS AND METHODS

Experimental Design

Seventy-two sheep were used in this study. The flock was divided into 9 groups of 8 sheep. Following pre-exposure (control) studies, the sheep tracheal lobe was exposed to

- 100 ml saline = Sa group (control);
- 100 mg latex beads in 100 ml saline = latex group;
- 100 mg graphite in 100 ml saline = graphite group;
- 100 mg SiC raw particles in 100 ml saline = SiCp group;
- 100 mg SiC ashed particles in 100 ml saline = SiCpa group;
- 100 mg Minusil-5 in 100 ml saline = Si group;
- 100 mg crocidolite fibers in 100 ml saline = Cro group;

NATO ASI Series, Vol. H30
Effects of Mineral Dusts on Cells
Edited by B.T. Mossman and R.O. Bégin

- 100 mg SiC raw fibers in 100 ml saline = SiCf group
- 100 mg SiC ashed fibers in 100 ml saline = SiCfa group

Exposure of the tracheal lobe was carried out via bronchoscopic infusion of the suspension in the lobe. The animals were studied prior to exposure and post-exposure at month 2, 4, 6, and 8 by BAL and by histopathological methods at month 8.

The minerals

The materials for exposure were obtained from the following sources: Latex beads from Sigma Chemical Co., St-Louis, MO, these particles are uniform in size with a mean diameter of 1 µ; Minusil-5 from Pennsylvania Glass Co., Pittsburg, PA, the silica particles are well characterized with 99.9% of diameter < 5 µ and 95% < 1 µ; crocidolite fibers from the Union Internationale Contre le Cancer (UICC). These fibers have a known diameter of 0.17 ± .01 µ with an average length of 3.9 ± 0.2 µ, 95% of fibers with a length < 10 µ, 82% < 5 µ.

All other samples were obtained from the Quebec SiC industry, prepared and characterized. Briefly, these materials were collected from the production sites in the Acheson furnaces of two Quebec SiC plants. The non-fibrous SiC was collected from the center of large lumps of produced materials. The SiC fibers were collected mainly from the outside part of the main cylindrical lump produced by the process. The graphite was extracted from the core of a fired Acheson furnace. The raw particulate and fibrous SiC were, as expected, contaminated with graphite flakes on surface, which at least in theory, could alter biological activity. To eliminate these contaminants, reference samples of fibrous and particulate SiC were ashed. Graphite particles were 98.8% < 5 µ, particulate SiC raw or ashed were 99.5% < 5 µ. For the SiC fibers, seventy percent of fibers were less than 5 µ, with some longer than 20 µ. The fibrous SiC raw or ashed were of an average of 0.27 µ ± 0.27 diameter with an average length of 6.8 ± 11.2 µ. These morphometric data were considered in the selection of the asbestos fiber crocidolite, for comparison in these experiments.

Assessment of lung reaction

To evaluate the disease process induced by exposure to these respirable minerals, we looked at lung lavage cellularity and biochemistry as biologic indicators of alveolitis. The severity of lung tissue damage was evaluated at autopsy, 8 months after exposure, by histopathology. To assess interstitial lung matrix changes we looked at the glycosaminoglycan accumulation in BAL fluid. We also measured the production of fibronectin by BAL cells in culture.

Lung lavage fluid was analyzed for the presence of molecules capable of enhancing fibroblast proliferation.

Histopathology

At month 8 of the study, all sheep were sacrificed and the lungs removed from the chest cavity. The tracheal lobe was identified and 9 samples of the lobe of each sheep were obtained each time for microscopic examination.

RESULTS

The particulates

In comparison to Sa group, all the particulate exposed groups had a slight and transient early increase in cellularity except for the Si group which had an early 500% increase in cellularity which decreased to 250% at month 4 but remained elevated to the end of experiment. This was largely due to increase in the macrophage population but increases in lymphocytes and neutrophils were also significant and sustained in the Si group. Similarly in the biochemical and cell culture analyses, only the Si group had significant increases in BAL lactate dehydrogenase, glycosaminoglycan, and increased production of fibronectin and fibroblast growth activity. The lung morphology of the sheep was normal in groups Sa and latex. The lung tissue of sheep in the graphite group, SiCp and SiCpa groups contained accumulation of particles in alveoli and interstitium without cellular reaction. In the graphite exposed sheep, the morphologic changes are reminescent of the early simple pneumoconiosis of coal workers (6). In the Si exposed sheep, the lung changes are characterized by a diffuse alveolitis with early nodular silicotic lesions as reported in our earlier studies. The pathological scores were 0 ± 0 for Sa group, latex group, graphite group, SiCp group and 2.9 ± 1.0 for the Si group ($p < 0.01$ for Si group vs others).

The fibers

Briefly, cellularity of BAL was increased in all the fiber groups following exposure with a larger attenuation for the SiCf group, this effect was pancellular and was also seen in the pattern of response of LDH overtime. Fibronectin production was significantly increased, again with some attenuation for the SiCf group, fibroblast growth activity was increased significantly in all fiber groups.

Pathologic analysis of lung tissue in the crocidolite exposed sheep revealed a peribronchiolar fibrosing alveolitis as previously reported in other asbestos exposure. In the fibrous carborundum exposed sheep, we found nodular lesions in the parenchyma which were not located around the bronchioles. These nodules were composed of multinucleated macrophages, monocytes and a few neutrophils and contained several carborundum fibers and "bodies". The intensity and profusion of lesions in the fiber groups were: 1.9 ± 0.25 for Cro group, 1.2 ± 0.21 for the SiCf and 1.6 ± 0.2 for the SiCfa group ($p > 0.05$ between the 3). The slightly lower score ($p > 0.05$) in the SiCf compared to SiCfa could suggest a partial inhibitory effect of the graphite on surface of the SiCf fibers.

DISCUSSION

In recent years, several epidemiological and clinical investigations have suggested that carborundum workers may have a specific occupational lung disease. This experimental study of the airborne dust particles and fibers in the carborundum industry provides significant new information on the pathogenesis of interstitial lung disease in workers of that industry. It documents that of all the non-fibrous particulate minerals encountered on site, only silica appears to be potentially responsible for some of the lung injury. In addition, this study documents clearly that fibrous SiC has significant biologic activity and can initiate a fibrosing lung disease. In SiC manufactures, SiC fiber inhalation can contribute to the genesis of an interstitial lung disease. Therefore, it is appropriate to recommend that surveillance of the work environment in that industry should include an assessment of airborne fiber levels.

REFERENCES

1. Dufresne, A., Lesage, J., Perrault, G.: Evaluation of Occupational Exposure to Mixed Dusts and Polycyclic Aromatic Hydrocarbons in Silicon Carbide Plants. Am. Ind. Hyg. Assoc. J. 48: 160-166 (1987).

2. Dufresne, A., Perrault, G., Sébastien, P., Adnot, A., Baril, M.: Morphology and Surface Characteristics of Particulates from Silicon Carbide Industries. Am. Ind. Hyg. Assoc. J. 49: in press (1988).

3. Durand, P., Bégin, R., Samson, L., Cantin, A., Massé, S., Dufresne, A., Perrault, G., Laflamme, J.: Silicon Carbide Pneumoconiosis: Current Exposure Levels do not Change the Radiographic Assessment of Disease. Submitted for publication, 1988.

4. Funahashi, A., Schlueter, D.P., Pintar, K., Siegesmund, K.A., Mandel, G.S , Mandel, N.S.: Pneumoconiosis in Workers Exposed to Silicon Carbide. Am. Rev. Respir. Dis. 129: 635-640 (1984).

5. Massé, S., Bégin, R., Cantin, A.: Pathology of Silicon Carbide Pneumoconiosis. Modern Pathology 1: 104-108 (1988).

6. Parkes, W.R. Occupational Lung Disease, 2nd Ed. Butterworth, London, U.K. (1982).

7. Peters, J.M., Smith, T.J., Bernstein, L., Wright, W.E., Hammond, S.K.: Pulmonary Effects of Exposures in Silicon Carbide Manufacturing. Br. J. Ind. Med. 41: 109-115 (1984).

INFLUENCE OF PHYSICOCHEMICAL PROPERTIES OF BERYLLIUM AND NICKEL COMPOUNDS ON CULTURED CELL TOXICITY*

G.L. Finch, A.L. Brooks, J.M. Benson, M.D. Hoover, J.A. Mewhinney, A.F. Eidson, R.G. Cuddihy, R.F. Henderson, C.H. Hobbs
Lovelace Inhalation Toxicology Research Institute
Albuquerque, NM 87185-5890

Introduction

Beryllium and nickel are toxic following deposition in the respiratory tract. Toxicity may be related to the bioavailability of these metals, which is in turn influenced by *in vivo* dissolution. This study was conducted to investigate the role of differing physical and chemical properties of beryllium and nickel compounds on their cytotoxicity to cell systems *in vitro*. We hypothesized that the two major factors influencing cytotoxicity are chemical form and specific surface area. Various forms of beryllium and nickel compounds were administered to cell cultures. The toxicity observed for particulate compounds was compared on the basis of particle mass and particle surface area. We found that for a given compound, toxicity was associated with the surface area of the compound available for interaction with cells.

Materials and Methods

Beryllium oxide particles were produced by nebulizing $Be(OH)_2$ and calcining the particles at either 500 or 1000°C for 16

NATO ASI Series, Vol. H30
Effects of Mineral Dusts on Cells
Edited by B.T. Mossman and R.O. Bégin
© Springer-Verlag Berlin Heidelberg 1989

hours (Hoover et al. 1988). Beryllium metal aerosols were fractionated by aerosolizing the metal powder (Brush-Wellman Corp., Elmore, OH) then passing the aerosol into a five-stage aerosol cyclone (Smith et al. 1979). Particles obtained from stages II and V of the cyclone were used. Nickel oxide particles were supplied by the Nickel Producers Environmental Research Association and were characterized by Sunderman et al. (1987); Ni_3S_2 was supplied by INCO, Ltd., Mississauga, Ontario. The specific surface area of the beryllium particles and the Ni_3S_2 was measured as described by Rothenberg et al. (1982).

Cultured Chinese hamster ovary cells (CHO) and rat lung epithelial cells (LEC) were used as described by Li et al. (1983). The culture medium consisted of Ham's F12 medium (Flow Laboratories, Inc., McLean, VA) supplemented with 10% heat-inactivated newborn calf serum (Hazelton Research Products, Inc., Lenexa, KS). Before exposure, cultures were grown to confluency for 24 hours in a humidified 37°C incubator containing an atmosphere of 5% CO_2/95% air.

Test compounds in culture medium were prepared fresh daily. For exposure, the medium containing appropriate test chemicals was ultrasonically agitated then added to cultures. Only LEC cultures were used for nickel exposures. Cells were cultured for 20 hours then harvested by trypsinization. Cells were subcultured into triplicate wells containing fresh medium. After 7 days, cultures were fixed, stained, and the number of colonies was counted. The relative colony-forming efficiency (RCE) was defined as the number of colonies formed in treated cultures expressed as a percentage of those formed in control cultures. The specific surface area of each particle sample was used to calculate the surface area concentration. Values for the mass and surface area concentrations that reduced the RCE by 50% (LC_{50}) were determined using a probit procedure (SAS Institute Inc., Cary, NC).

Results

Properties of the beryllium and nickel compounds used are shown in Table 1. Particles were of respirable size. Figure 1 shows the effect of BeO and Be metal particles on CHO cells. Toxicity plotted as a function of mass concentration is shown at left, and as a function of surface area concentration at right. On a mass basis, the order of toxicity was 500°C-BeO > $BeSO_4$ > 1000°C-BeO ~ Be-V metal > Be-II metal, and on a surface area basis, the order was Be-V ~ Be-II > 500°C-BeO ~ 1000°C-BeO.

TABLE 1. Properties of Compounds Used in Toxicity Studies

Compound	Preparation	Diameter[a] (µm)	Specific Surface Area (m^2/g)
$BeSO_4$	Soluble Salt	NA[b]	NA
BeO	Calcined at 500°C	1.8 (1.9)	146 or 189[c]
	Calcined at 1000°C	1.9 (1.8)	35.8
Be Metal	Cyclone Fraction II	2.4 (1.9)	7.6
	Cyclone Fraction V	1.0 (1.5)	20.4
$NiSO_4$	Soluble Salt	NA	NA
Ni_3S_2	Industrial Source	2.8	1.8
NiO	Calcined at 650°C	2 to 5	41.5
	Calcined at 1045°C	2 to 5	< 0.2

[a] Geometric standard deviation given in parentheses. Value given is aerodynamic diameter for BeO and Ni_3S_2, and physical diameter for Be metal and NiO.
[b] NA = Not applicable.
[c] Two separate preparations of 500°C-BeO were used.

Alterations in the RCE of LEC cells are shown in Figure 2 for beryllium compounds, and values for LC_{50} are shown in Table 2. Just as for the CHO cultures, conversion to surface area concentration for each beryllium particle type caused values for LC_{50} to converge, and the 95% confidence intervals overlapped in each case. For nickel compounds (Fig. 3), the order of toxicity was Ni_3S_2 > $NiSO_4$ > NiO. Conversion to surface area concentration demonstrated that Ni_3S_2 was much more toxic than

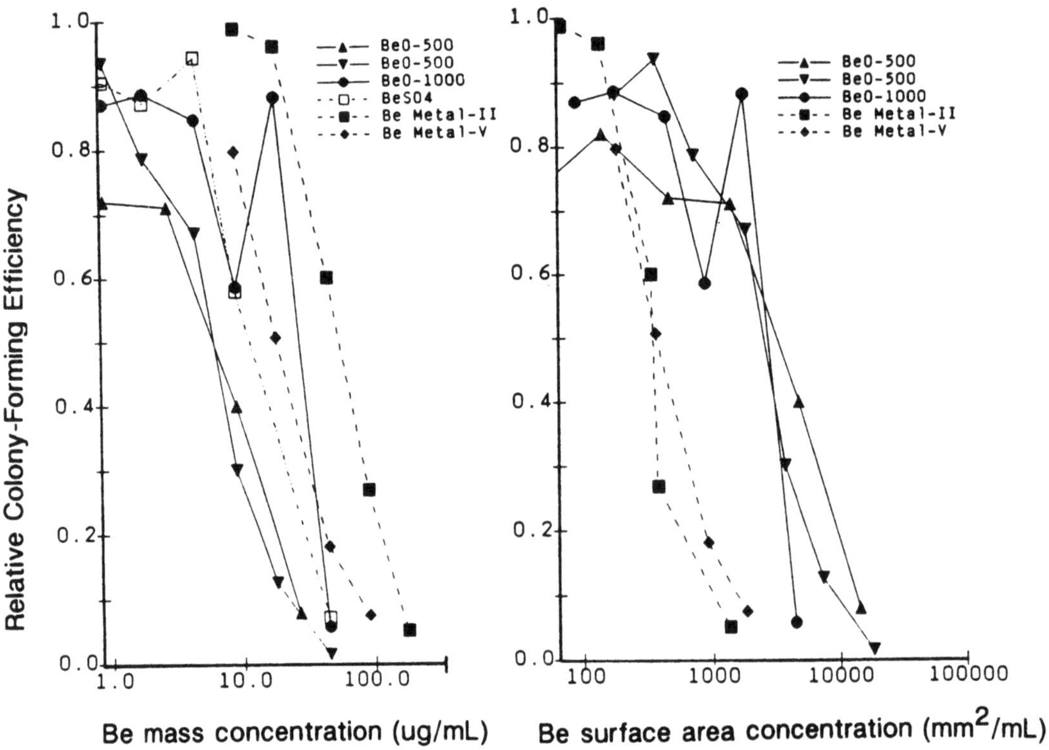

Fig. 1. Reduction in CHO cell relative colony-forming efficiency (RCE) as a function of BeO or beryllium metal.
Left: RCE vs. the beryllium mass concentration.
Right: RCE vs. the beryllium surface area concentration.

either NiO preparation. The RCE could only be plotted against mass concentration for the sulfate salts.

Discussion

We observed that CHO cell cultures were less sensitive to beryllium than were LEC cultures. Differences may have been related to differing rates of particle uptake or differing susceptibilities to beryllium in our assay system.

For both cell types, we found that the toxicity of beryllium

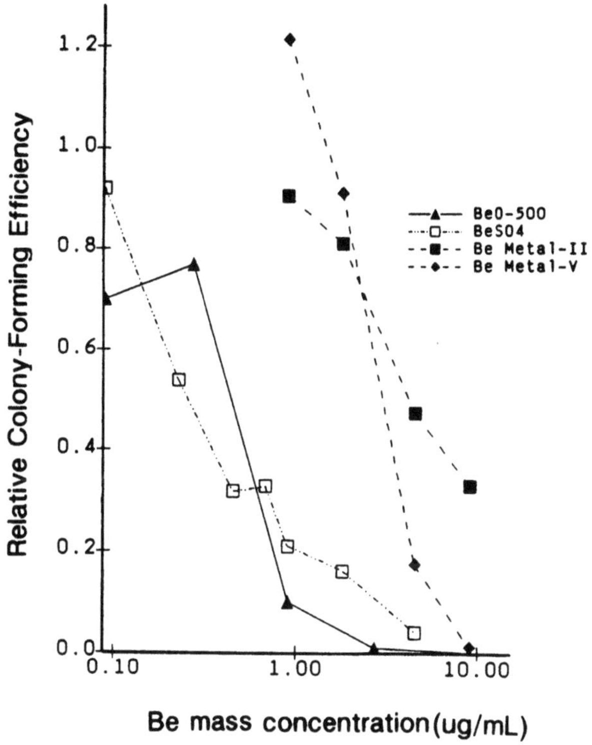

Fig. 2. Reduction in LEC cell relative colony-forming efficiency (RCE) as a function of the mass concentration of administered BeO or beryllium metal.

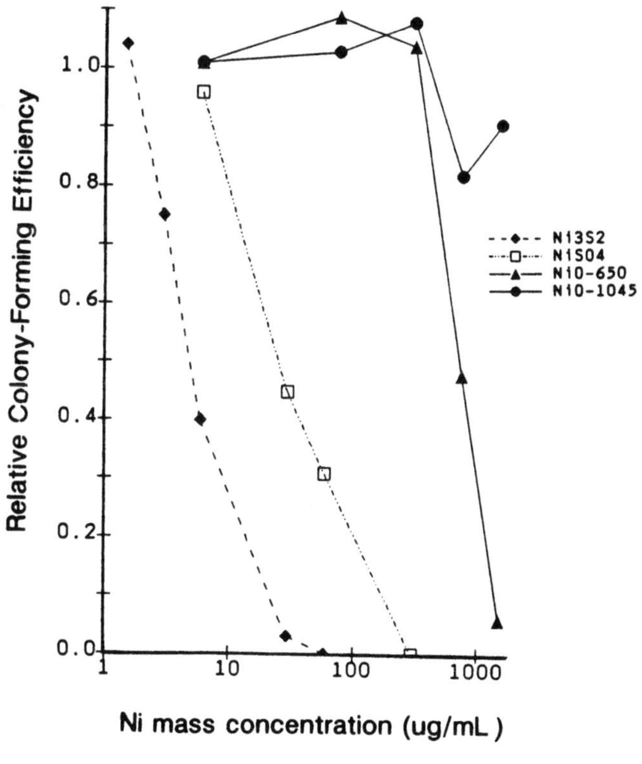

Fig. 3. Reduction in LEC cell relative colony-forming efficiency (RCE) as a function of the mass concentration of administered nickel compounds.

TABLE 2. Cytotoxicity of Be and Ni Compounds to Cultured Cells

Cell Type	Compound	LC_{50} µg/mL (±95% CI)[a]	LC_{50} mm^2/mL (±95% CI)[a]
CHO	$BeSO_4$	9.8 (6.5-16)	NA[b]
	BeO-500°C	3.4 (1.2-12)	1800 (630-6300)
	BeO-1000°C	19 (17-22)	1900 (1700-2200)
	Be-II	56 (50-62)	430 (380-480)
	Be-V	19 (16-22)	390 (340-450)
LEC	$BeSO_4$	0.3 (0.2-0.5)	NA
	BeO-500°C	0.3 (0.1-0.6)	130 (90-340)
	Be-II	4.8 (4.1-5.9)	37 (31-45)
	Be-V	3.1 (2.8-3.4)	63 (57-69)
	$NiSO_4$	30 (14-56)	NA
	Ni_3S_2	5.0 (4.3-5.8)	12 (10-14)
	NiO-650°C	720 (580-880)	30000 (24000-37000)
	NiO-1045°C	> 1400	> 280

[a] Concentration required to reduce relative colony-forming efficiency by 50%. Values of the 95% confidence interval obtained from the probit fit are given in parentheses.
[b] NA = Not applicable.

metal was greater than that of BeO when expressed on the basis of surface area concentration. In support of this observation, we previously found that beryllium metal was slightly more soluble than BeO in a non-biological dissolution system when the effects of specific surface area were removed (Finch et al. 1988a). We also observed that the conversion of mass to surface area concentration yielded a similar ranking of toxicity caused by physically different forms of the same beryllium particulate compound. Previous studies of canine alveolar macrophages demonstrated that 500°C-BeO was more cytotoxic than 1000°C-BeO (Finch et al. 1988b). A reanalysis of those data showed that differences in LC_{50} decreased when data were converted from a mass to a surface area concentration basis.

For nickel, the ranking of LEC cytotoxicity was Ni_3S_2 > $NiSO_4$ > NiO. Conversion by specific surface area demonstrated that Ni_3S_2 was much more toxic than either oxide. This observation is supported by Fisher et al. (1984) who observed that particulate Ni_3S_2 was at least ten-fold times more toxic to cultured bovine alveolar macrophages than solubilized Ni_3S_2 or $NiCl_2$. In addition, Kuehn et al. (1982) indicated that Ni_3S_2 did not conform to an inverse association between dissolution rate and uptake of nickel compounds by rat peritoneal macrophages.

In addition to influencing the clearance rate of material from the lung, the chemical form of administered toxic metals is a crucial determinant of the expression of toxicity in cultured cell systems. Although beryllium ion is poorly incorporated into cultured cells (Hart and Pittman 1980), we found that $BeSO_4$ was the second most toxic beryllium compound tested here, and was more toxic than BeO to cultured canine alveolar macrophages (Finch et al. 1988b). Although a small fraction of administered ions may be internalized, probably by pinocytosis or phagocytosis of colloidal precipitates, they may be in a more bioavailable form compared with insoluble particles.

A plausible mechanism of toxicity due to relatively insoluble particles involves phagocytosis followed by intracellular dissolution within acidic phagolysosomes. Because specific surface area influences the dissolution rate of insoluble particles (Mercer 1967), it follows that once internalized by cells, particles of identical chemical form should initially dissolve at rates dependent on the surface area presented. Once dissolved, metallic ions could form complexes and cause toxic effects to the cell. This hypothesis accounts for the finding that equal masses of particle samples that are chemically identical can result in differing levels of toxicity.

In conclusion, we have demonstrated that chemically identical particulate beryllium compounds with differing physical properties induce toxicity to CHO or LEC cell cultures at different mass concentrations but similar surface area concentrations. In contrast, Ni_3S_2 was more toxic than would be predicted on the basis of surface area alone. These results underscore the importance of characterizing physicochemical properties of toxicants when designing and interpreting toxicologic studies.

Acknowledgements. These studies were supported by the U.S. Department of Energy under Contract DE-AC04-76EV01013 in facilities fully accredited by the American Association for the Accreditation of Laboratory Animal Care. We also acknowledge the assistance of M.J. Waltman, M.K. McLeod, B.D. Ritchey, C.D. Kahle, L.F. Blair, and S.J. Rothenberg.

References

Finch GL, Mewhinney JA, Eidson AF, Hoover MD, Rothenberg SJ (1988a) In vitro dissolution characteristics of beryllium oxide and beryllium metal aerosols. J Aerosol Sci 19:333-342.
Finch GL, Verburg RJ, Mewhinney JA, Eidson AF, Hoover MD (1988b) The effect of beryllium compound solubility on in vitro canine alveolar macrophage cytotoxicity. Toxicol. Lett. 41:97-105
Fisher GL, Chrisp CE, McNeill KL, McNeill DA, Democko C, Finch GL (1984) Mechanistic evaluations of the pulmonary toxicology of nickel subsulfide. In: MacFarland H et al. (eds) Applied Toxicology of Petroleum Hydrocarbons, vol VI. Princeton Sci Publ, Princeton NJ, p 49.
Hart BA, Pittman DG (1980) The uptake of beryllium by the alveolar macrophage. J Reticuloendothel Soc 27:49-58.
Hoover MD, Eidson AF, Mewhinney JA, Finch GL, Greenspan BJ, Cornell CA (1988) Generation and characterization of respirable beryllium oxide aerosols for toxicity studies. Aerosol Sci Technol 9:83-92.
Kuehn K, Fraser CB, Sunderman FW Jr. (1982) Phagocytosis of particulate nickel compounds by rat peritoneal macrophages in vitro. Carcinogenesis 3:321-326.
Li A, Hahn FF, Zamora PO, Shimizu RW, Henderson RF, Brooks AL, Richards R (1983) Characterization of a lung epithelial cell strain with potential applications in toxicological studies. Toxicol 27:257-272.
Mercer TT (1967) On the role of particle size in the dissolution of lung burdens. Hlth Phys 13:1211-1221.
Rothenberg SJ, DeNee PB, Cheng YS, Hanson RL, Yeh HC, Eidson AF (1982) Methods for the measurement of surface areas of aerosols by adsorption. Adv Coll Int Sci 15:223-249.
Smith WB, Wilson RR, Harris DB (1979) A five-stage cyclone system for in situ sampling. Envirn Sci Technol 13:1387-1392.
Sunderman FW, Jr, Hopfer SM, Knight JA, McCully KS, Cecutti, AG, Thornhill PG, Conway K, Miller C, Patierno, Costa M (1987) Physicochemical characteristics and biological effects of nickel oxides. Carcinogenesis 8:305-313.

Comparative Study of the Effect of Various Fibres on the Secretion of Plasminogen Activator by Murine Peritoneal Macrophages

D. Godelaine, and H. Beaufay
International Institute of Cellular and Molecular Pathology
UCL 75.39
Avenue Hippocrate, 75
B-1200 Brussels
Belgium

Introduction

Urokinase-type plasminogen activator (PA) is produced by many cell types, including macrophages. This neutral protease converts the abundant extracellular zymogen plasminogen into active plasmin, another neutral protease. PA is thought to play an important role in regulating proteolytic events that occur in various normal conditions, but also in pathological processes involving tissue destruction and cell migration as, for instance, inflammatory reactions and invasive growth of cancer cells (for a review, see Danø et al. 1985).

Involvement of PA in the development of asbestos-induced inflammatory lesions has already been suggested by the work of J. Hamilton. Chrysotile B, when injected in the peritoneal cavity of mice, elicits macrophages which produce high levels of PA (Hamilton et al. 1976). In vitro addition of these fibres to cultured mouse peritoneal macrophages also stimulates PA production (Hamilton et al. 1976 ; Hamilton 1983).

We have undertaken a comparative study of the <u>in vitro</u> effect of various dusts on the PA production by mouse peritoneal macrophages with the aim of examining whether this biological activity could help discriminating between harmful and innocuous dusts. Under examination were mineral fibres (chrysotile A, crocidolite, wollastonite, rockwool and glass), organic fibres (cellulose), and mineral particles (rutile and quartz).

NATO ASI Series, Vol. H30
Effects of Mineral Dusts on Cells
Edited by B. T. Mossman and R. O. Bégin

Materials and Methods

Peritoneal cells

Peritoneal cells were obtained from NMRI mice, 6-7 weeks old, grown in pathogen-free conditions in a local animal house. Cells referred to as "LPS-activated" were obtained from mice which had received an intraperitoneal injection of 30 µg of Lipopolysaccharide W from S. typhosa (Difco) 4 days before sacrifice. Cells obtained from normal (i.e. not injected) mice are referred to as "resident".

Dusts

Samples of UICC chrysotile A and crocidolite were provided by Dr. J.C. Wagner (MRC Pneumoconiosis Unit, Penarth, GB). Fractions of aerodynamic diameter not greater than 7 µm were prepared by Professor J. Helsen (Depart. Metaalkunde, Katholieke Universiteit Leuven, Heverlee, Belgium) from the following samples : wollastonite (NYAD G of Nyco), rockwool (Lapinus), fibrous glass (Cemfil III of Pilkington), and cellulose (Prothermac) ; the first three samples were finely grinded before fractionation. Particles of rutile (2-4 µm) and quartz (DQ 12, < 5 µm) were kindly provided by Professor Dr. K. Robock (Institute for Applied Fibrous Dust Research, Neuss, FRG).

Measurement of PA production

PA production was estimated from the plasminogen-dependent fibrinolytic activity of macrophages cultured on an insoluble coat of tritiated fibrin. To this purpose, peritoneal cells (2 x 10^6) in 1 ml of Eagle-Dulbecco medium supplemented with 5 % of foetal calf serum (FCS) and 100 µg/ml of soybean trypsin inhibitor were plated on Nunclon 24 well-dishes coated with [^{3}H]fribrin (40 µg/2 cm^2) and incubated in a 10 % CO_2-90 % air atmosphere. At 20 hours after plating, non-adherent cells were removed by washing three times with serum-free medium. Adherent cells (in duplicate) were placed again in 1 ml of medium supplemented with FCS and trypsin inhibitor, but also in the presence of dusts carefully distributed in the medium by sonication for 2 min. After 2 hours incubation, the cells were washed three times to remove dusts and trypsin inhibitor, placed in 2 ml of dust-free medium supplemented with 5 % of acidified FCS and returned to the incubator. The time-course of fibrinolysis was estimated by withdrawing aliquots of medium every 24 hours over a period of 7 days, and counting the radioactivity in a liquid scintillation counter (Beckman LS 2800). Results are given in % of fibrin degraded. The radioactivity corresponding to 100 % degradation was determined by incubation in the presence of 50 µg of trypsin. The plasminogen dependence of fibrinolysis was examined by measuring the activity in the presence of acidified FCS which had been depleted of plasminogen by chromatography on Lysyl-Sepharose (Deutsch and Mertz 1970).

Results

Figure 1 shows the results of a preliminary experiment in which mouse peritoneal macrophages, either resident (Fig. 1a) or LPS-activated (Fig. 1b), were exposed <u>in vitro</u> to UICC chrysotile A fibres at various doses. Both types of macrophages induced increasing fibrinolysis after contact with 10 or 100 µg of chrysotile, but there was a longer lag period for resident macrophages. With 100 µg of fibres, there was some cell rounding and detachment after 3-4 days ; therefore asbestos cytotoxicity possibly explains why fibrinolysis levelled off at that time. Cultures exposed to lower doses of asbestos appeared not to be affected. The fibrinolytic activity was predominantly plasminogen-dependent, and may thus be attributed to the release of PA by the cells. These results are in good agreement with those of Hamilton (Hamilton et al. 1976 ; Hamilton 1983).

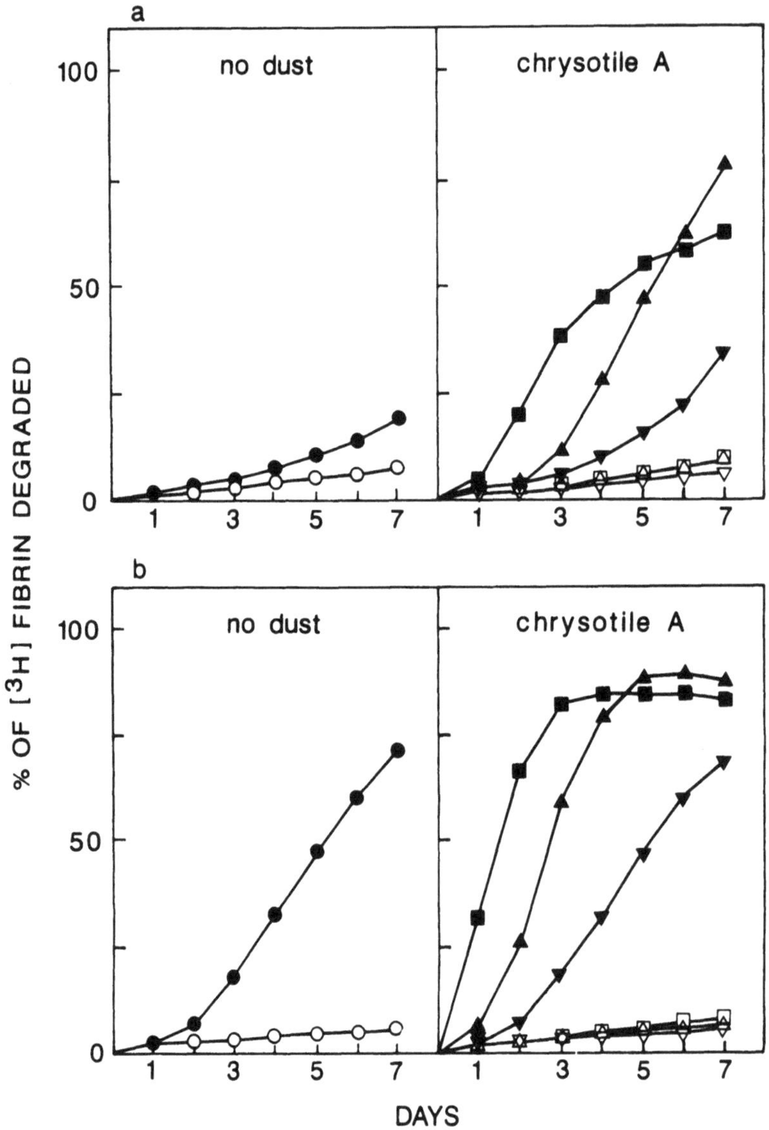

Fig. 1. Degradation of fibrin by macrophages exposed in vitro to chrysotile. Mouse peritoneal macrophages, either resident (a) or LPS-activated (b), plated on [^{3}H]fibrin-coated dishes, were incubated for 2 hours in the absence of dust (○ , ●) or in the presence of chrysotile A at 1 (▽ , ▼), 10 (△ , ▲) or 100 (□ , ■) µg/ml. Fibrin degradation was allowed to proceed for 7 days either in the presence of acidified FCS (closed symbols), or in the presence of the same serum depleted of plasminogen (open symbols).

Figure 2 shows the results of an experiment in which the PA activity of LPS-activated macrophages was compared after in vitro exposure to various respirable dusts generated from fibres of chrysotile A, crocidolite, wollastonite, rockwool, or glass, and to particles of rutile or quartz. From the comparison of the fibrous samples, two patterns emerge : strong stimulation of PA release by chrysotile, crocidolite, wollastonite and cellulose, as 10 µg of these materials are enough to degrade 40 % of the fibrin in less than 3-4 days ; slight stimulation by rockwool and glass as 100 µg are needed to reach a similar result. Rutile particles behave like weak stimulators. Quartz particles, in contrast, are not able to stimulate fibrinolysis. The fact that fibrin degradation by cells exposed to 100 µg of quartz is below the level of untriggered cells results from the high cytotoxicity of that material, as evidenced by the nearly complete release of the cytoplasmic lactate dehydrogenase from cells after 24 hours (data not shown).

Discussion

In vitro, asbestos fibres are a potent inducer of PA production by resident as well as activated macrophages. This property, originally reported by Hamilton (Hamilton et al. 1976), is consistent with a role for PA in the inflammatory and tissue-destructive processes that result from exposure to asbestos. Therefore, the secretion of PA by macrophages triggered in vitro was envisaged as a possible approach to evaluate the potential hazards of substitution fibres.

However, our results show no tight correlation between known pathogenicity and in vitro fibrinolysis. On the one hand, organic dusts generated from cellulose fibres were even more active than chrysotile A and crocidolite, contrasting with the absence of specific pathology related to cellulose. The in

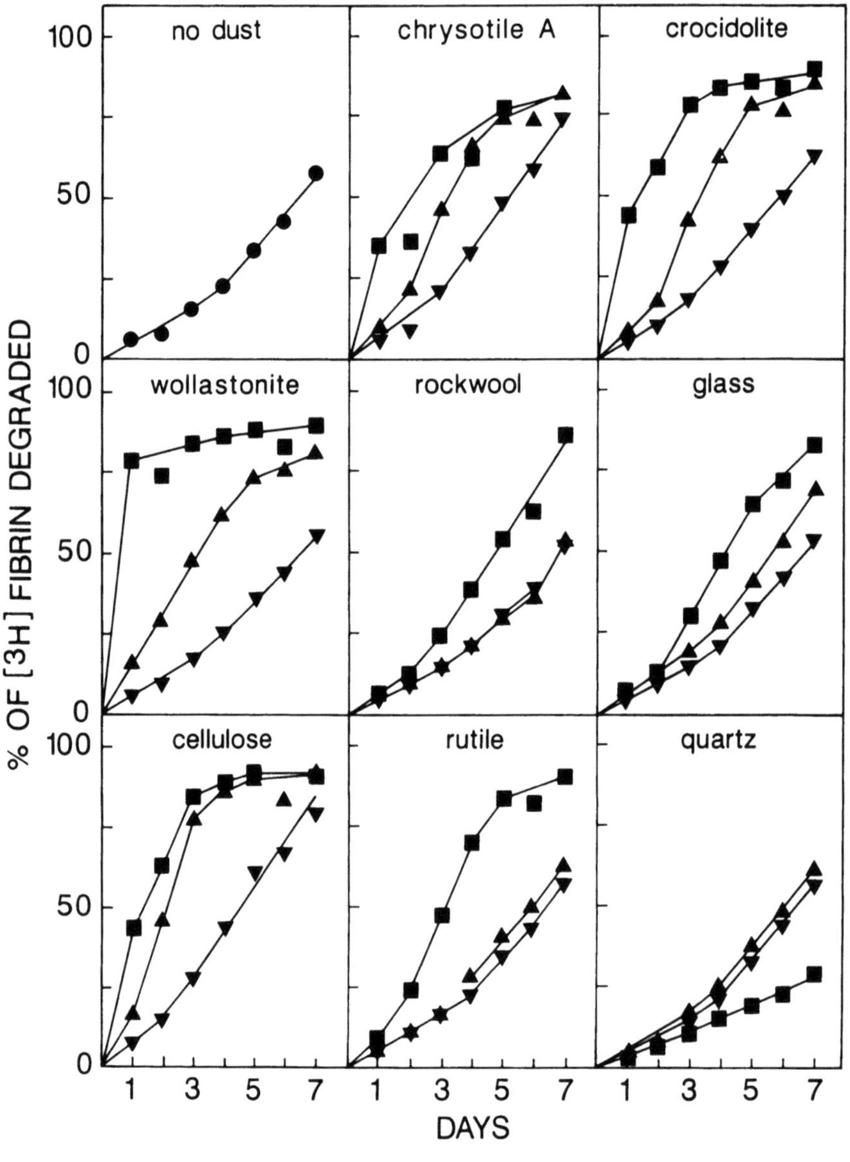

Fig. 2. Degradation of fibrin by macrophages exposed in vitro to various dusts. LPS-activated mouse peritoneal macrophages plated on [^{3}H]fibrin-coated dishes, were incubated for 2 hours in the absence of dust (o), or in the presence of the dust samples described under "Materials and Methods" at 1 (▼), 10 (▲) or 100 (■) µg/ml. Fibrin degradation was allowed to proceed for 7 days in the presence of acidified FCS.

vitro fibrinolytic activity of this material has been confirmed on another sample made from kraft cellulose (data not shown). On the other hand, quartz particles, although notoriously fibrogenic, do not promote PA release in vitro. Quartz particles, even at low dose, strongly stimulate cultured macrophages to secrete prostaglandins E_2 (PGE_2), which, by feedback regulation, inhibit production of PA (Hamilton 1981 ; Drapier and Petit 1984). Thus, conceivably, PGE_2 produced by quartz-exposed macrophages could be responsible for the lack of PA secretion by these cells. However, addition of indomethacin, a potent inhibitor of PGE_2 biosynthesis, did not allow quartz-exposed macrophages to produce PA (Godelaine and Beaufay 1988). Inhibition of PA activity by a specific protein has been documented (Chapman and Stone 1985 ; Wohlwend et al. 1987) ; the release of such a component by cells cultured in the presence of quartz remains to be examined.

Obviously, the lack of correlation between established pathogenicity and in vitro fibrinolysis prevents us from drawing any predictive conclusion about the materials examined. Without questioning the in vitro approach, our results underline the necessity to take into account a wide spectrum of cellular responses when addressing such a question.

Acknowledgements

This research was supported by the Belgian Institut pour l'Encouragement de la Recherche Scientifique dans l'Industrie et l'Agriculture (Convention n° 4396).

References

Chapman HA, Stone OL (1985) Characterization of a macrophage-derived plasminogen-activator inhibitor. Biochem J 230:109-116.

Danø K., Andreasen PA, Grøndahl-Hansen J, Kristensen P, Nielsen LS and Skriver L (1985) Plasminogen activators, tissue degradation, and cancer. Adv Cancer Res 44:139-266.
Deutsch DG, Mertz ET (1970) Plasminogen : purification from human plasma by affinity chromatography. Science 170:1095-1096.
Drapier J-C, Petit J-F (1984) Involvement of prostaglandins in LPS-mediated regulation of plasminogen activator synthesis by inflammatory macrophages. Int J Immunopharmacol 6:345-350.
Godelaine D, Beaufay H (1988) Effect of asbestos on the secretion of plasminogen activator by murine peritoneal macrophages in culture. Arch. internat. Physiol. Biochim. 96:B37.
Hamilton J, Vassalli J-D, Reich A (1976) Macrophage plasminogen activator : induction by asbestos is blocked by anti-inflammatory steroïds. J Exp Med 144:1689-1694.
Hamilton JA (1981) Regulation of prostaglandin and plasminogen activator production by mouse peritoneal macrophages. J Reticuloendothel Soc 30:115-128.
Hamilton JA (1983) Suppression of the neutral protease activity of macrophages treated with asbestos in vitro. Environ Health Perspect 51:103-108.
Wohlwend A, Belin D, Vassalli J-D (1987) Plasminogen activator-specific inhibitors produced by human monocyte-macrophages. J Exp Med 165:320-339.

Release of Interleukin-1 and Tumor Necrosis Factor by Rat Alveolar Macrophages after In Vivo or In Vitro Exposure to Mineral Dusts

K.E. Driscoll, R.C. Lindenschmidt, J.K. Maurer and J. M. Higgins
Human and Environmental Safety Division
The Procter & Gamble Company
Cincinnati, OH 45239

Introduction

An important role in the pathogenesis of chronic fibrotic lung diseases has been ascribed to the alveolar macrophage (AM). This cell, depending on its state of activation, has the capacity to regulate various immunologic and inflammatory responses through the elaboration of biologically active cytokines. Two major macrophage-derived cytokines are interleukin-1 (IL-1) and tumor necrosis factor (TNF). IL-1, has been suggested as an AM-derived factor that is active in fibrogenic responses to mineral dusts (Schmidt et al., 1984). Since TNF, like IL-1, has pro-inflammatory activities, and recent reports indicate it can modulate fibroblast growth (Vilcek et al., 1986), it too may play a role in the development of pulmonary inflammation and fibrosis following mineral dust exposure.

The present investigation was undertaken to better elucidate the role of AM activation and cytokine release in the development of fibrotic lung disease resulting from mineral dust exposure. We examined the ability of silica (SiO_2), a fibrogenic material, and titanium dioxide (TiO_2), a nonfibrogenic dust, to activate AM IL-1 and TNF secretion. In vivo and in vitro AM exposures were used to evaluate the role of direct particle:AM interactions in activating cytokine release.

Materials and Methods

Animals and Exposure Protocol. For in vivo dust exposures, specific pathogen-free (SPF) male, Fischer 344 (F344) rats weighing 200-220g, were intratracheally instilled with sterile saline (control) or saline suspensions of SiO_2 (Min-U-Sil; Pennsylvania Glass Sand Corp., Pittsburgh, PA) or TiO_2 (Fisher Scientific, Fair Lawn, NJ). The dusts were administered at a dose of 5.0 mg/100g bw and a dosing volume of 0.1 ml/100g bw. Six rats/treatment were sacrificed at 7, 14, and 28 days after exposure for bronchoalveolar lavage and analysis of the free lung cell populations. Five rats/treatment were sacrificed 28 days after exposure for light

NATO ASI Series, Vol. H30
Effects of Mineral Dusts on Cells
Edited by B. T. Mossman and R. O. Bégin

microscopic evaluations of hematoxylin and eosin stained and Masson's trichrome stained lung sections.

In vitro exposures were performed on AM obtained from SPF, male F344 rats. The AM were cultured in serum-free media, as described below, in the presence of various concentrations of heat-sterilized SiO_2 or TiO_2. After 24 hr of in vitro exposure the macrophage conditioned media (MCM) was harvested and analysed for IL-1 and TNF.

Bronchoalveolar Lavage. Bronchoalveolar lavage was performed on animals sacrificed by a lethal i.p. injection of sodium pentabarbital (5.0 mg/100g bw). The animals were exsanguinated via the abdominal aorta and the trachea exposed and cannulated with sterile plastic tubing. Using a 30 ml syringe connected to the tracheal cannula the respiratory tract was lavaged, in situ, 6 times with 8 ml of calcium- and magnesium-free phosphate buffered (PBS; pH=7.2) saline. The thorax was massaged gently during the lavage procedure to enhance cell yield. The bronchoalveolar lavage fluid (BALF) from each rat was centrifuged and the cell pellet resuspended in RPMI 1640 culture media (GIBCO, Grand Island, NY) containing 2 mg/ml bovine serum albumin (BSA). Total cell number and cell viability was determined by hemocytometer counting and trypan blue exclusion, respectively. Cell differentials were performed on cytocentrifuge preparations stained with Diff Quik™ (Sigma, St. Louis, MO).

Alveolar Macrophage Culture. Alveolar macrophages were cultured at a concentration of 1×10^6 AM/ml in 35 mm culture dishes. The cells were allowed to attach for 1 hr in a humidified incubator (37°C, 5% CO_2) after which the adherent monolayers were rinsed vigorously with RPMI 1640 media. When the rinsed AM monolayers were from in vivo exposures they were cultured (37°C, 5% CO_2) for 24 hr in RPMI 1640 containing 2 mg/ml BSA. When the monolayers were of naive cells for in vitro exposures they were cultured for 24 hr in RPMI 1640 containing 2 mg/ml BSA and 10 to 1000 µg/ml of SiO_2 or TiO_2. In some cases naive cells were incubated with lipopolysaccharide (LPS) as a positive control stimulus of AM IL-1 secretion.

Cytokine Analysis. Macrophage conditioned media from both in vivo and in vitro exposed cells were analysed for IL-1 and TNF activity. Interleukin-1 activity was determined by its ability to stimulate C3H/HeJ murine thymocyte proliferation in the presence of suboptimal concentrations of concanvalin A (Mizel, 1981). TNF activity was determined by its ability to lyse L929 fibroblasts (Flick and Gifford, 1984). Units of cytokine activity were determined by probit analysis and comparison to IL-1 or TNF standard preparations.

Statistical Analysis. Differences between treatments were evaluated using a one-way ANOVA followed by pairwise comparisons using the Newman-Keuls test. Statistical significance was considered at $p<0.05$.

Results

Cell Number and Viability. Intratracheal instillation of either SiO_2 or TiO_2 caused significant changes in the numbers and types of cells obtained by bronchoalveolar lavage (Table 1). The predominant cell type in the BALF of controls was the AM while dust exposed rats also had increased numbers of neutrophils and lymphocytes. SiO_2 elicited a more pronounced increase in cell numbers compared to TiO2. SiO2-induced increases in AM and lymphocyte numbers were progressive while after TiO2 exposure they returned toward control levels. In all cases the viability of the cells was greater than 93%.

Table 1. Bronchoalveolar Lavage Fluid Cell Numbers[a]

SacrificeDay: Treatment	Macrophages	Neutrophils	Lymphocytes
Day 7:			
saline	75.0 ± 2.0	0.3 ± 0.2	0.6 ± 0.2
TiO_2	125.1 ± 21.2[b]	61.1 ± 12.9[b]	16.5 ± 3.8[b]
SiO_2	118.2 ± 58.6[b]	140.0 ± 34.2[c]	37.3 ± 9.6[c]
Day 14:			
saline	75.2 ± 3.0	0.1 ± 0.1	1.4 ± 0.3
TiO_2	112.5 ± 9.4[b]	63.0 ± 10.5[b]	15.4 ± 3.0[b]
SiO_2	213.5 ± 8.9[c]	273.3 ± 28.1[c]	55.4 ± 7.2[c]
Day 28:			
saline	72.1 ± 5.0	0.7 ± 0.3	0.6 ± 0.2
TiO_2	98.0 ± 5.4	44.5 ± 6.9[b]	12.4 ± 1.1[b]
SiO_2	218.1 ± 21.5[c]	212.7 ± 60.7[c]	78.6 ± 8.1[c]

[a] data represent X±SE; cell numbers are $\times 10^5$
[b] significantly different from saline group mean; $p<0.05$.
[c] significantly different from saline and TiO_2 group mean; $p<0.05$.

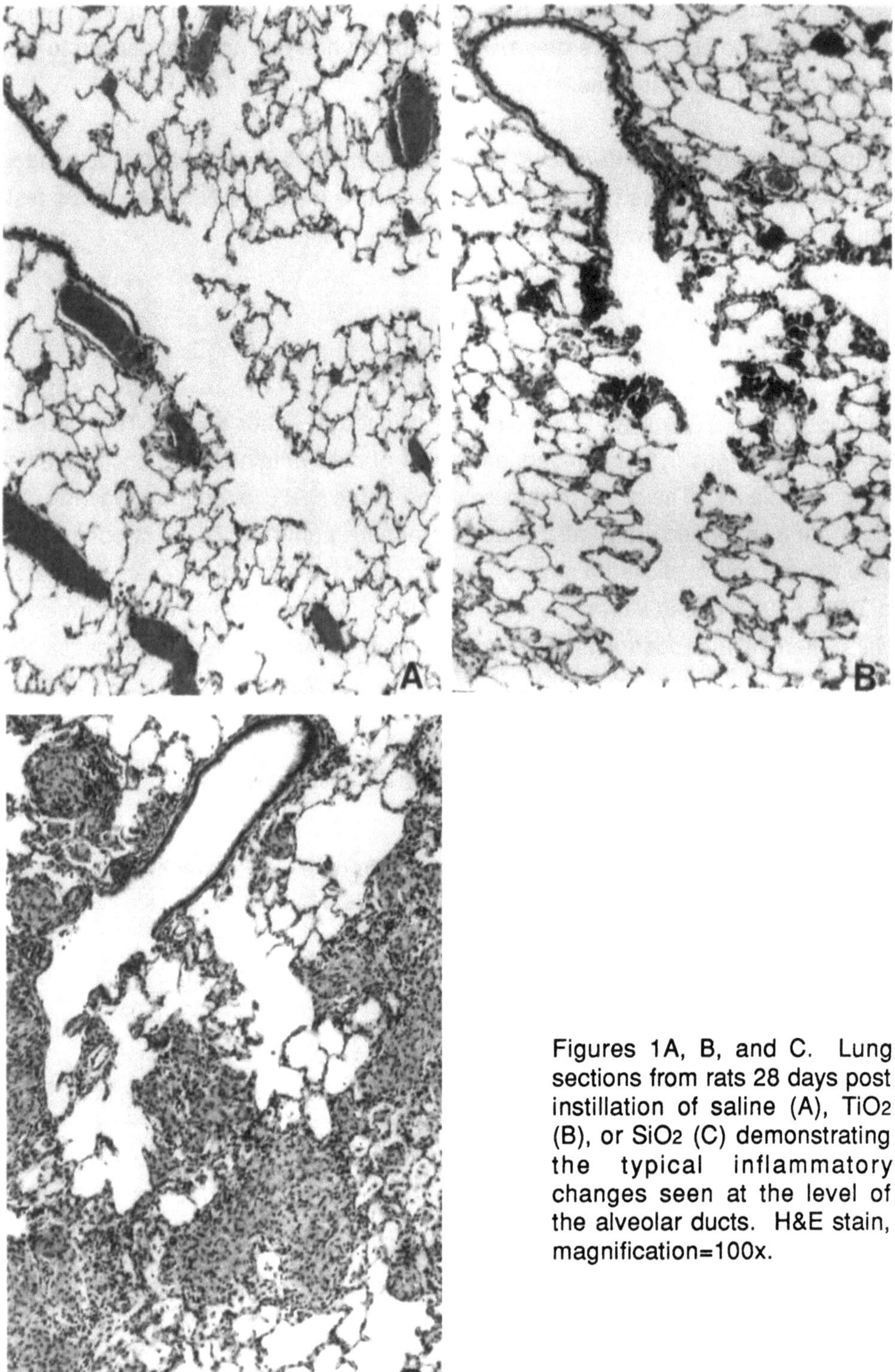

Figures 1A, B, and C. Lung sections from rats 28 days post instillation of saline (A), TiO_2 (B), or SiO_2 (C) demonstrating the typical inflammatory changes seen at the level of the alveolar ducts. H&E stain, magnification=100x.

Histopathology. Control lungs were normal in appearance (Figure 1A). With TiO_2 there were particle ladened macrophages at the alveolar duct level and scattered minimal interstitial inflammation (Figure 1B). SiO_2 (Figure 1C) resulted in prominant granulomas, the accumulation of neutrophils and foamy macrophages, necrosis of principally macrophages, and interstitial inflammation. Minimally increased collagen was observed with SiO_2 using Masson's Trichrome stain.

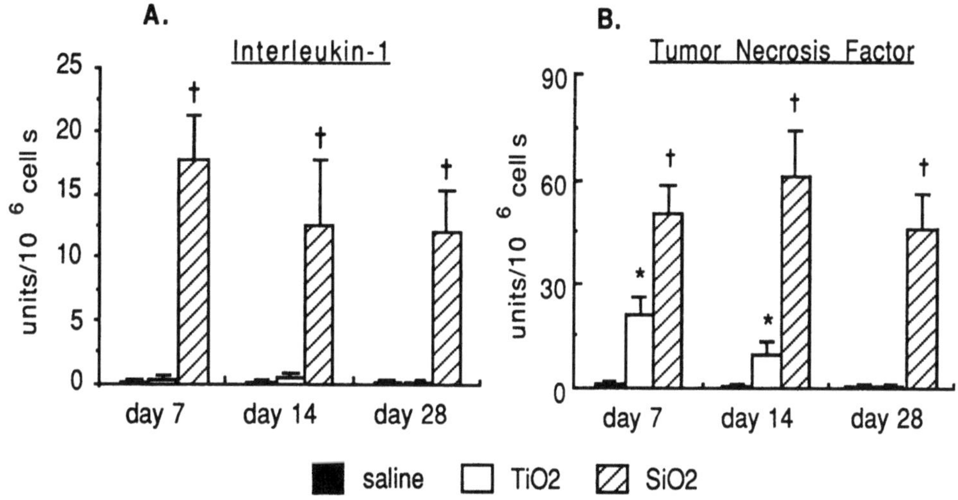

Figures 2A and B. Release of IL-1 and TNF by AM from rats intratracheally instilled with saline, TiO_2, and SiO_2. Results represent the X±SE of 6 animals sacrificed at 7, 14 and 28 days post treatment. (*) denotes a significant difference from the saline control group; $p<0.05$. (†) denotes a significant difference from the saline and TiO_2 groups; $p<0.05$.

Cytokine Production. Instillation of SiO_2 resulted in AM which spontaneously released IL-1 and TNF (Figure 2A and B); a response which persisted throughout the 28 days. TiO_2 did not stimulate an IL-1 response and TNF was increased on days 7 and 14 but, the levels detected were significantly less than that seen with SiO_2.

In vitro treatment of AM with either dust did not stimulate IL-1 release by AM (Figure 3) while exposure to LPS (a macrophage activating agent) elicited the release of IL-1. TNF increased in a dose-related fashion with SiO_2 but not TiO_2 (Figure 4).

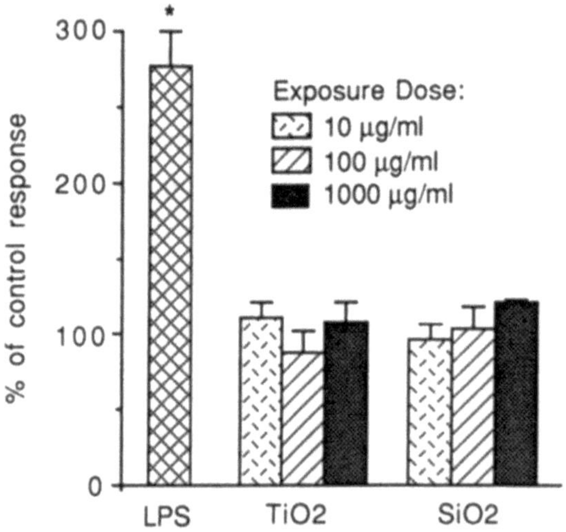

Figure 3. Release of IL-1 after *in vitro* exposure to SiO_2 or TiO_2. Results are presented as a % of the nonexposed control group response. All values represent the X±SE of 3 experiments. LPS= lipopolysaccharide at 3µg/ml. (*) denotes significantly different from nonexposed control AM; $p<0.05$.

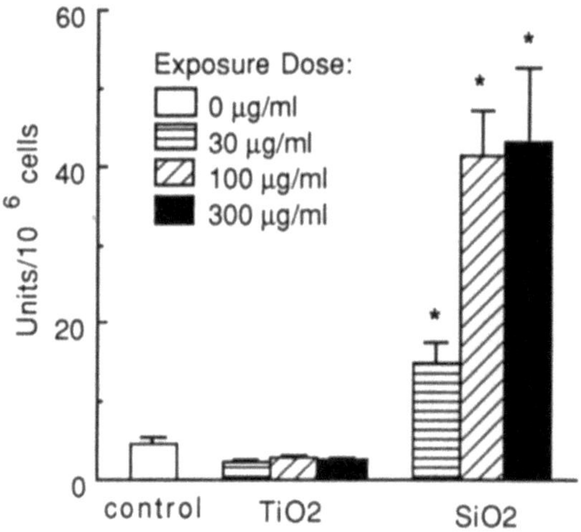

Figure 4. Release of TNF after *in vitro* exposure to SiO_2 or TiO_2. All values represent the X ± SE of 3 experiments. (*) denotes a significant difference from control and TiO_2 group means; $p<0.05$.

Discussion

Intratracheal instillation of either dust elicited a pulmonary inflammatory response. However, SiO_2 caused a significantly greater increase in cell numbers than TiO_2, and the changes in numbers of AM and lymphocytes after SiO_2 were progressive in nature while after TiO_2 they were transient. These BALF changes reflected marked differences in the lung histopathology after treatment with the two dusts. SiO_2 resulted in a chronic-active inflammatory response, with the accumulation of neutrophils, macrophages and lymphocytes, as well as, the formation of granulomas and increased collagen deposition. In contrast, TiO_2 resulted in minimal inflammation, with primarily accumulation of macrophages.

There were significant differences in the effects of SiO_2 versus TiO_2 on AM cytokine release. Instillation of SiO_2 activated AM to release increased quantities of IL-1, a response not observed with TiO_2. This observation confirms and extends those of Oghiso and Kubuta (1986) who reported increases in IL-1 release by AM from some but not all rats inhaling ~135 mg/M^3 SiO_2, 1 hr/d for 6 days. In vivo exposed AM released increased levels of TNF with SiO_2 and TiO_2, however, the degree of response was significantly greater and more perisistent with SiO_2. These findings indicate that instillation exposure to the fibrogenic dust SiO_2 is more effective in activating AM than the nuisance-type dust, TiO_2.

To determine the role of direct particle:cell interaction in stimulating AM to release IL-1 and TNF, normal AM were exposed in vitro to increasing concentrations of the two dusts. SiO_2 but not TiO_2 was sufficient to activate TNF secretion, however, neither dust stimulated IL-1 release. This absence of an IL-1 response in vitro was not due to the culture conditions since in vitro treatment with LPS stimulated AM to release of IL-1. These in vitro findings suggest that the direct effect of the SiO_2 on the AM may be responsible, at least in part, for the TNF response seen after in vivo SiO_2 exposure. However, the IL-1 response after SiO_2 instillation and the TNF release with TiO_2 in vivo can not be explained simply by a direct activating effect of the dust on the AM. These results suggest that factors, in addition to direct particle:cell interactions, are involved in vivo in activating AM secretory activity after dust exposure and they illustrate the potential difficulty in interpreting results solely from in vitro studies.

Expansion of the lung fibroblast population and increased production of collagen are important components of chronic interstitial lung disease. Interleukin-1 has been shown to augment the proliferative response of fibroblasts to growth factors such as platelet-derived growth factor (Estes et al., 1985). Additionally, other investigators (Schmidt et al., 1984; Libby et al., 1985) have reported that IL-1 may itself act as one of the fibroblast growth promoting factors released by mononuclear phagocytes. Recently, Vilcek et al. (1986) reported that TNF, while cytotoxic for many transformed cell lines, can stimulate the proliferation of normal fibroblasts. Thus, the release of IL-1 and/or TNF by AM activated after SiO_2 exposure represents a potential mechanism of increased fibroblast proliferation in silicosis.

The present study demonstrated that intratracheal instillation of SiO_2, at a dose that causes granulomas and increased collagen, activates AM to release IL-1 and TNF. In contrast, TiO_2, has no or a limited effect on macrophage IL-1 and TNF release, respectively. The effects of SiO_2 instillation on AM cytokine release appear to be only in part the result of a direct effect of the particle on the cell with other in vivo factors likely playing an important role in the activation process, particularly with respect to IL-1 release. Our findings further suggest a role for AM activation and associated cytokine production in the pathogenesis of silicosis.

References

Estes JE, Pledger WJ, Gillespie GY (1984) Macrophage-derived growth factor for fibroblasts and interleukin-1 are distinct entities. J Leuk Biol 35:115-129

Libby P, Wyler DJ, Janicka MW, Dinarello CA (1984) Differential effects of human interleukin-1 on growth of human fibroblasts and vascular smooth muscle cells. Arteriosclerosis 5:186-191

Mizel S B (1984) Production and quantitation of lymphocyte activating factor (interleukin-1). In: Herscowitz, Holden, Bellanti, Ghaffar (eds) Manual of macrophage methodology. Marcel Dekker, New York, p 407

Oghiso Y, Kubota Y (1986) Enhanced interleukin-1 production by alveolar macrophages and increases in Ia-positive lung cells in silica-exposed rats. Microbiol Immunol 30:1189-1198

Schmidt JA, Oliver CN, Lepe-Zuniga JL, Green I, Gery I (1984) Silica-stimulated monocytes release fibroblast proliferation factors identical ti interleukin-1. J Clin Invest 73:1462-1472

Vilcek J, Palombella VT, Henriksen-DeStefano D, Swenson C, Feinman r, Hirai M, Tsujimoto M (1986) Fibroblast growth enhancing activity of tumor necrosis factor and its relationship to other polypeptide growth factors. J Exp Med 163:632-643

The Effect of Aramid and Metaphosphate Fibers on Macrophage Viability and Function

Candace S. Wheeler and Charles D. Garner
Biomedical Science Department
General Motors Research Laboratories
Warren, MI 48090-9058

ABSTRACT

This study was designed to evaluate the cytotoxicity of a number of asbestos substitute fibers on primary rat alveolar macrophages (AMs) and a mouse macrophage-like cell line (J774A.1) in vitro. Following 24- and 48-hour incubations, the cells were observed under phase contrast microscopy for changes in cell morphology and cell viability was assessed using trypan blue exclusion. A viability index (VI) was calculated to reflect changes in both cell number and cell viability. All of the materials studied produced a concentration-dependent reduction in VI. In addition, both Kevlar and calcium sodium metaphosphate (CSMP) fibers produced a concentration-dependent release of lactate dehydrogenase (LDH) from primary AMs. Finally, no chemotaxis was observed in response to either Kevlar or CSMP fiber alone, but both fibers stimulated measurable concentration-dependent chemotaxis when incubated with AMs. These results suggest that all of the materials were cytotoxic to AMs in vitro in the form and at the concentrations (25, 250 μg/mL) used in this study and can affect AM function. What effect these materials have on the human health remains to be determined.

INTRODUCTION

Substitute materials have been developed as replacements for asbestos but the potential health impact of these materials is largely unknown. The impact of these materials on human health will depend primarily on the deposition and clearance of the material in the lung and the chemical and physical properties (composition, size, shape, and durability) of the fibers themselves. Alveolar macrophages (AMs) are the primary cell type to encounter inhaled materials in the lung and are responsible for their effective removal. Adverse effects on the activity of these cells can,

NATO ASI Series, Vol. H30
Effects of Mineral Dusts on Cells
Edited by B.T. Mossman and R.O. Bégin

therefore, lead to increased retention of inhaled materials, impairment of the defense capacity of the lung, and an increased susceptibility to disease. The purpose of this study was to evaluate the relative cytotoxicity of a number of asbestos fiber substitutes on primary rat AMs and a mouse macrophage-like cell line, J774A.1, in vitro. In addition, the effect of these materials on AM chemotaxis was assessed as an indicator of cell function.

EXPERIMENTAL

Test Materials: UICC chrysotile, crocidolite, and amosite were used as positive controls. The calcium sodium metaphosphate™ fibers ($[CaNa(PO_3)_3]_n$, CSMP, fine, Monsanto) were bent fibers which differed greatly in length and diameter while Fiberfrax® (milled ceramic fiber, Sohio) composed mainly of Al_2O_3 and SiO_2, had uniform, straight blunt-ended fibers approximately 14 µm long and 2-3 µm in diameter. Kevlar® (special inhalation grade aramid fiber, DuPont) was composed of highly interwoven fibers and fibrils that differed greatly in size and shape.

Cell Preparations: Primary AMs were obtained by bronchopulmonary lavage of male Fischer 344 rats weighing 200-250g using Hanks' balanced salt solution prepared without calcium and magnesium. AMs were collected by a brief centrifugation, diluted in Dulbecco's MEM supplemented with 10% fetal bovine serum, 2 mM M-glutamine, 0.1 mM MEM non-essential amino acids, 100 units/mL penicillin, and 100 µg/mL streptomycin, and cultivated at 37° C in an atmosphere of 5% CO_2. A macrophage-like cell line, J774A.1, was obtained from the American Type Culture Collection and cultivated as outlined above.

Fiber Exposure: 1.0×10^5 adherent cells were exposed to 0, 2.5, 25, or 250 µg fiber/mL of medium for 24 or 48 hours. Exposures were performed on a weight per volume basis and did not take into consideration differences in the size of the materials or their relative densities.

Analysis: Following exposure, changes in cell morphology were assessed using phase contrast photomicroscopy. Cell counts were made and viability measured using trypan blue exclusion. A viability index was calculated to reflect changes in both of these parameters.

Membrane Integrity: 1.0 x 10^6 adherent cells/1 mL were incubated with 0, 2.5, 25, or 250 µg fibers/mL medium for 24 or 48 hours. Lactate dehydrogenase activity was assessed spectrophotometrically by following the decrease in NADH absorbance at 340 nm using pyruvate as a substrate for both the medium and cell lysate.

Chemotaxis: AM chemotaxis was assessed using a modification of the Boyden method (1962) by placing 0.2 mL of medium containing the material to be tested in the lower of a two-compartment chamber either with or without AMs and adding 0.2 mL of cell suspension to the upper compartment. A 5.0 µm pore polycarbonate filter and a 0.45 µm pore cellulose filter were used to separate the two compartments. The filters were removed after four hours, fixed, stained, and mounted on glass slides. N-formyl-methionyl-phenylalanine and zymosan-activated serum were used as positive controls while medium alone was used as a negative control. Chemotaxis was quantified by counting the number of cells that had migrated through the filter in 20 random microscopic fields on duplicate filters.

RESULTS AND CONCLUSIONS

In this study, simple but quantitative markers of cellular integrity were used to compare and evaluate the relative cytotoxicity of a broad spectrum of fibrous and particulate materials in a mammalian cell system. Cell morphology was not significantly altered after 24 or 48 hour incubations with any of the materials studied. All cells remained an intact adherent monolayer engaged in active phagocytosis with a number of cells containing engulfed fibers and some large fibers surrounded by more than one cell.

Table 1

Cytotoxicity of Asbestos and Nonasbestos Fibers on J774A.1 Cells In Vitro

Fiber Concentration	0.25 µg/ml	2.5 µg/ml	25 µg/ml	250 µg/ml
Crocidolite	85.2	87.0	44.1	16.0
Chrysotile	79.2	75.7	46.7	NA
Amosite	83.0	80.8	42.1	15.2
Fiberfrax	81.7	76.2	65.9	61.1
Metaphosphate	83.8	83.0	60.3	21.4
Kevlar	NA	92.9	65.1	62.1

All of the fibers studied produced a concentration-dependent decrease in VI when incubated with either cell type for 24 or 48 hours. In the J774A.1 cells, this decrease was significant at the two highest concentrations for chrysotile, crocidolite, and amosite and the three asbestos substitute fibers (Table 1).

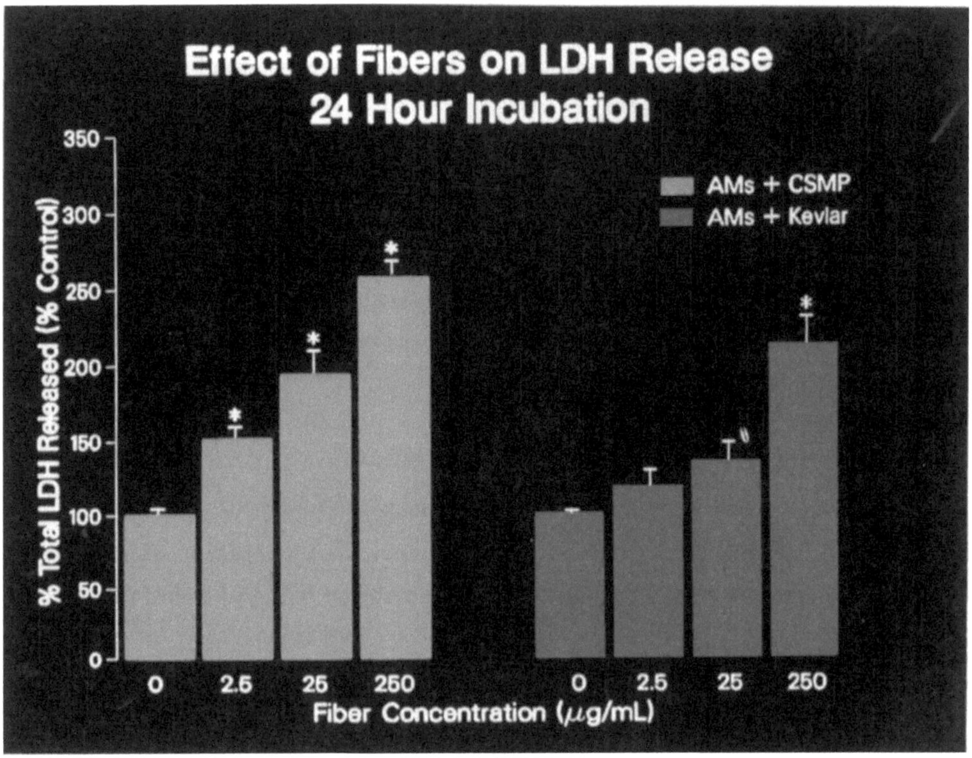

Figure 1 Incubation of AMs with Kevlar and CSMP fibers for 24 and 48 hours produced a significant concentration-dependent increase in LDH activity in the medium.

These results suggest that all of the materials studied were cytotoxic to AMs in vitro in the form and at the concentrations used in this study. At equivalent concentrations (μg/ml), the relative cytotoxicity to J774A.1 cells after a 24 hour incubation was amosite, crocidolite, and chrysotile > CSMP > Kevlar and Fiberfrax.

Both Kevlar and CSMP fibers stimulated measurable dose-dependent chemotaxis when incubated with AMs but were not directly chemotactic (Figure 2). These results suggest that these fibrous materials can effect AM function and support earlier studies by this laboratory (Wheeler and Garner, 1986a,b) which suggest that the interaction of particles with AMs is necessary to stimulate the production and release a factor chemotactic for additional AMs and is independent of the type of particle involved.

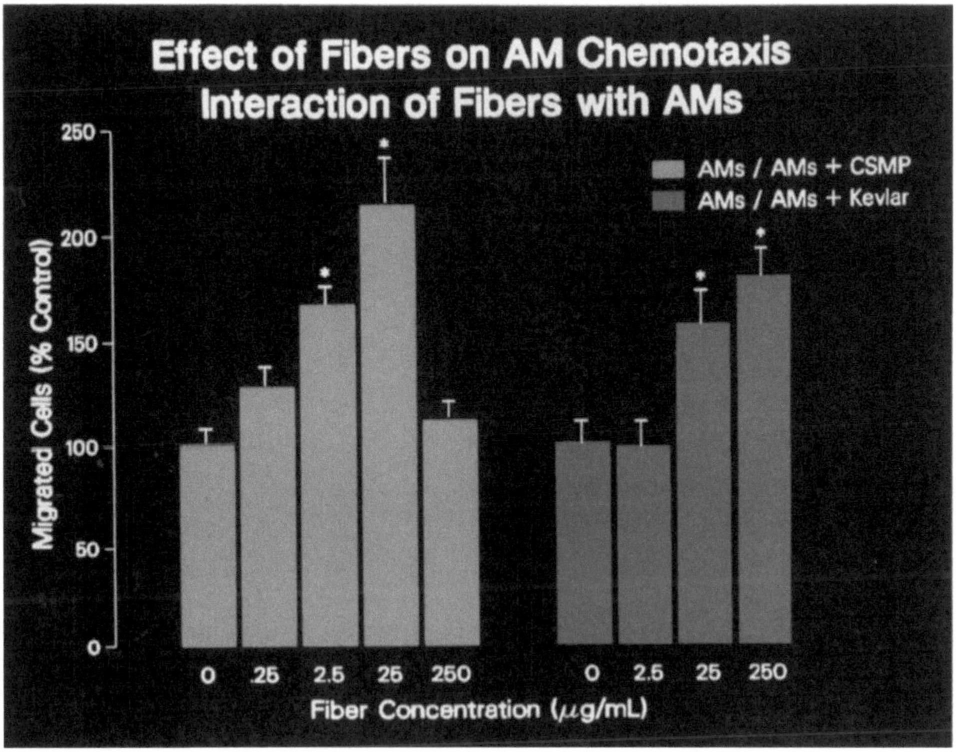

Figure 2 Both CSMP and Kevlar when incubated with AMs stimulate the release of a chemotactic factor and produce a dose-dependent increase in AM chemotaxis.

This study represents an important first step in the analysis of the potential health effects of these materials. Studies are under way to determine the mechanism of action of these fibrous materials and the effect of fiber size and shape on the relative toxicity of the fibers.

REFERENCES

Boyden S (1962) The Chemotactic Effect of Mixtures of Antibody and Antigen on Polymorphonuclear Leukocytes. J Exp Med 115:453-466

D'Arcy JB, Schreck RM, Siak JS (1986) Health Effects Analysis of Wear-Generated Particles from Semi-Metallic Brake Materials. GMR-5611

Wheeler CS, Garner CG (1986a) The Role of Phagocytosis in Alveolar Macrophage Chemotaxis. GMR-5311

Wheeler CS, Garner CG (1986b) The Effect of Diesel Particulate on Alveolar Macrophage and Polymorphonuclear Leukocyte Chemotaxis. GMR-5325

EFFECTS OF NICOTINAMIDE ON THE CYTOTOXICITY OF MINERAL DUSTS TOWARDS PULMONARY ALVEOLAR MACROPHAGES.

D. Nadeau, D. A. Lane, D. Paradis and L. Fouquette
Laboratoire de Caractérisation de l'Amiante
Département de Biologie
Faculté des Sciences
Université de Sherbrooke
Sherbrooke, Qc
Canada J1K 2R1

Introduction

It is well acknowledged that the inhalation of particles and/ or fibres can be linked to different pulmonary pathologies. In the deep lung, the pulmonary alveolar macrophages (PAM) play a major role in the defense of the lung against the toxicity of respirable dusts (Green, 1970). During the respiratory burst, the oxygen radical species associated with the phagocytosis of foreign matters by PAM have been implicated in the mechanism of toxic injury by mineral dusts such as asbestos fibres (Mossman et al., 1987). Recently, we demonstrated that nicotinamide (NAM) protected PAM against the <u>in vitro</u> toxic effects induced by chrysotile asbestos fibres (Nadeau and Lane, 1988a). Our results indicated that, most probably, the major mechanism of protection of NAM did not involve the poly(ADP-ribosylation) system. We have also shown that the energy metabolism of the PAM monolayers was significantly altered by the presence of NAM in the culture medium (Nadeau and Lane,1988b).

In view of our initial observations, the objective of this study was to determine if the NAM could also protect PAM

NATO ASI Series, Vol. H30
Effects of Mineral Dusts on Cells
Edited by B.T. Mossman and R.O. Bégin

monolayers against the cytotoxicity induced by mineral dusts other than chrysotile asbestos. For that purpose, we exposed PAM monolayers to UICC crocidolite asbestos fibres, and to quartz (DQ12) and silica particles. For comparison purposes, very short and UICC B chrysotile fibres were also included in the study. The response of PAM towards those mineral dusts were evaluated with our usual cytotoxicity parameters, that is the extracellular liberations of lactate dehydrogenase (LDH) and ß-galactosidase (ß-GAL), and the intracellular levels of ATP and NAD^+.

Material and Methods

Mineral Dusts

The short chrysotile fibres ($\bar{L}$= 0.917 µm; $\bar{D}$= 0.08 µm) were isolated as described before (Nadeau et al., 1987). The UICC chrysotile (type B,Canadian) and crocidolite fibres, the quartz DQ12 and the silica particles were kindly provided by l'Union Internationale Contre le Cancer, Dr. Klauss Robock and Dr. G.W. Gibbs, respectively.

Cell Isolation and Culture

As described before (Nadeau et al., 1987), PAM were isolated by bronchoalveolar lavages from male Long-Evans rats with an Hepes-phosphate buffered saline. The macrophages were seeded in multiwells (4 x 10^5 cells/well; ≈2 cm^2/well, 24-well Linbro culture plates) with 1.5 ml of a modified MEM medium containing 4% foetal bovine serum. Then, after a 2 h incubation at 37°C to establish the cell monolayer, the culture medium was removed to eliminate nonadherent cells. After the addition of fresh medium, supplemented or not with 20 mM NAM, the monolayers were exposed to the mineral dusts for 18 h at 37°C. Thereafter, the different cytotoxicity parameters were measured.

Extracellular Enzyme Releases

The LDH activity was measured by following the rate of transformation of NAD^+ into NADH, in the presence of lactic acid. The β-GAL assay rely on the quantification of the amount of p-nitrophenol released from the substrate p-nitrophenyl-β-D-galactopyrannoside (Nadeau et al.,1987).

ATP and NAD^+ Cell Contents

After an extraction of the PAM monolayers with DMSO, the ATP cell contents were quantified with a bioluminescence method [luciferine and luciferase] (Nadeau et al., 1987). For the NAD^+ determinations, after an acidic extraction, the samples were neutralized. Then, in the presence of LDH and lactic acid, the NAD^+ is converted into NADH. Thereafter, the luminescence was measured [NADP(H)-FMN oxidoreductase and luciferase] (Nadeau and Lane, 1988a; Lane et Nadeau, 1988).

Results and Discussion

Very Short Chrysotile Fibres

For the PAM monolayers exposed to the very short chrysotile fibres, the presence of 20 mM NAM considerably attenuated the extracellular releases of LDH (Fig. 1a) and β-GAL (Fig. 1b), even at the highest dose studied (500 μg). The protective ef-

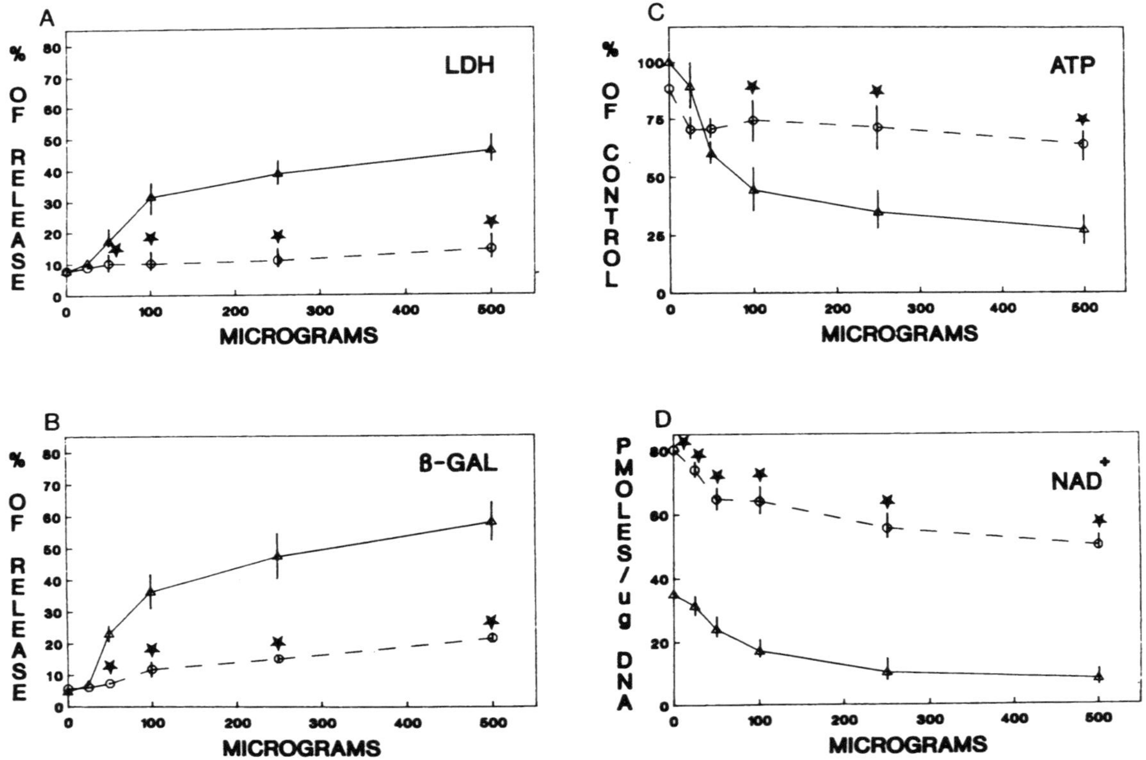

Fig. 1. Effects of the addition of 20 mM NAM on the cytotoxic response of PAM monolayers exposed to very short chrysotile fibres. The results are expressed as the mean ± the standard error (S.E.) of 3 experiments. △——△ no NAM; ○----- ○ 20 mM NAM. The stars (★) indicate that the addition of NAM produced a statistically significant effect ($p \leq 0.05$, nonparametric Kruskal-Wallis ANOVA coupled to the Conover's t test for multiple comparisons; Nadeau et al., 1988d).

fect of NAM was confirmed by the prevention of a "no return" depletion of the ATP pools (Fig. 1c). As described before (Nadeau and Lane, 1988a), the exposure of the PAM monolayers to the short asbestos fibres caused a depletion of the NAD^+ pools in the presence, or in the absence of NAM (Fig. 1d). However, in the presence of NAM, the NAD^+ pools of the cell monolayers exposed to the highest dose of the short chrysotile fibres always remained higher than those of unexposed cell monolayers incubated without NAM. Thus, when PAM were exposed for 18 h to doses up to 500 µg of very short chrysotile fibres, the presence of 20 mM NAM prevented the cytotoxic responses of the cells. Despite the cytotoxicity of the UICC B chrysotile fibres was weak, a protection by the vitamin was also observed with these longer fibres (data not shown).

UICC Crocidolite Fibres

The effects of NAM on the cytotoxicity induced by the exposure of PAM monolayers to crocidolite are presented in Table 1. In the presence of NAM, the cytotoxicity was lower as indicated by the significant reductions in the extracellular leakage of LDH and ß-GAL, and the slower depletion of the ATP pools. However, contrary to the chrysotile fibres, the protection was only partial. Effectively, in the presence of the vitamin supplement, there was still significant extracellular releases of the marker enzymes, and marked depletions of the ATP pools at the highest doses studied (e.g. $\approx$ 50% at $\geq$ 250 µg).

Quartz DQ12 and Silica Particles

The effects observed with these mineral dusts were very similar (Table 1). Effectively, these two particles were very toxic to the PAM monolayers as shown by the extensive extracellular releases of LDH and ß-GAL, and by the massive deple-

TABLE 1. EFFECTS OF THE ADDITION OF 20 mM NAM ON THE *IN VITRO* BIOLOGICAL RESPONSE OF PAM MONOLAYERS EXPOSED TO DIFFERENT MINERAL DUSTS.

EXPERIMENTAL GROUP	DOSE (μg)	EXTRACELLULAR RELEASE (% of total cell content) LDH		β-GAL		ATP CELL CONTENTS (% of control)		NAD^+ CONTENTS (pmoles/μg DNA)	
		- NAM	+ NAM	- NAM	+ NAM	- NAM	+ NAM	- NAM	+ NAM
CONTROL	-	8 ± 2[a]	8 ± 2	5 ± 1	6 ± 1	100 ± 1	89 ± 6	24 ± 2	58 ± 3[c]
UICC CROCIDOLITE FIBRES	25	20 ± 3	11 ± 2[c]	14 ± 1	8 ± 1[c]	71 ± 7	71 ± 5	N.D.[b]	N.D.
	50	27 ± 3	12 ± 3[c]	18 ± 1	10 ± 1[c]	63 ± 9	61 ± 9	N.D.	N.D.
	100	39 ± 2	20 ± 1[c]	29 ± 2	13 ± 1[c]	47 ± 2	64 ± 8[c]	N.D.	N.D.
	250	60 ± 3	29 ± 3[c]	43 ± 2	21 ± 2[c]	26 ± 4	52 ± 7[c]	N.D.	N.D.
	500	68 ± 3	31 ± 3[c]	49 ± 4	26 ± 3[c]	19 ± 3	52 ± 9[c]	N.D.	N.D.
QUARTZ DQ12 PARTICLES	5	9 ± 2	9 ± 1	5 ± 2	4 ± 2	99 ± 9	84 ± 8	26 ± 2	54 ± 3[c]
	10	15 ± 1	16 ± 1	8 ± 2	7 ± 1	94 ± 8	78 ± 9	22 ± 1	52 ± 2[c]
	25	37 ± 3	30 ± 4	19 ± 1	15 ± 1	56 ± 5	46 ± 7	15 ± 1	41 ± 2[c]
	50	64 ± 3	47 ± 4[c]	37 ± 2	24 ± 2	23 ± 6	29 ± 5	9 ± 1	30 ± 2[c]
	100	67 ± 3	51 ± 8	42 ± 5	26 ± 3	11 ± 2	21 ± 4	4 ± 1	16 ± 1[c]
	250	78 ± 6	60 ± 7	49 ± 6	33 ± 2	7 ± 1	12 ± 2	N.D.	N.D.
SILICA PARTICLES	5	10 ± 1	10 ± 1	7 ± 2	4 ± 2	106 ± 7	96 ± 2	26 ± 2	58 ± 4[c]
	10	13 ± 1	13 ± 1	9 ± 3	5 ± 2	100 ± 7	84 ± 9[c]	27 ± 2	54 ± 3[c]
	25	25 ± 1	23 ± 2	12 ± 1	11 ± 1	77 ± 2	68 ± 7	24 ± 2	50 ± 4[c]
	50	43 ± 3	40 ± 3	21 ± 2	16 ± 1	54 ± 5	53 ± 9	22 ± 1	37 ± 2[c]
	100	63 ± 8	43 ± 9	30 ± 5	18 ± 3	24 ± 4	34 ± 1	11 ± 1	26 ± 2[c]
	250	73 ± 4	57 ± 11	40 ± 4	26 ± 3	9 ± 2	18 ± 1	5 ± 1	15 ± 1[c]

[a]Mean ± S.E. (n = 3 to 5)　[c] $p \leq 0.05$ (nonparametric Kruskal-Wallis ANOVA coupled to the Conover's t test for multiple comparisons; Nadeau et al., 1988d)

[b]N.D. : Not determined

tions of the cellular ATP and NAD+ pools. With the 20 mM NAM supplement, there was no protective effect of the vitamin with the exception of small reductions in the release of ß-GAL at doses higher than 100 µg. Even if the NAD+ pools were increased in the presence of NAM, the cellular depletions caused by these particles were massive (e.g. ≈ 50% at 100 µg).

Conclusion

To summarize, we observed major differences in the protective effect of NAM on the cytotoxicity induced by mineral dusts. There was an almost complete protection with the chrysotile fibres, a partial protection with the crocidolite fibres, and no significant protection with the quartz and the silica particles. Apparently, when PAM are exposed for 18 h to these mineral dusts, the protective effect of NAM does not seem to be clearly related to the magnitude of a given cellular response. Hence, one can notice that the exposure to 250 µg of very short chrysotile fibres, to 100 µg of crocidolite fibres, and to 25 µg of quartz particles caused similar extracellular releases in LDH (≈ 40%). However, the protection by NAM was almost complete with the short asbestos fibres (release of ≈10%), while no protective effect of the vitamin was observed with the quartz particles.

Kinetic studies revealed striking differences in the cytotoxic responses induced by very short chrysotile and crocidolite fibres, and quartz particles towards PAM (Nadeau et al., 1988c; to be published elsewhere). While the biological responses of the cell monolayers were quite linear over an 18 h-exposure to the short asbestos fibres, the maxima were observed sooner with the crocidolite fibres and more so with the quartz DQ12 particles. It appears then that the efficiency of the vitamin in protecting PAM monolayers against the toxicity of mineral dusts is inversely related to the time at which the cytotoxic response of a given dust becomes significant.

It is therefore postulated that the magnitude of the protective effect of NAM against the toxicity of mineral fibres and particles towards PAM, could be related in part to differences in the level of production of reactive oxygen radicals species associated with the phagocytosis of mineral dusts by PAM(This study was supported by l'Institut de l'Amiante, Montréal, Québec, Canada; Grant #19, Project LCA).

References

Green GM (1970) Amberson lecture: in defense of the lung. Amer Rev Respir Dis, 102:691-700

Lane DA, Nadeau D (1988) Determination of pyridine nucleotides contents of cell monolayers by bioluminescence. J Biochem Biophys Meth, 17:107-118

Mossman BT, Marsh JP, Shatos MA, Doherty J, Gilbert R, Hill S (1987) Implication of active oxygen species as second messengers of asbestos toxicity. Drug Chem Toxicol, 10: 157-180

Nadeau D, Fouquette-Couture L, Paradis D, Khorami J, Lane D., Dunnigan J (1987) Cytotoxicity of respirable dusts from industrial minerals: comparison of two naturally occuring and two man-made silicates. Drug Chem Toxicol, 10:49-86

Nadeau D, Lane DA (1988a) On the cytotoxicity of chrysotile asbestos fibers towards pulmonary alveolar macrophages. I. Effects of inhibitors of the ADP-ribosyl transferase. Cell Biol Toxicol, 4:13-30

Nadeau D, Lane DA (1988b) On the cytotoxicity of chrysotile asbestos fibres towards pulmonary alveolar macrophages. II. Effects of nicotinamide on the cell metabolism. Toxicol Appl Pharmacol,(Accepted for publication)

Nadeau D, Lane DA, Paradis D, Fouquette L (1988c) Protection by nicotinamide of pulmonary alveolar macrophages (PAM) exposed in vitro to chrysotile asbestos. Proc Can Fed Soc Biol, 31:136

Nadeau D, Paradis D, Fouquette L, Dunnigan J (1988d) In vitro toxicology data on respirable particles and fibres and its evaluation for statistical significance. In: Goldberg AM (ed) Progress in in vitro toxicology. Alternative Methods in Toxicology, vol 6. Mary Ann Liebert, Inc., Publishers, New York, p 237-251

LONG-TERM EFFECT OF 3METHYLCHOLANTHRENE-COATED ASBESTOS FIBRES ON THE IMMUNE STATUS OF F1(NZB*C57Bl/6) MICE

A. Lange, B. Pietrzyca, A. Moniewska, J. Rabczynski*

Institute of Immunology and Experimental Therapy
53-114 Wrocław, Czerska 12
Poland

Introduction

There are several pieces of evidence that asbestos is a carcinogenic substance. Asbestos exposure also affects the immune system, what may be relevant for carcinogenic process. The immunological outcome of exposure depends in part on the genetically dependent responsiveness of the host (Matej et al. 1978). Among environmental factors polycyclic aromatic hydrocarbons (PAH) increases the risk of asbestos-related cancer (Mossman et al. 1983). This effect also depends on the inherited ability of the host to metabolize PAH to ultimate carcinogens (Kouri et al. 1980). Both asbestos and PAH augment their carcinogenic potential, however, this effect is not seen when asbestos-related immunopathology is evaluated (Lange 1980, Wagner 1981). Therefore, study on the interaction between PAH and asbestos fibres should address the question of the effect on the immune system. To make this work possible an appriopriate model has to be employed to provide a susceptible system for the effect of asbestos fibres and/or PAH on the immunity. F1(NZB*C57Bl/6) mouse hybrid was used. Due to the genetic trait inherited from NZB maternal strain, these mice show the immunological type of responsiveness (Theofilopoulos and Dixon 1981) which resmebles the human situation (HLA B8 antigen associated humoral hypereactivity) known to be a risk factor of asbestos-related immunopathology (Matej et al. 1978). F1(NZB*C57Bl/6) are also prone to the AHH inducement due to the

*Department of the Morbid Anatomy, Medical School, Wrocław

NATO ASI Series, Vol. H30
Effects of Mineral Dusts on Cells
Edited by B. T. Mossman and R. O. Bégin

trait inherited from C57Bl/6 paternal strain (Dzugaj et al. 1985). 3methylcholanthrene (3MC) as an immunocytotoxic substance was chosen because it impaires T cell function of C57Bl/6 mice and in addition alters the percent distribution of T-helper and T-supresor cells in favor of the suppressors (Alfred et al. 1983).
In this work the effect of 3MC coated chrysotile on the naturally occuring immunological abnormalities in F1(NZB*C57Bl/6) mice was investigated.

Materials and Methods

Mice. F1(NZB*C57Bl/6) hybrid mice were raised in our institute from parental strains (NZB/Iiw and C57Bl/6JBoyIiw) and used for experiments at 8-12 weeks of age.

Injected material. Aqueous suspension of chrysotile (A/PBS) (UICC chrysotile A), 3MC-coated chrysotile (3MC/A/PBS), linseed oil 3MC solution (3MC/OIL), were all prepared as described in the accompanying paper (Pietrzyca et al. see this volume).

Design of experiments. Five groups of mice were injected i.p. in 4 days intervals with two doses (listed respectively) of aqueous chrysotile suspension (2.5 mg, 1.5 mg), 3MC (500 ug, 300 ug)-coated chrysotile (2.5 mg, 1.5 mg), 3MC oil solution (500 ug, 300 ug) and as controls with PBS and linseed oil. After 14 weeks of observation all mice were sacrificed and necropsized as well as mice found dead between 0 - 14 weeks after the injection. Gross anatomical observation of the peritoneal tissue was done. The spleens were weighted and the kidneys taken for pathomorphology.

Antisera: against mouse gamma-globulin from rabbit came from Behring (Marburg,FRG), against mouse IgM from goat (kindly donated by Dr. E. Lisowska), against mouse IgG FITC conjugated came from Heintel (Vienna, Austria), Orion-Diagnostica (Helsinki, Finland), Hybritech (San Diego, USA), against rabbit gamma globulin FITC conjugated came from Behring (Marburg, FRG), against mouse gamma globulins FITC conjugated came from Difco (Detroit, USA) and from Amersham (Buckinghamshire, UK).

Immunology. Serum IgM concentrations were measured by radial immunodiffusion in agar. Sera were added to wells cut in agar (Difco, Detroit,USA) impregnated with monospecific IgM antibody. NZB mouse prototype serum in the range of dilution from 1:1 to 1:8 were used for preparing the standard curve. The plates were incubated for 48 hrs in $4^{o}C$ and the rings diameters were measured. The results were expressed as a fraction of a prototype serum. Antinuclear antibodies (ANA) were evaluated in sera with the use of indirect immunofluorescence technique (IF) as described elsewhere (Roitt and Doniach 1969) employing rat liver as an antigenic substrate and FITC conjugated anti-mouse gammaglobulin serum.

Tissue sections. Paraffin sections were routinely stained with hematoxylin and eosin; cryostat sections of mice kidneys were stained for the presence of immunoglobulin deposits with the use of IF technique. Percentage of the glomeruli showing unequivocal immunoglobulin (Ig) deposits was calculated from the enumeration of at least 20 glomeruli.

Results

ANA appear spontaneously in NZB mice and also in some their crosses (Theofilopoulos et al. 1985) including female F1(NZB*C57Bl/6) mice (not shown). The incidence of ANA in F1(NZB*C57Bl/6) mice injected with PBS varied from 45% to 83% in 4 different experiments. This incidence was lowered by i.p. exposure of the mice to 3MC-coated chrysotile (Table 1).

Kidney lesions. In three experiments immuno-pathomorpology of kidneys was evaluated. In two out of three experiments percentage of glomeruli with immunoglobulin deposits was lower in mice injected with 3MC-coated chrysotile as compared to the PBS and 3MC oil solution injected mice. In the third experiment there was no apparent differences between the groups (Table 2). To substantiate the results of the IF the paraffin sections of the kidneys were evaluated after hematoxylin eosin staining. Nine out of 16 mice injected with 3MC oil solution showed glomeruli abnormalities but only 2 out of 16 mice injected with 3MC-coated chrysotile (data from two experiments).

Table 1. ANA incidence (%) in female F1(NZB*C57Bl/6) mice determined at 14 weeks after the second injection

Exp.	Intraperitoneal exposure				
	PBS	A/PBS	3MC/A/PBS	L.OIL	3MC/OIL
1	57 (4/7)	44 (4/9)	0* (0/6)	n.d.	44 (4/9)
2	45 (5/11)	n.d.	0* (0/8)	20 (3/15)	30 (3/10)
3	78 (11/14)	50 (7/14)	43 (3/7)	77 (10/13)	55 (6/11)
4	83 (5/6)	67 (4/6)	33 (2/6)	17 (1/6)	50 (3/6)

* statistically significant ($p<0.05$, chi^2-test) as compared to PBS control group; in brackets number of ANA positive /total number of animals

Serum IgM levels and spleen enlargement. Lower values of serum IgM were found in the group of mice injected with 3MC-coated chrysotile as compared to any other injection regime groups (Table 3). Similarly, weights of the spleens were also decreased by the i.p. injection with 3MC-coated chrysotile . This tendency was seen in all experiments except one in females

(Table 4) and in all three experiments in males (not shown).

Table 2. Percentage of glomeruli with Ig deposits in female F1(NZB*C57Bl/6) mice in the kidneys (14 weeks after the 2nd injection)

Exp.	Intraperitoneal exposure				
	PBS	A/PBS	3MC/A/PBS	L. OIL	3MC/OIL
1	11.5±15.5* (8)	n.d.	3.9±9.1 (12)	8.6±10.2 (14)	6.8±10.8 (15)
2	25.4±25.8 (13)	26.8±26.2 (14)	14.9±17.8 (7)	34.6±21.5 (14)	18.6±25.5 (11)
3	15.6±12.6 (10)	11.7±16.6 (9)	13.3±24.6 (6)	7.6±7.7 (10)	15.7±13.4 (7)

* mean±SD; in brackets number of animals

Discussion

It was shown that asbestos workers who smoke have show less pronounced immunological abnormalities (Lange 1980, Wagner et al.1979). Therefore, the question arose whether lack of asbestos-related immunological abnormalities in smokers reflects more severe abnormality which impaires "normal" response of the host to the asbestos exposure. As a result the symptoms known as asbestos-related are alievated. The latter possibility could be predicted due to the known data on the ability of PAH to block the immune system function (Luster and Blank 1987). It is known, that benzo(a)pyrene (BP) disturbs interleukins generation and macrophages are one of the principle targets of BP (Lyte and Bick 1986; Lyte at al.1987; Myers et al. 1987). Generation of the monokines takes place at the begining of the immune response. It is also suspected that the interaction between macropahages and asbestos fibres constitutes the very primary step in the chain of reactions which leads to the lung fibrosis and immunological abnormalities (Pernis and Vigliani 1982).

Table 3. Serum IgM level in female F1(NZB*C57Bl/6) mice determined at 14 weeks after the 2nd injection.

Exp.	Intraperitoneal exposure				
	PBS	A/PBS	3MC/A/PBS	L.OIL	3MC/OIL
1	77±5* (7)	74±11 (8)	44±16[aa] (6)	n.d.	76±9 (9)
2	66±10 (11)	n.d.	38±14[a] (8)	69±18 (15)	67±15 (9)
3	107±12 (14)	90±14 (14)	71±16[ab] (7)	91±11 (15)	90±13 (12)
4	39±11 (4)	51±7 (5)	26±6[aa] (6)	56±14 (6)	50±9 (6)

* expressed in units, 100 U represents the IgM concentration in a pool of NZB mice sera; mean±SD;in brackets number of animals statistically significant ([a] $p<0.001$, b $p<0.05$, Student's t-test as compared to the groups injected with PBS and A/PBS, respectively

Table 4. Weight of the spleens in female F1(NZB*C57Bl/6) mice (14 weeks after the 2nd injection)

Exp.	Intraperitoneal exposure				
	PBS	A/PBS	3MC/A/PBS	L.OIL	3MC/OIL
1	0.09±0.03* (10)	0.12±0.04 (10)	0.04±0.01[a] (8)	n.d.	0.08±0.03 (10)
2	0.12±0.04 (11)	n.d.	0.06±0.02[a] (12)	0.08±0.02 (16)	0.09±0.02 (15)
3	0.12±0.05 (14)	0.11±0.04 (14)	0.15±0.12 (9)	0.09±0.03 (14)	0.13±0.07 (12)

* in grams, mean±SD ;in brackets number of animals
[a] statistically significant ($p<0.001$, Student's t-test) as compared to PBS control

3MC-coated chrysotile while injected into the female F1(NZB*C57Bl/6) mice caused a decline in serum IgM level and in weight of the spleens. Also the incidence of ANA and glomeruli lesions which depend on the autoantibodies formation in NZB mice (Theofilopoulos and Dixon 1981) was partly declined.

The mechansim of this suppresive effect of the combined exposure to PAH and chrysotile is not apparent from the present work. However, at least two possibilities have to be discussed. (1) 3MC-coated chrysotile exert a toxic effect in a general term what resulted in the waisting syndrome. This could cause the presence of malnutrition and anemia. From the work of others it is known that low calorie diet diminishes autoimmunity in NZB/NZW mice (Anderson et al. 1985), and this mechanism might be responsible for the modulatory effect of 3MC-coated chrysotile on the autoimmune syndrome of the mice under the study. (2) 3MC-coated asbestos fibres residing in the peritoneum (after injection) or those floating in the lymphatic system might be taken up by macrophages and brought by those cells to the spleen and lymph nodes. This could result in the accumulation of 3MC in form of "hot spots". These "hot spots" of 3MC could produce a (geno)toxic hit which caused the anemia (Pietrzyca et al. see this volume) and could switch off highly activated clones of lymphocytes in F1 (NZB*C57Bl/6) mice. Somatic mutation may lead to the acquisition or to the lost of the primary antibodies specificity (Davidson et al. 1987). It has been documented that N-Methyl-N-nitrosourea in susceptible C57Bl/6 mice can induce a specific pre-mutagenic base modification which leads to autoimmune disease (Harris et al.1983). The mutation may operate in two directions. Therefore, it is possible that the expansion of B cells, what is reflected by an elevation of serum IgM level and ANA production, could be modified by an exposure to 3MC-coated chrysotile via genotoxic hit. Indeed, it was found, in the in vitro system, that presentation of PAH in context of asbestos is critical for enhanced alkylation (Mossman et al. 1983).

Independently on the mechanism of the modification of the immune system, this was seen only in those F1 (NZB*C57Bl/6) mice which were injected with 3MC-coated asbestos fibres, but not with any of these two agents given separately. Asbestos fibres did not simply provide a general accesibility of 3MC because when the latter compound was injected with oil solvent no apparent toxic effect was seen. Therefore, asbestos fibres facilitate rather accumulation of 3MC in some target organs. And this may have important consequences including carcinogenesis, and the immune systems function alterations.

References

Alfred LJ, Wojdani A, Nieto M, Perez R, Yoshida G (1983) A chemical carcinogen, 3-methylcholanthrene, alters T-cell function and induces T-suppressor cells in a mouse model system. Immunology 50:207-213

Anderson DJ, Watson ALM, Yunis EJ (1985) Environmental and genetic factors that influence immunity and longevity in mice. In: Woodhead AD, Blackett AD, Hollaender A (eds) Molecular biology of aging, Plenum Press, New York and London, p 231

Davidson A, Shefner R, Diamond B (1987) Studies on the role of somatic mutation in the generation anti-DNA antibodies. In: Dukor P, Hanson LA, Kallos P, Shakib F, Trnka Z, Waksman BH (eds) Monographs in allergy, S.Karger, Basel, p 31

Dzugaj A, Szyba K, Lange A (1985) Adsorption of 3-methylcholanthrene onto chrysotile fibres enhances aryl hydrocarbon hydroxylase inducement in mice. In: Beck EG , Bignon J (eds) In vitro effects of mineral dusts, NATO ASI Series, Vol.G3, Springer-Verlag Berlin Heidelberg,p 467

Harris G, Lawley PD, Asbery LJ, Chandler PM, Jones MG (1983) Autoimmune haemolytic disease in mice after exposure to a methylating carcinogen. Immunology 49:439-449

Kouri RE, Schechtman LM, Nebert DW (1980) Metabolism of chemical carcinogens. In: Kouri RE (ed) Genetic differences in chemical carcinogenesis, CRC Press, Boca Raton, p 21

Lange A (1980) An epidemiological survey of immunological abnormalities in asbestos workers. II. Serum immunoglobulin levels. Environ Res 22:176-183

Luster MI, Blank J (1987) Molecular and cellular basis of chemically induced immunotoxicity. Ann Rev Pharmacol Toxicol 27:23-49

Lyte M, Bick PH (1986) Modulation of interleukin-1 production by macrophages following benzo(a)pyrene exposure. Int J Immunopharmac 8:377-381

Lyte M, Blanton RH, Myers MJ, Bick PH (1987) Effect of in vivo

administration of the carcinogen benzo(a)pyrene on interleukin-1 and interleukin-3 production. Int J Immunopharmac 9:307-312
Matej H, Lange A, Garncarek D, Smolik R, Roszak E (1978) HLA and antinuclear antibody incidence in asbestos workers. Arch Immunol Ther Exp 26:201-206
Mossman B, Light W (1983) Asbestos: mechanism of toxicity and carcinogenicity in the respiratory tract. Ann Rev Pharmacol Toxicol 23:595-615
Myers MJ, Schook LB, Bick PH (1987) Mechanisms of benzo(a)pyrene-induced modulation of antigen presentation. J Pharm Exp Ther 242:339-404
Pernis B, Vigliani EC (1982) The role of macrophages and immunocytes in the pathogenesis of pulmonary diseases due to mineral dust. Am J Ind Med 3:133-137
Roitt IM, Doniach D (1969) WHO Mannual on autoimmune serology, WHO,Geneva.
Theofilopoulos AN, Dixon FJ (1981) Etiopathogenesis of murine SLE. Immunological Rev 55:179-216
Theofilopoulos AN, Prud'homme GJ, Dixon FJ (1985) Autoimmune aspects of systemic lupus erythematosus. In: Cruse JM, Lewis RE (eds) Autoimmunity: basic concepts;systemic and selected organ-specific diseases. Karger Basel, p 190
Wagner MMF, Campbell MJ, Edwards RE (1979) Sequential immunological studies on an asbestos-exposed population. I. Factors affecting peripheral blood leucocytes and T lymphocytes. Clin Exp Immunol 38:323-331

PRODUCTION OF GROWTH FACTORS BY RAT PLEURAL MESOTHELIAL CELLS IN VIVO OR IN VITRO TRANSFORMED BY CHRYSOTILE FIBERS.

A. Van der Meeren*, G. Clement**, J. Bignon*, D. Barritault**, and M. C. Jaurand*

* INSERM U 139. CHU Henri-Mondor. 94010 Créteil Cedex, France

** Laboratoire de Biotechnologie des Cellules Eucaryotes. Université Paris-Val de Marne 94010 Créteil Cedex, France

INTRODUCTION.

Mesothelial cells are target cells in asbestos-related carcinogenesis. Chrysotile fibers are known to induce mesothelioma in vivo (Wagner et al. 1969; Monchaux et al. 1981) and mesothelial cell transformation in vitro (Paterour et al. 1985).

Since growth factor production is strongly involved in the carcinogenic process (Sporn et al. 1980, Bradshaw 1986), it was important to study the growth factor production by the transformed rat pleural mesothelial cell lines. In addition, it was of particular interest to compare the effects of different growth factors on both normal and transformed cell growth.

We have reported elsewhere (Van der Meeren et al. 1988; Levy et al., this meeting) the effect of conditioned medium from chrysotile-transformed mesothelial cells on mesothelial cell growth. The characterization of growth factors contained in these conditioned media is described in this paper. The growth factors found in the conditioned media were tested for their ability to stimulate DNA synthesis in normal and transformed cells.

NATO ASI Series, Vol. H30
Effects of Mineral Dusts on Cells
Edited by B. T. Mossman and R. O. Bégin

MATERIALS AND METHODS.

Cell cultures. Normal rat pleural mesothelial cells (RPMC) were obtained by a previously described method (Jaurand et al. 1981). RPMC were routinely cultured in HAM F10 medium (Biopro) supplemented with 10% fetal calf serum (FCS), (Flow Laboratories), 2mM glutamine (Gibco), 50 μg/ml streptomycin (Biomérieux), 100 U/ml penicillin (Biomérieux) and 10 mM HEPES (Eurobio).

In vivo transformed rat pleural mesothelial cells (RPMC-AV1) were obtained from a histologically confirmed mesothelioma. This mesothelioma was induced by the intrapleural injection of chrysotile asbestos fibers as described elsewhere (Jaurand et al. 1986).

In vitro transformed RPMC (RPMC-AT1) were obtained after weekly treatment of normal mesothelial cells with 2 μg/ml chrysotile for 48 hours for 42 weeks.

These two transformed cell lines were tumorigenic in nude mice (Saint-Etienne et al., this meeting). The mesothelioma cell lines were maintained in RPMI 1640 medium (Biopro) supplemented with 10% FCS, 2 mM glutamine and 10 mM HEPES.

Conditioned media and growth factors. Conditioned serum-free media were collected every 48 hrs and concentrated as described previously (Van der Meeren et al. 1988). Control medium consisted in serum-free RPMI 1640 concentrated according to the same procedure.

Platelet derived growth factor (PDGF) and epidermal growth factor (EGF) were purchased from Collaborative Research. Fibroblast growth factors (FGFs acid and basic forms) were a gift from D. Barritault.

Determination of DNA synthesis. To determine ^{3}H-thymidine incorporation, cells were plated in their respective growth media in 24 microtest well plates (Falcon). When confluence was reached, medium was replaced with serum-free medium to stop cell growth. Twenty four hours later, cultures were incubated with conditioned media or growth factors for an additional 24 hours. ^{3}H-thymidine (0.5 μCi/well) was added for the last four hours.

Growth factor characterization. Ion exchange chromatography was performed in order to separate EGF from PDGF. Both S sepharose and G sepharose columns were used, and the growth factors were eluted with 0.1 M phosphate, 0.5 M NaCl.

Affinity chromatography using heparin sepharose column was carried out on conditioned media in order to partially purify FGF. The elution buffer was 1.15 M phosphate/NaCl to elute aFGF and 2.15 M phosphate/NaCl for bFGF elution. All fractions obtained from these purification steps were tested for their biological effects on normal confluent RPMC.

Radioreceptor assays. After the ion exchange chromatography step, the conditioned media were tested for their ability to compete with radiolabeled PDGF and EGF.

PDGF radioreceptor assay : confluent mouse embryo cells were incubated with serum-free medium for 24 hours and then treated with increasing concentrations of unlabeled PDGF or conditioned media samples. ^{125}I-PDGF was added for 2 hours at 4°C and the membrane bound activity determined with a gamma counter.

EGF radioreceptor assay : 1.35 ng/ml A 431 membranes, 10^5 cpm ^{125}I-EGF, and several concentrations of unlabeled EGF or conditioned medium were incubated for 1 hour and then spun down 15 min at 3000×g. Membrane bound radioactivity was determined as described above.

RESULTS

Effects of conditioned media on RPMC growth. When concentrated conditioned medium was added to confluence normal cells, a decrease in ^{3}H-thymidine incorporation was observed in a dose-dependent fashion, when compared to the addition of control medium. This inhibitory effect was obtained both with in vivo transformed cells (RPMC-AV1) conditioned medium and in vitro transformed cells (RPMC-AT1) conditioned medium (Table 1).

Characterization of growth factors in conditioned media. The partial purification using ion exchange chromatography and the following step of the radio receptor-assays demonstrated the presence of PDGF-like molecules in both RPMC-AV1 AND RPMC-AT1 conditioned media. The concentration found were respectively 106 pg/ml and 126 pg/ml.

TABLE 1 - Effect of in vivo and in vitro transformed cells conditioned medium on normal cells. 3H thymidine incorporation in cultures treated with conditioned medium (% of control medium)

Conditioned medium from:	μl added per well	^{3}H-thymidine incorporation: Exp # 1	Exp # 2
In vivo transformed cells	5	100.2	43.1
	10	94.9	68.7
	50	73.3	30.0
	100	53.8	19.3
In vitro transformed cells	5	88.6	26.9
	10	39.7	27.3
	50	7.5	13.8
	100	10.9	6.0

Normal confluent cells were placed for 24 hours in serum-free medium. Then, concentrated conditioned medium or control medium (RPMI concentrated according to the same procedure as conditioned medium) were added to the cells for 24 hours. ^{3}H-thymidine incorporation was determined as described in Material and Methods.

TABLE 2 - Concentrations of growth factors obtained in conditioned media from transformed mesothelial cells.

	PDGF-like	EGF-like	FGF
Conditioned medium from in vivo transformed cells	106 pg/ml	40 pg/ml	0
Conditioned medium from in vitro transformed cells	126 pg/ml	ND*	0

* not determined

EGF-like molecules were also found in RPMC-AV1 conditioned medium at a concentration of 40 pg/ml (Table 2). FGF acidic and basic forms were found neither in RPMC-AV1 nor in RPMC-AT1 conditioned media.

Effects of growth factors on normal and transformed mesothelial cell DNA synthesis. As is shown in Table 3, EGF was tested on normal RPMC in a range of concentration from 0.05 to 10^3 ng/ml. A dose-dependent inhibitory effect of EGF was clearly observed from 0.1 ng/ml to 10^3 ng/ml. At this high concentration, ^{3}H thymidine incorporation was only 40% of the control value obtained without EGF.

On confluent in vivo and in vitro transformed cells, EGF was tested from 1 ng/ml to 10^5 ng/ml. RPMC-AV1 cells showed an increase in DNA synthesis when 100 ng/ml of EGF was added. When 10^5 ng/ml of EGF was added, ^{3}H-thymidine incorporation was 6 times higher than in control cells.

RPMC-AT1 cells seemed more sensitive to EGF since a stimulation of DNA synthesis was obtained with 5 ng/ml of EGF.

PDGF was tested from 1 to 5 units/ml, and it was not shown to have any effect on normal or transformed mesothelial cells.

TABLE 3 - Effect of growth factors on DNA synthesis of normal and transformed mesothelial cells.

Cells tested	Normal cells		Transformed cells	
Growth factors	EGF	PDGF	EGF	PDGF
Range of concentration tested	0.05 - 10^3 ng/ml	1 - 5 stimulation units	1 - 10^5 ng/ml	1 - 5 stimulation units
Effects on DNA synthesis	Inhibitory effect	No effect	Mitogenic effect	No effect

DISCUSSION.

Conditioned media from in vivo (RPMC-AV1) and in vitro (RPMC-AT1) transformed mesothelial cells contain factors which inhibit normal mesothelial cell DNA synthesis. This inhibitory effect is dose dependent and reversible (Van der Meeren et al. 1988). In addition a G1 blocking was observed by flow cytometry with RPMC-AV1 conditioned media (Levy et al., this meeting).

In a previous paper we showed an autocrine growth stimulation with RPMC-AV1 cells. In addition, we demonstrated a stimulatory effect on chinese hamster lung fibroblasts DNA synthesis with either conditioned media (data not shown). In order to investigate which growth factor could be responsible for these inhibitory and stimulatory effects, we performed a partial purification of the conditioned media. PDGF-like molecules were found in both in vivo or in vitro transformed cell conditioned medium. In vivo transformed cells produced EGF-like molecules.

No FGF, basic or acid form, was found in conditioned media of either type of transformed cells. FGFs are a large family of polypeptides having an affinity for heparin. These growth factors are ubiquitous and are secreted by many tissues such as eye (Courty et al. 1986), brain (Schreiber et al. 1985) and hypothalamus (Klagsbrun et al. 1985) as well as by endothelial cells in culture (Burgess et al. 1985). These factors do not seem to be implicated in the tumorigenic process. Our results confirm this hypothesis. However absence of production of FGF might be a specificity of RPMC. Since EGF and PDGF-like molecules were secreted by transformed cells, it seemed important to determine the effect of these growth factors on mesothelial cell DNA synthesis.

The obtained results raise important considerations. First, since EGF has an inhibitory action on normal RPMC, it could be responsible for the inhibitory effect of conditioned media. In this way, EGF might be important for the growth regulation of rat pleural mesothelial cells.

PDGF has no effect on either normal or transformed cell DNA synthesis. However, PDGF-like molecules are produced by in vivo and in vitro transformed cells. Some similar results were obtained by another group using human mesothelial cells (Gerwin et al. 1987); the mesothelioma cells did not respond to PDGF but produced PDGF-like molecules in their conditioned medium. PDGF is strongly mitogenic for many mesenchymal cells (Antoniades et al. 1979, Clemmons et al. 1983), but it seems that neither epithelial cells nor endothelial cells have PDGF receptors (Sitaras et al. 1987). Production of PDGF-like molecules might be of importance for growth of submesothelial fibroblasts. Many studies showed the importance of PDGF in the carcinogenic process. Further, PDGF is frequently involved in the autocrine growth of tumor cells (Lens et al. 1986). The behavior of normal mesothelial cells seems different according to their origin: rat cells behave differently than human cells do. Normal rat pleural mesothelial cells are inhibited by EGF and not affected by the presence of PDGF. However, normal human mesothelial cells are stimulated by EGF (La Rocca et al. 1985, Gabrielson et al. 1988) and by PDGF (Gerwin et al. 1987).

The results reported here strongly suggest a role for growth factors in the mechanisms of mesothelial carcinogenesis. The autocrine growth of transformed cells and the inhibition of normal cells by transformed cell conditioned medium might result in a growth advantage to the tumorigenic cells.

References.

Antoniades HN, Scher CD, and Stiles CD (1979) Purification of human platelet derived growth factor. Proc Natl Acad Sci USA 76:1809-1813

Bradshaw TK (1986) Cell transformation : the role of oncogenes and growth factors. Mutagenesis 1:91-97

Burgess W, Mehlman T, Friesel R, Johnson W and Maciag T (1985) Multiple forms of endothelial cell growth factor. J Biol Chem 260:11389- 11392

Clemmons D (1983) Age dependent production of a competence factor by human fibroblasts. J Cell Physiol 114:61-67

Courty J, Chevallier B, Moenner M, Loret C, Lagerte O, Bohlen P, Courtois Y, Barritault D (1986) Evidence for FGF-like growth factor in adult bovine retina : analogies with EDGF-1. Biochem Biophys Res Comm 136:102-108

Gabrielson EW, Gerwin BI, Harris CC, Roberts AB, Sporn MB and Lechner JF (1988) Stimulation of DNA synthesis in cultured primary human mesothelial cells by specific growth factors. FASEB J 2:2717- 2721

Gerwin BI, Lechner JF, Reddel RR, Roberts AB, Robbins KC, Gabrielson EW and Harris CC (1987) Comparison of production of transforming growth factor ß and platelet derived growth factor by normal human mesothelial cells and mesothelioma cell lines. Cancer Res 47:6180- 6184

Jaurand MC, Bernaudin JF, Renier A, Kaplan H, Bignon J (1981) Rat pleural mesothelial cells in culture. In Vitro 17:98-108

Jaurand MC, Renier A, Van der Meeren A, Magne L, Pinchon MC and Bignon J (1986) In vitro growth characteristics of rat mesothelioma cells in culture. Biol Cell 57:249-256

Klagsborn M and Shirg Y (1985) Heparin affinity of anionic and cationic capillary endothelial cell growth factors: analysis of hypothalamus derived growth factors and fibroblast growth factors. Proc Natl Acad Sci USA 82: 805- 809

La Rocca P and Rheinwald JG (1985) Anchorage independent growth of normal human mesothelial cells : a sensitive bioassay for EGF which discloses the absence of this factor in fetal calf serum. In Vitro Cellular et Dev Biol 21:67-72

Lens PF, Altera B, Nuser (1986) Expression of c-sis and platelet-derived growth factor in in vitro transformed glioma cells from rat brain tissue transplacentally treated with ethylnitrosourea. Mol and Cell Biol 6: 3537-3540

Levy F, Van der Meeren A, Renier A, Katz A, Bignon J and Jaurand MC. (1988) First studies on cell cycles of rat pleural mesothelial cells submitted to direct and indirect hit by chrysotile fibres. 4th Int Workshop on Effects of Mineral Dusts on Cells, Sherbrooke, Canada 21- 23 Sept

Monchaux G, Bignon J, Jaurand MC, Lafuma J, Sebastien P, Masse R, Hirsch A and Goni J (1981) Mesotheliomas in rats following inoculation with acid-leached chrysotile asbestos and other mineral fibres. Carcinog 2:229-236

Paterour MJ, Bignon J and Jaurand MC (1985) In vitro transformation of rat pleural mesothelial cells by chrysotile fibres and/or benzo(a) pyrene. Carcinog 6:523-529

Saint-Etienne L, Paterour MJ, Nebut M, Fleury J, Kheuang L, Bignon J and Jaurand MC (1988) Occurence and morphology of tumors induced in nude mice transplanted with pleural mesothelial cells in vitro treated with chrysotile fibres or benzo 3-4 pyrene. 4th Int Workshop on Effects of Mineral Dusts on Cells, Sherbrooke, Canada 21-23 Sept

Schreiber A, Kenny S, Kowalski J, Thomas K, Gimenez-Galleco G, Los Candelarc M, Di Salvo J, Barritault D, Courty J, Courtois Y, Moenner M, Loret C, Burges W, Mehlman T, Firesel R, Johnsson W and Maciag J (1985) A unique family of endothelial cell polypeptide mitogens : the antigenic and receptor cross reactivity of bovine endothelial cell growth factor, brain derived acidic fibroblast growth factor and eye derived growth factor II. J Cell Biol 101:1623-1626

Sitaras MN, Sariban E, Pantazis P, Zetter B and Antoniades HN (1987) Human iliac artery endothelial cells express both genes encoding the chains of platelet derived-growth factor (PDGF) and synthesize PDGF like mitogen. J Cell Physiol 132:376-380

Sporn MB and Todaro GJ (1980) Autocrine secretion and malignant transformation of cells. New Engl J Med 303:878-880

Van der Meeren A, Levy F, Bignon J and Jaurand MC (1988) Growth of normal and neoplastic rat pleural mesothelial cells in the presence of conditioned medium from neoplastic mesothelial cells. Biol Cell 62:293-296 Wagner JC and Berry G (1969) Mesotheliomas in rats following inoculation with asbestos. Brit J Cancer 23:567-581

MUTAGENICITY OF ASBESTOS AND ERIONITE IN A SALMONELLA TESTER STRAIN OF TA102

Eiji Yano, Makoto Shimizu*, Naoko Urano, Peter H. Evans+
Department of Public Health
Teikyo University School of Medicine
Kaga, Itabashi-ku
Tokyo 173
JAPAN

* School of Health Sciences, Kyorin University, Tokyo, Japan
+ MRC Dunn Nutrition Unit, Milton Road, Cambridge, England

Introduction

Asbestos is a significant exception among carcinogens with respect to the finding that most chemical carcinogens have been identified as mutagens in the Ames test (Chamberlain and Tarmy, 1977). Indeed, apart from ploidy changes in mammalian chromosomes (Sincock et al. 1975) and sister chromatid exchange (SCE) (Livingston et al. 1980; Babu et al. 1981) asbestos has produced negative results in most of the short-term genotoxicity tests for the detection of genetic effects. As a consequence, Hesterberg and Barret (1985) have postulated an epigenetic mechanism of asbestos carcinogenicity.

However erionite, a fibrous zeolite which has been identified as a causative agent for the endemic outbreak of malignant mesothelioma in Turkey (Baris et al.1981), has shown positive results in several in vitro studies. Poole et al. (1983) showed that erionite caused an increase in morphological transformation and unsheduled DNA (UDS) repair synthesis in C3H10T$_{1/2}$ cells and in A549 cells. In addition, Kelsey et al. (1986) have demonstrated that erionite causes an increase in SCE in synchronous CHO cells. Although the increases were marginal, these findings suggest that erionite may initiate cancer through direct modification of DNA.

NATO ASI Series, Vol. H30
Effects of Mineral Dusts on Cells
Edited by B. T. Mossman and R. O. Bégin

In order to obtain further evidence for this possibility, we have examined the mutagenicity of erionite in the Ames test (Yano et al. in preparation). As a result, only TA 98 with the S9 mix showed a small increase of the revertant colonies, but no increase was observed without the S9 mix. The other Salmonella typhimurium tester strains of TA1537, TA100, and TA1535 and a E. coli strain (WP2 uvr) showed virtually no reaction. We also examined crocidolite and TiO_2 for comparison and they demonstrated no increase of revertant colonies regardless of the addition of the S9 mix. In the experiments of simultaneous addi-tion of erionite and chemical carcinogens, enhancement of the increase of revertant colonies of TA98 induced by benzo[a]pyrene (BP) and 2-aminoanthracene (2AA) was observed and the effect of erionite on the increase of the revertants by chemical carcinogens appeared multiplicative rather than additive.

Several studies on the pathogenic mechanism of how mineral fibers cause cell damage have focused on the role of reactive oxygen metabolites. We have demonstrated that asbestos and erionite stimulate luminol-dependent chemiluminescence using human polymorphonuclearcytes (Evans et al. 1987). The formation of lipid hydroperoxides and production of malondialdehyde by crocidolite in phagocytic cells has also been demonstrated (Yano, 1988).

Among the Salmonella tester strains of the Ames test, TA102 has been established to detect oxidative mutagens (Levin et al. 1982). In the present study, the possible participation of reactive oxygen metabolites in the mutagenicity of erionite was therefore examined using the TA102 strain.

Materials and methods

Mineral dusts

Erionite fibers from Rome, Oregon (U.S.A.) prepared in the MRC Pneumoconiosis Unit, U.K. were used as a test material. Detailed information about erionite fiber has been described elsewhere (Poole et al., 1983; Wagner et al., 1985). Samples of UICC standard crocidolite asbestos and titanium dioxide (TiO_2 BHD Chemicals ltd., Poole, England) were used respectively for positive and negative controls. For erionite and crocidolite, the assays were carried out at concentrations of 10µg to 5000µg /plate. TiO_2 was examined up to 500µg/plate because of the difficulties in counting the number of colonies at the higher concentrations.

Bacterial mutation test

In all experiments, the pre-incubation method (Nagao et al. 1977), which is a modification of the method of Ames et al. (1975), was adopted using the strain of Salmonella typhimurium TA102. The S9 mix was obtained from Oriental Kobo Co. (Tokyo) and experiments were performed with and without the S9 mix. Dust samples suspended in distilled water were homogenized by ultrasonication and preincubated for 30 minutes with bacteria and added to the test plates. After 70 hour's incubation the number of colonies were counted. At least 3 plates were examined for each assay. N-ethyl-N'-nitro-N-nitrosoguanidine (ENNG) and mitomycin-C (MMC) were used as positive controls in addition to negative control of distilled water to monitor the bacteria and the S9 mix.

Test of synergistic effect

The interaction between erionite and a known chemical carcinogen was examined using 2.5, 5, and 10µg/plate of BP.

Results

The mutagenicity assay of erionite, crocidolite, and TiO_2 using TA102 is illustrated in Table 1. To be considered positive in mutagenicity test, it generally requires the increase of revertants to the level of more than twice above the baseline level and apart from the positive controls of ENNG and MMC, no samples showed a distinct increase in the number of revertant colonies at any dose range. However, a very slight tendency of increase of revertants can be noted in higher concentrations of crocidolite without the S9 mix.

Table 1. Mutagenicities of mineral dusts on TA102

Compound	Dose (µg/plate)	Crocidolite		Erionite		TiO_2	
		S-9(-)	S-9(+)	S-9(-)	S-9(+)	S-9(-)	S-9(+)
Test minerals	0	240	285	243	254	208	223
	5					236	223
	10	241	293	253	301	252	226
	50	313	333	205	277	227	260
	100	280	308	259	278	220	228
	500	288	296	248	293	214	215
	1000	313	275	270	277	199	273
	5000	314	272	249	316		
Positive controls							
ENNG	1	1953		1579		1612	
MMC	25		1625		1579		2134

The values represent the mean number of revertants/plate of 3 or more plates.

The assay of mutagenicity of erionite with BP was carried out to examine the interaction with known carcinogen. The S9 mix was used throughout the experiments. As shown in Fig. 1, the number of revertants increased with the concentration of erionite when examined with BP of 2.5, 5.0, and 10 µg/plate and up to the highest dose of 5000µg/plate of erionite, the number of revertants increased without a growth inhibition.

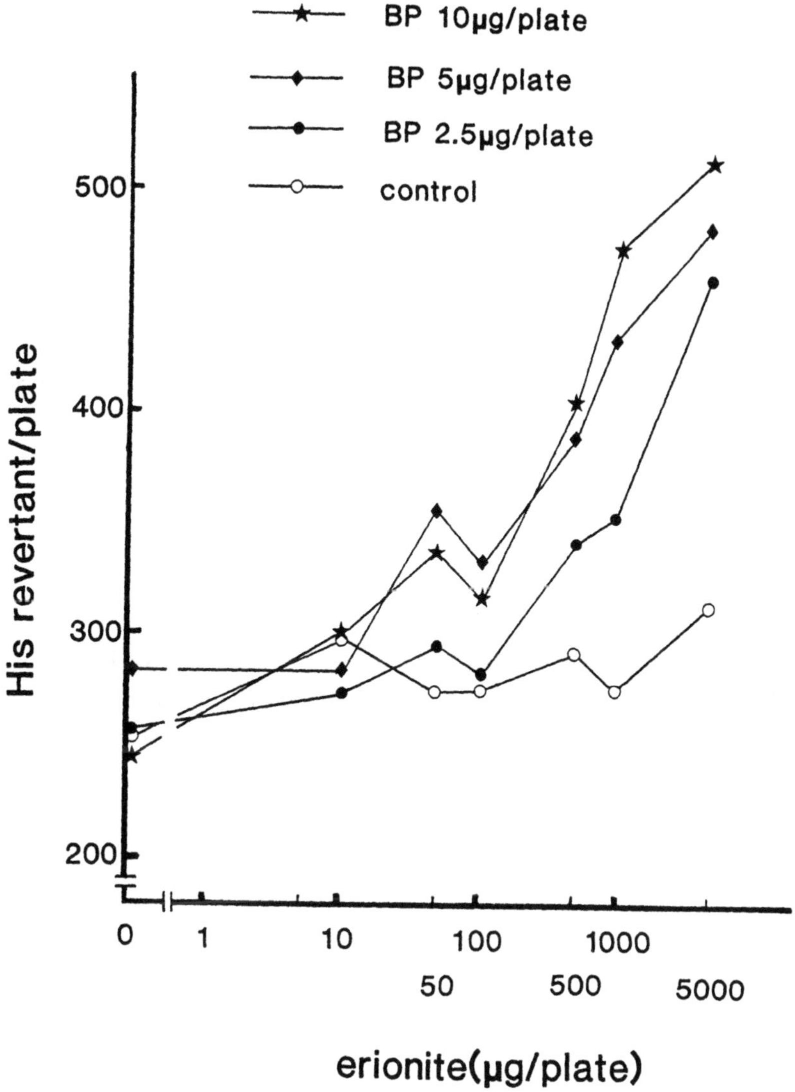

Fig. 1. Combined effects of erionite and various doses of benzo[a]pyrene (BP) with the S9 mix. Each point represent the mean of the triplicated experiment. The number of revertant colonies increased with the dose of erionite and BP.

Discussion

In the first bacterial mutation test study of Chamberlain and Tarmy (1977) no mutagenic activity was found to be associated with asbestos or glass fibers. Only one exception has been the report by Cleveland (1984) who found a positive results for asbestos, however, hydrocarbon contaminants were postulated as causative agents in this study.

In the present study, the results of the assay with mineral dust alone were virtually negative corresponding to most of the previous reports. With BP, the number of revertants increased with the dose of erionite, while erionite alone did not increase revertants up to the concentration of 1000 ug/ml. Considering the non-additive increase of the revertants by erionite and BP, the contaminated chemical carcinogens are not likely as the cause of erionite mutagenicity.

As the Salmonella strain of TA102 is sensitive to oxidative mutagens, the increase, although marginal, in the number of revertants by crocidolite is noteworthy. One of the present authors (Yano,1988), has recently reported that crocidolite produces lipid peroxidation in cultured phagocytic cells. Fenton type Herber-Weiss reaction was postulated for the mechanism of the lipid peroxidation and this reaction is associated with the production of hydroperoxide. The reaction of crocidolite in TA102 supports the postulated mechanism of the lipid peroxidation by crocidolite.

Comparing the results of the present study and those of our previous one with classical strains of the Ames test (Yano et al. in preparation) several differences between crocidolite and erionite may be noted with regard to the responder strain of Salmon-ella and the necessity of the S9 mix. As summarized in Table 2, a slight increase of the revertants of TA102 was induced by crocidolite without the S9 mix while erionite did not induce the increase of revertant regardless of the addition of the S9 mix. In TA98, a slight increase of revertants was observed not by crocidolite but by erionite if the S9 mix was added. Both in TA98 and TA102, the synergistic effect of erionite and BP required the S9 mix. This may suggest the different mechanism of the mutagenicity between crocidolite and erionite and the former may primarily associated with the oxidative reaction. This speculation is supported by the fact that malondialdehyde produced by crocidolite from phagocytic cells was much greater than that by erionite (Yano, 1988).

The increase of the revertant colonies caused by mineral fibers shown in the present study was so slight as to indicate that erionite is not directly mutagen. However, the findings provide some further insight into the potensial carcinogenic mechanism of of mineral fibers.

Table 2. Summary of the Ames test of mineral dusts on TA98 and TA102

Compound		Crocidolite		Erionite		TiO_2	
		S-9(-)	S-9(+)	S-9(-)	S-9(+)	S-9(-)	S-9(+)
Tester strain	TA98	-	-	-	±	-	-
	TA102	±	-	-	-	-	-

-:no increase of revertants compared to the blank
±:slight (less than 2X) increase of revertants

References

Ames BN, McCann J, Yamasaki E (1975) Method for detecting carcinogens and mutagens with the Salmonella/mammalian-microsome mutagenicity test. Mutat Res 31:347-364

Babu KA, Lakkard BC, Nigam SK, Bhatt DK, Karnik AB, Thakore KN, Kashyap SK Chatterjee SK (1980) In vitro cytological and cytogenetic effects of an Indian variety of chrysotile asbestos. Environ Res 21:416-422

Baris YI, Simonato L, Artivinli M, Saracci R, Skidmore JW (1981) Malignant mesothelioma and radiological abnormalities in two villages in central Turkey. Lancet i:984-987

Chamberlain M Tarmy EM (1977) Asbestos and glass fibers in bacterial mutation tests. Mutat Res 43:159-164

Cleveland MG (1984) Mutagenesis of Escherichia coli (CSH50) by Asbestos (41954). Proc Soc Exp Biol Med 177:343-346

Evans PH, Campbell AK, Yano E, Goodman B (1987) Phagocytic oxidant stress and antioxidant interactions in the pneumoconioses and dust-induced tumorigenic lung disease. In: Rice-Evans C (ed) Free radicals, oxidant stress and drug action. Richelieu Press, London, p213-235

Hesterberg TW, Barrett JC (1985) Induction by asbestos fibers of anaphase abnormalities: mechanism for aneuploidy induction and possibly carcinogenesis. Carcinogenesis 6:473-475

Kelsey KT, Yano E, Liber HL, Little JB (1986) The in vitro genetic effects of fibrous erionite and crocidolite asbestos. Br J Cancer 54:107-114

Levin DE, Hollstein M, Christman MF, Schweiers EA, Ames BN (1982) A new Salmonella strain (TA102) with A-T base pairs at the site of mutation detects oxidative mutagens. Proc Nat Acad Sci USA 79:7445-7449

Livingston GK, Rom WN, Morris MV (1980) Asbestos-induced sister chromatid exchanges in cultured Chinese hamster ovarian fibroblast cells. J Environ Pathol Toxicol 4: 373-382

Nagao M, Yahagi T, Seino Y, Sugimura T, Ito N (1977) Mutagenicities of quinoline and its derivatives. Mutat Res 42:335-342

Poole A, Brown RC, Turver CJ, Skidmore JW, Griffiths DM (1983) In vitro genotoxic activities of fibrous erionite. Br J Cancer 47:697-705

Sincock A, Seabright M (1975) Induction of chromosome changes in Chinese hamster cells by exposure to asbestos fibres. Nature 257:56-58

Wagner JC, Skidmore JW, Hill RJ, Griffiths DM (1985) Erionite exposure and mesotheliomas in rats. Br J Cancer 51:727-730

Yano E (1988) Mineral fiber-induced malondialdehyde formation and effects of oxidant scavengers in phagocytic cells. Int Arch Occup Environ Health 61:19-23

Induction of Sister Chromatid Exchanges by Fibrous Dusts Alone and in Combination with Other Xenobiotics in Chinese Hamster Cells

Anna B. Fischer
Institute of Hygiene
Justus-Liebig-University of Giessen
Federal Republic of Germany

Introduction

The naturally occuring mineral asbestos is known to constitute an occupational as well as environmental health hazard leading to lung fibrosis and two types of cancer: mesothelioma and pulmonary carcinoma. Smoking potentiates the risk of lung cancer. The responsible components of cigarette smoke have so far not been identified, though polycyclic aromatic hydrocarbons have been suspected.

In mammalian cell cultures in vitro asbestos fibres but also Man-Made Mineral Fibres are cytotoxic. In addition, morphologic cell transformation has been demonstrated for chrysotile, crocidolite, amosite, and erionite (Oshimura et al. 1984, Kelsey et al. 1986). These fibres also produce cytogenetic damage such as above all changes of chromosome numbers and ploidy as well as structural alterations (gaps, breaks and exchanges) (Kenne et al. 1986, Lavappa et al. 1975, Sincock and Seabright 1975). However, studies on the induction of point mutations have usually been negative with the exception of the investigations by Huang et al. who reported weakly positive effects of amosite, chrysotile and crocidolite (Huang et al. 1978, Huang 1979). Studies using sister chromatid exchanges (SCE) as a mutagenic endpoint have to date yielded conflicting results. While three working groups found no effects in human lymphocytes and lymphoblastoid cells, C3H mouse cells, and Chinese hamster cells V79, chrysotile, crocidolite, and erionite were positive in Chinese hamster ovary cells (Casey 1983, Hei et al. 1985, Price-Jones et al. 1980, Livingston et al. 1980, Babu et

NATO ASI Series, Vol. H30
Effects of Mineral Dusts on Cells
Edited by B. T. Mossman and R. O. Bégin

al. 1981). In rat pleural mesothelial cells SCE were produced by crocidolite but not by attapulgite (Achard et al. 1987).

The present study had the purpose to reexamine the induction of SCE by asbestos fibres and to see if a synergistic effect of other environmental mutagens could be detected. The following mutagenic xenobiotics were employed: methanesulfonic acid, ethyl ester (EMS) as a model mutagen, benzo(a)pyrene (BaP) as an important aromatic constituent of cigarette smoke, and cadmium, because this heavy metal also occurs in cigarettes and has recently been shown to induce lung cancer in rats (Takenaka et al. 1983).

Materials and Methods

Cells. Chinese hamster peritoneal cells, line B14F28, with a mean chromosome number of 22 - 23 were cultured in MEM supplemented with nonessential amino acids and 5% Newborn Calf Serum. Fibres. Asbestos and glass fibres sifted by Spurny were employed (Table I). Mutagens. Benzo(a)pyrene and ethylmethanesulfonate were purchased from Sigma and $CdCl_2$ from Merck. Exposure. Fibres were added to 0.8 * 10^6 cells/53 cm^2 flask during inoculation. EMS and $CdCl_2$ were added at the same time. One d later the cells were washed with phosphate buffered saline (PBS) and 7.5 µg/ml BrdU added in new growth medium. Cultures were terminated 24 - 26 h later, i.e. after 2 cell cycles. B(a)P-exposure took place between the first and second halves of the incubation; together with S9-mix (Fresenius, Bad Homburg) cells were shaken gently for 1 h at 37°C, washed with PBS and refed with medium containing BrdU. Preparation of slides and SCE stain. The methodology of Perry and Wolf (1974) was followed. Scoring and statistical analysis. Thirty metaphases with 20 - 24 chromosomes were examined for SCE and the results expressed as SCE/metaphase. For the assessmenht of ploidy 100 metaphases/sample were evaluated. The Kruskal-Wallis test was used to compare the results obtained with treated and control cultures.

Results

Exposure of B14F28 cells to chrysotile, crocidolite, and amosite as well as glass fibres resulted in a high incidence of polyploid cells. While the respective controls contained 5 - 9% polyploid metaphase, exposure to 150 µg of CFF, KFF, Aff, and GFF resulted in 35, 22, 20 and 33% of polyploidy. In addition, all fibres induced a non-spectacular, but significant increase in SCE (Figs. 1-3). No clear dose-response effect could be seen. Application of different fractions varying in fibre length (Table I) which were dosed gravimetrically also showed no consistent relationship between fibre size and SCE incidence (Fig. 4).

Combined exposure of the fibres with the other mutagens yielded the following results: In three tests where cells were simultaneously treated with 150 µg/10^6 cells CFF or KFF and rising concentrations of EMS, there was no synergism detectable (Fig. 5). Combination of CFF and KFF with BaP also proved negative (Fig. 6), as did the combined treatment with asbestos and $CdCl_2$ (Fig. 7).

<u>Table I:</u> Distribution of fibre length, mean diameters, and fibre counts/ng of the different asbestos and glass fibres

	Percentage shorter than stated length (µm)									
	< 2	3	5	8	10	20	25	50	Mean diameter (µm)	Fibres per ng
Chrysotile										
F (modified UICC)	64.0	-	83.8	-	97.0	-	98.5	99.7	0.19	7210
FF (fine)	80.6	95.1	98.7	99±1	-	-	-	-	0.16	12 500
fF (ultrafine)	-	99±1	-	-	-	-	-	-	0.13	25 000
Crocidolite										
F (modified UICC)	86.7	-	97.0	-	99.3	-	99.9	100.0	0.27	3900
FF (fine)	-	93.4	98.9	-	-	-	-	-	0.23	7000
fF (ultrafine)	-	97.3±1	-	-	-	-	-	-	0.21	8400
Amosite										
F (modified UICC)	79.5	-	94.2	-	98.6	-	99.7	99.8	0.51	710
FF (fine)	-	72.1	92.8	-	99±1	-	-	-	0.47	900
fF (ultrafine)	-	99±1	-	-	-	-	-	-	0.23	8500
Glass fibres										
MN 104	-	72.4	88.3	-	99.1	-	-	-	0.32	520
MN 100	-	-	97.9	-	99.8	-	-	-	0.23	740
G_{FF} (fine)	-	68.9	85.6	-	95.4	99±1	-	-	0.43	850
G_{fF} (ultrafine)	-	99±1	-	-	-	-	-	-	0.19	10 000

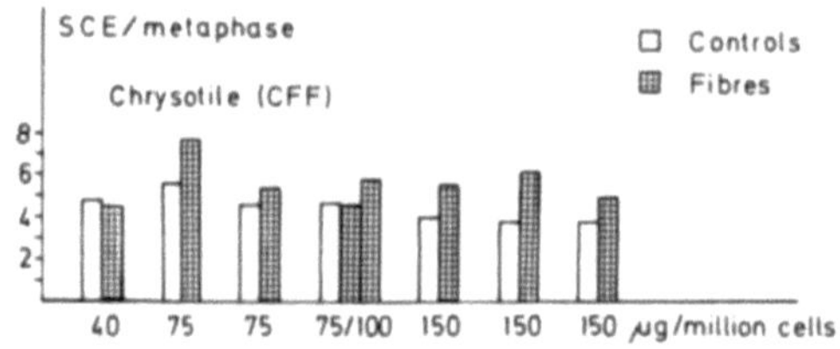

Fig. 1: Induction of SCE by different concentrations of chrysotile fibres

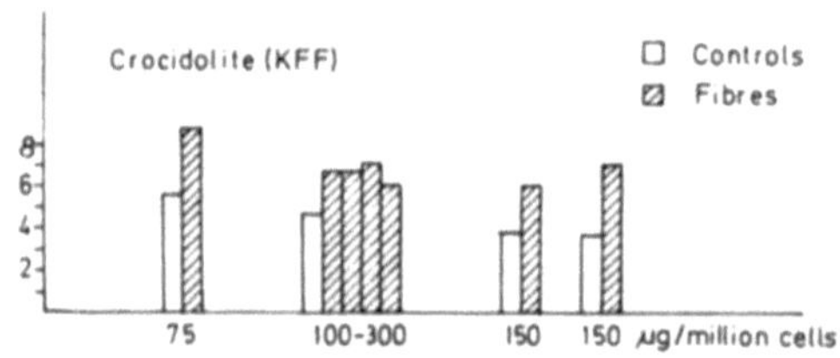

Fig. 2: Induction of SCE by different concentrations of crocidolite fibres

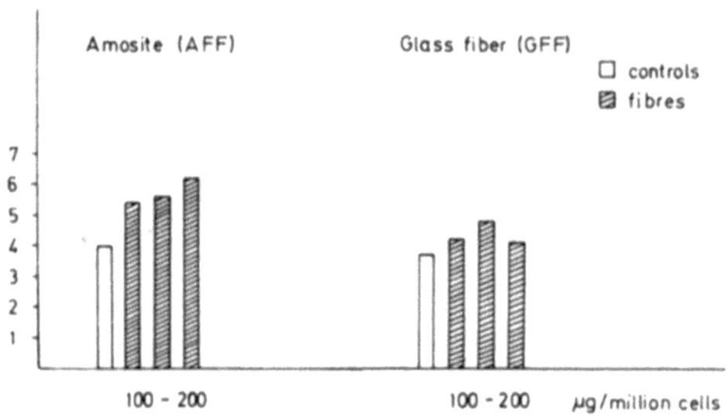

Fig. 3: Induction of SCE by different concentrations of amosite and glass fibres

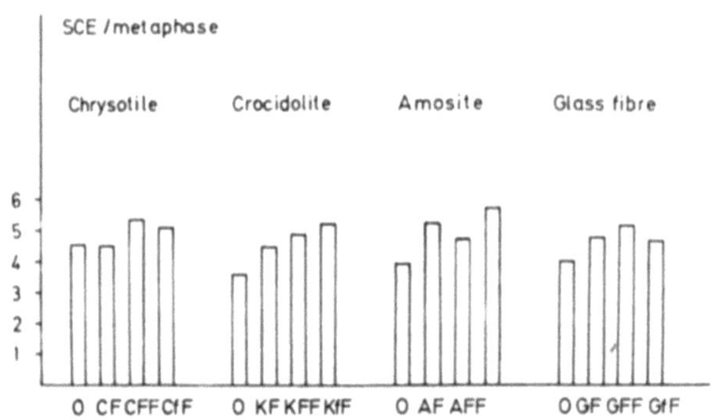

Fig. 4: Induction of SCE by different fractions of asbestos and glass fibres dosed at a concentration of 125 µg/10^6 cells

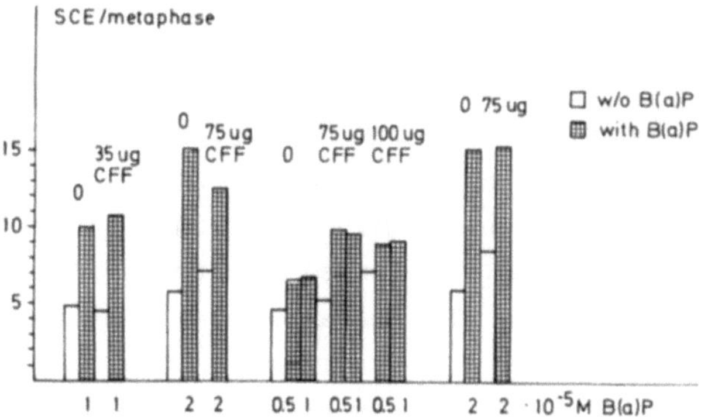

Fig. 5: Induction of SCE by ethylmethanesulfonate in combination with chrysotile and crocidolite

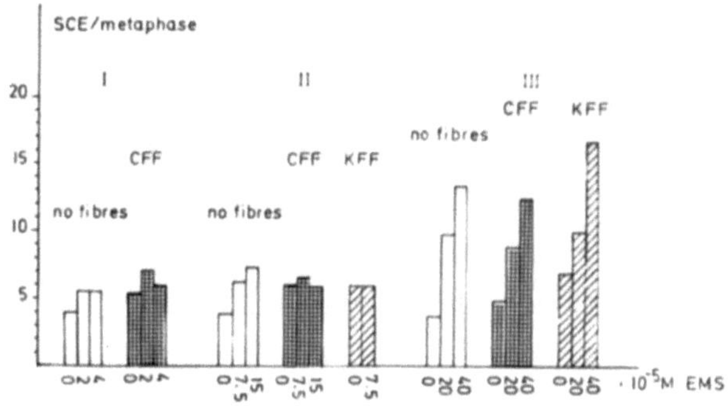

Fig. 6: Induction of SCE by benzo(a)pyrene in combination with chrysotile and crocidolite

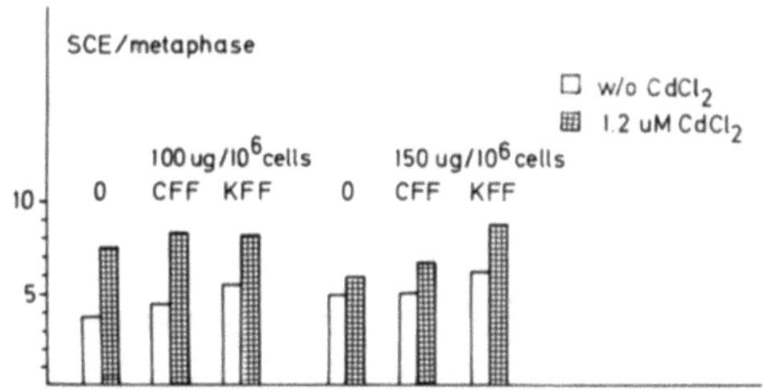

Fig. 7: Induction of SCE by cadmium chloride in combination with chrysotile and crocidolite

Discussion

The present study demonstrates a weak but significant mutagenic effect of asbestos in terms of SCE confirming the results of three previous working groups (Kelsey et al. 1986, Livingston 1986, Babu et al. 1981). It is of interest that glass fibre also caused an elevation of SCE incidence confirming again that Man-Made Mineral Fibres are not free from mutagenic risk. The frequently reported fibre effects on ploidy were also clearly seen in our cells (Sincock and Seabright 1975, Babu et al. 1981, Livingston 1986, Kelsey et al. 1986).

The importance of fibre dimensions for the biological effects in vivo and in vitro have been stressed. Long fibres are more potent in inducing pleural mesothelioma than short ones (Stanton et al. 1977). In vitro there was an excellent correlation between fibre size and cytotoxicity (Tilkes and Beck 1980, Hesterberg et al. 1987). Cytogenetic effects such as the incidence of micronuclei and cell transformation were also correlated to fibre length (Hesterberg et al. 1987). Achard et al. (1987) speculated that the failure to induce SCE in rat mesothelial cells by attapulgite in contrast to crocidolite was due to the short fibre size of their attapulgite sample. In the present study the short fibre fractions "fF" were as effective as the two longer fractions "FF" and "F" to produce SCE following gravimetric dosage. This might be explained by the greater quantity of the ultrafine fibres "fF" per weight unit. Experiments with numerical dosage could clarify this point. On the other hand the incidence of SCE induced by treatment with the various fibre fractions and concentrations is relatively low and the scattering of the number of SCE per metaphase comparatively high so that slight differences between the experimental groups are likely to be obscured.

Combination experiments with chrysotile or crocidolite and other mutagens did not result in greater than expected SCE numbers. Benzo(a)pyrene, EMS, and $CdCl_2$ though in themselves effective inducers of SCE showed no synergistic effect with the fibres. These results should be discussed in the context of experimental investigations showing synergistic effects. A synergism between BaP and asbestos was noted with respect to the morphological transformation of Syrian hamster embryo cells. (DiPaolo et al. 1983). Recently Kimizuka et al. (1987) demonstrated a co-carcinogenic effect of chrysotile asbestos and BaP. The interaction between a nitrosamine, Cd and crocidolite was observed in rats. Animals receiving intratracheal instillation of all components showed an apparent synergism, though on the other hand the combination of Cd and crocidolite alone caused no increase in bronchial carcinomas (Harrison and Heath 1986).

In addition to the fact that the differing results concerning the interaction of asbestos and BaP were obtained using different test methods *in vivo* and *in vitro*, discrepancies may be explained by the finding that the induction of SCE is probably not the most important genotoxic effect of asbestos fibres. Other effects such as chromosome aberrations including alterations of ploidy are obviously more significant, possibly also for the production of cell transformation and cancer induction.

Summary

In the present study the asbestos fibres chrysotile, crocidolite and amosite as well as also glass fibres produced weak but significant increases in SCE rates, while ethylmethanesulfonate, benzo(a)pyrene and cadmium chloride applied singly caused clear dose-dependent elevations. The combination of crocydolite and chrysotile with the three other mutagens did not result in greater than additive SCE levels. It can be concluded that the genetic events producing SCE are probably less important than other genotoxic effects for the induction of heritable changes leading to the appearance of cancer cells.

References

Achard S, Perderiset M, Jaurand M-J (1987) Sister chromatid exchanges in rat pleural mesothelial cells treated with crocidolite, attapulgite, or benzo 3-4 pyrene. Brit J of Ind Med 44:281-283

Babu KA, Lakkad BC, Nigam SK, Bhatt DK, Karnik AB, Thakore KN, Kashyap NK, Chatterjee SK (1980) In vitro cytological and cytogenetic effects of an Indian variety of chrysotile asbestos. Environ Res 21:416- 422

Casey G (1983) Sister-chromatid exchange and cell kinetics in CHO-K1 cells, human fibroblasts and lymphoblastoid cells exposed in vitro to asbestos and glass fibre. Mutation Res 116:369-377

DiPaolo JA, De Marinis AJ, Doniger J (1983) Asbestos and benzo(a)pyrene synergism in the transformation of Syrian hamster embryo cells. Pharmacology 27: 65-73
Harrison PTC, Heath JC (1986) Apparent synergy in lung carcinogenesis: interactions between N-nitrosoheptamethyleneimine, particulate cadmium and crocidolite asbestos fibers in rats. Carcinogenesis 7:1903-1908
Hesterberg TW, Butterick C, Oshimura M, Brody AR, Barett JC (1986) Role of phagocytosis in Syrian hamster cell transformation and cytogenetic effects induced by asbestos and short and long glass fibres. Cancer Res 46:5795-5802
Huang SL (1979) Amosite, chrysotile and crocidolite asbestos are mutagenic in Chinese hamster lung cells. Mutation Res 68:265-274
Huang SL, Saggioro D, Michelmann H, Malling HV (1978) Genetic effects of crocidolite asbestos in Chinese hamster lung cells. Mutation Res 57:225-232
Kelsey KT, Yano E, Liber HL, Little JB (1986) The in vitro genetic effects of fibrous erionite and crocidolite asbestos. Br J Cancer 54:107-114
Kenne K, Ljungquist S, Ringertz NR (1986) Effects of asbestos fibers on cell division, cell survival, and formation of thioguanine-resistant mutants in Chinese hamster ovary cells. Environ Res 39:448-464
Kimizuka G, Ohwada H, Hayashi Y (1987) Co-carcinogenic effect of asbestos and benzo(a)pyrene in the lung of hamster. Acta Pathol Jpn 37:465-474
Lavappa KS, Fu MM, Epstein SS (1975) Cytogenic studies on chrysotile asbestos, Environ Res 10:165-173
Livingston GK, Rom WN, Morris MV (1980) Asbestos-induced sister chromatid exchanges in cultured Chinese hamster ovarian fibroblasts cells. J Environ Pathol Toxicol 4:373-382
Oshimura M, Hesterberg TW, Tsutsui T, Barrett JC (1984) Correlation of asbestos induced cytogenetic effects with cell transformation of Syrian hamster embryo cells in culture. Cancer Res 44: 5017-5022
Perry P, Wolff S (1974) New Giemsa method for the differential staining of sister chromatids. Nature 251:156-158
Price-Jones MJ, Gubbings G, Chamberlain M (1980) The genetic effects of crocidolite asbestos; comparison of chromosome abnormalities and sister-chromatid exchanges. Mutation Res 79:331-336
Sincock A, Seabright M (1975) Induction of chromosome changes in Chinese hamster cells by exposure to asbestos fibres. Nature 257:56-58
Stanton MF; Layard M, Tegeris A, Miller E, May M, Kent E (1977) Carcinogenicity of fibrous glass: pleural response in the rat in relation to fiber dimension. J Natl Cancer Inst 58:587-597
Takenaka S, Oldiges H, König H, Hochrainer D, Oberdörster G (1983) Carcinogenicity of cadmium chloride aerosols in Wistar rats. J Natl Cancer Inst 70:367-373
Tilkes F, Beck EG (1980) Comparison of length-dependent cytotoxicity of inhalable asbestos and man-made mineral fibres. In: Wagner JC, Davis W (eds)(1980) Biological effects of mineral fibres. International Agency for Research on Cancer, 475

PLEURAL AND PARENCHYMAL RESPONSES IN RATS TO SHORT-TERM INHALATION EXPOSURES TO ERIONITE, CROCIDOLITE, CHRYSOTILE, SILICA AND TITANIUM DIOXIDE*

N. F. Johnson[1] and J. C. Wagner[2]

[1]Inhalation Toxicology Research Institute
Lovelace Biomedical and Environmental Research Institute
P. O. Box 5890
Albuquerque, NM 87185 USA

[2]59 Combe Valley Road
Weymouth, Dorset
DT3 6N4, UK

Introduction

The ultimate assay for determining the pathogenicity of respirable fibrous materials is a rodent 2-year inhalation lifetime study. Under these circumstances, asbestos exposure gives rise to lesions in rats which are similar to those seen with human occupational exposures, i.e., interstitial fibrosis, primary lung cancer, and mesothelioma (Wagner *et al*., 1974). This assay presents the fibrous material in a physiological manner and in relevant doses. However, the assay is costly and time consuming, and it would be impractical to test all the fibrous materials available commercially or being developed for commercial exploitation in this manner. Alternative assays are *in vitro* cell systems and implantation or inoculation studies in animals. These approaches have limitations. *In vitro* cell systems monitor cell/fiber interactions over a short period, thus neglecting the important effects of residence in lung tissue and interactions with host factors such as the immune system. These cell systems do provide a practical method for alerting one to the overall potential pathogenic nature of the material, but it is frequently difficult to relate this activity in terms of fibrogenicity or carcinogenicity. Inoculation or implantation studies in animals involve

NATO ASI Series, Vol. H30
Effects of Mineral Dusts on Cells
Edited by B. T. Mossman and R. O. Bégin

administration of extremely high nonphysiological doses of material to the mesothelium of the peritoneal or pleural cavities or to the epithelium of the conducting airways. These studies have produced false positive results with materials such as glass fibers, where the animal inhalation studies and human epidemiological studies indicate no excess of either lung or mesothelial tumors (Wagner et al., 1984; Simonato et al., 1987). Silica, a nonfibrous mineral, causes a high incidence of tumors following intraperitoneal inoculation (Pott et al., 1987); however, no mesothelial tumors occur after intrapleural inoculation (Wagner et al., 1973) and silica exposure is not associated with mesotheliomas in humans.

This study was undertaken as a preliminary investigation to determine whether short-term inhalation studies are of value in predicting the long-term consequences of fibrous dust inhalation. The materials used in this study have been previously tested by in vitro cell assays and by long-term inhalation studies (Wagner et al., 1974, 1984, 1985, 1987; Brown et al., 1983; Davies, 1980; Poole et al., 1983; Chamberlain et al., 1978).

Materials and Methods

SPF Fischer 344 rats were exposed at age 6 wk to an atmosphere of UICC (Union International Contre le Cancer) crocidolite, fibrous erionite (Rome, Oregon), UICC chrysotile, Min-U-Sil (Pennsylvania Glass and Sand Co.), and titanium dioxide (BDH Poole) or filtered room air.

The details of the aerosol generation, exposure chambers and aerosol monitoring have been published previously (Wagner et al., 1974). In brief, groups of 3 rats were exposed to a dust atmosphere of 10 mg/m^3 for 7 hr/day, 5 days/wk for 1, 2, 4, 8 and 12 wk, and then sacrificed. A further group of 3 animals were exposed for 12 wk and held for 12 months before sacrifice. An equal number of unexposed control rats were used at each exposure period. The animals were exposed by whole-body exposure in a Timbrell chamber. The dust was generated by a Timbrell generator, and the respirable dust concentration was measured using size-selective gravimetric dust samples.

Following the appropriate exposure period, the animals were sacrificed by inhalation of ether. The lungs were inflated with phosphate-buffered glutaraldehyde at constant pressure, removed and immersed in fixative. Tissue was taken from the same location in the left lung of all animals. Transverse sections from upper, mid, and lower lung were taken for processing for conventional paraffin wax histology. Adjacent sections were taken and processed for electron microscopy. The paraffin-embedded tissues were sectioned and stained with hematoxylin and eosin or Masson's trichrome.

Tissue from the mid lung was used to count the various cell types present in the lung parenchyma of animals exposed for 1 and 12 wk, and for 12 wk followed by 12 months rest and of the appropriate controls. The details of this morphometric analysis have been previously published (Johnson, 1987).

Results

The characteristics of the inhaled dusts are given in Table 1. UICC chrysotile contained a greater number of fibers than either crocidolite or erionite samples. Detailed fiber size analysis of these particular materials have been previously published (Wagner et al., 1974, 1985).

The control lungs displayed no apparent ultrastructural abnormalities, and the distribution of the various cell types in the lung parenchyma was similar to previous results (Johnson, 1987).

The earliest exposure periods to crocidolite, erionite, chrysotile and silica (1 and 2 wk) were associated with a focal perivascular and alveolar lymphocytic infiltration. This response was most marked in the erionite-exposed animals and was not evident in the 40- and 60-day-exposed animals. The morphometric analysis failed to reveal this lymphocytic infiltration because the areas selected for analysis were away from major blood vessels. Apart from these changes, the lung parenchyma was unremarkable until 12 months after the 12 wk exposure.

RESPIRABLE DUST CHARACTERISTICS

	CROCIDOLITE	ERIONITE	CHRYSOTILE	SILICA	TITANIUM DIOXIDE
SOURCE	UICC	(Rome)Oregon	UICC-B	Min-U-Sil 5*	BDH
MEAN CONCENTRATION	10mg/m^3	10mg/m^3	10mg/m^3	10mg/m^3	10mg/m^3
EXPOSURE PERIOD	7hr/day 5day/wk	7hr/day 5day/wk	7hr/day 5day/wk	7hr/day 5day/wk	7hr/day 5day/wk
FIBERS/ml>5 μm	1630	354	3567	—	—
FIBERS/ml>5 μm % RESPIRABLE FRACTION	52.7	44.0	90.5	—	—
FIBERS/ml>5 μm LENGTH <0.2 μm DIAMETER	78	51	317	—	—
FIBERS/ml>5 μm LENGTH <0.2 μm DIAMETER % RESPIRABLE FRACTION	4.8	14.3	89	—	—

*Mean Geometric size 2.0 ± 0.2 μm

Table 1. Table showing the characteristics of the various dust samples used.

Alveolar macrophages contained variable numbers of dust particles throughout all the exposure periods, including exposures to titanium dioxide. Dust-laden macrophages were largely confined to the airspaces. The morphometric analysis of fiber- and silica-exposed animals revealed a significant increase in the absolute number of alveolar macrophages present after 12 wk exposure and after the recovery period. However, no significant change was noted in the differential cell count of alveolar-free cells, with the exception of cells from the silica-exposed animals. In all cases except for silica-exposed animals, the percentage of alveolar macrophages obtained was in excess of 96%. In the silica-exposed animals this value was 89%. The remainder of the cells were polymorphonuclear leukocytes and lymphocytes. A few dust-laden cells were found in the interstitium after 12 wk dusting. Following 12 months rest from the latter exposure period, accumulations of dust-laden macrophages were found at the alveolar duct-terminal bronchiolar region. This region was also associated with the presence of interstitial dust-laden cells, polymorphonuclear leukocytes, fibroblasts and with increased amounts of collagen evident by both light and electron microscopy in the fiber- and silica-exposed animals.

Type II cell hyperplasia was seen after all fiber and silica exposure periods, however, this was only evident by detailed numerical evaluation. The increase in type II cells was not accompanied by any reduction in the number of type I cells. In addition to the hyperplastic changes seen in the type II alveolar cells, there were areas of bronchiolar/alveolar hyperplasia associated with the recovery period from the 12 wk exposure (Table 2).

LUNG AND PLEURAL CHANGES

12 Weeks Exposures and 12 Months Rest

	CONTROL	CROCIDOLITE	ERIONITE	CHRYSOTILE	SILICA	TITANIUM DIOXIDE
Alveolar Macrophages	-	+	+	+	+	-
Interstitial Macrophages	-	+	+	+	+	-
Type I Cell Changes	-	-	-	-	-	-
Type II Cell Changes	-	+	+	+	++	-
Interstitial Fibrosis	-	+	+	+	++	-
Pleural Fibrosis	-	+/-	++	+/-	++	+
Bronchiolar / Alveolar Hyperplasia	-	+[1]	+[1]	+[2]	++	-
Mesothelioma	-	-	++[3]	-	-	-

+ - Positive Response ++ - Marked Positive Response

1 - One of three animals 2 - Two of three animals 3 - Three of three animals

Table 2. Table showing lung and pleural changes after 12 wk exposure and 12 months rest. The individual dusts produced various lesions.

Pleural changes were evident by light microscopy after 12 wk exposure with erionite, silica and titanium dioxide, but not with crocidolite or chrysotile. This change involved collagen deposition adjacent to the alveolar wall side of the sub-mesothelial connective tissue layer. Similar changes could be seen by electron microscopy after 8 wk exposure. Twelve weeks exposure followed by 12 months rest was associated with further changes. Crocidolite and chrysotile exposure produced mild changes typified by collagen deposition adjacent to the alveolar wall of the subpleural connective tissue zone. In contrast, erionite- and silica-exposed animals displayed a spectrum of marked pleural changes (Table 2). Pleural fibrosis was widespread and the fibrotic lesions were either covered by squamoid or hyperplastic mesothelial cells. All of the erionite-exposed animals possessed multiple focal mesotheliomas. In one exposed animal, the tumor had spread and involved the pericardium. The focal tumors were either well-described and composed largely of epithelial-like cells or more diffuse and mainly composed of sarcomatous cells. The tumors were covered with a complete layer of normal squamoid or hyperplastic mesothelial cells. Electron microscopy confirmed the light microscopy findings of both epithelial and sarcomatous tumor cells.

Discussion

This study has shown that short-term inhalation of crocidolite, erionite, chrysotile, or silica induces a spectrum of pathological changes including pleural fibrosis, mesothelioma, and interstitial fibrosis. The individual dusts have differing pulmonary effects. Silica results in nodular interstitial fibrosis and pleural fibrosis, erionite produced mesotheliomas and pleural fibrosis, while crocidolite exposure resulted in mild interstitial changes.

The fibrotic activity of the dusts following short-term inhalation mirrored that of previous long-term inhalation studies (Wagner et al., 1974, 1984; Holland et al., 1984). Apart from the occurrence of mesotheliomas with erionite, the neoplastic potential of the materials was not revealed. Crocidolite, chrysotile and silica have been shown to induce lung tumors following long-term inhalation (Wagner *et al.*, 1974; Holland *et al.*, 1984). This suggests that short-term inhalation studies may be suitable for identifying fibrogenic materials, but not for identifying their neoplastic potential.

The *in vitro* assays conducted with the same materials revealed the overall biological activity of the samples but did not mirror or reflect the wide spectrum of lesions seen in the exposed animals (Chamberlain *et al.*, 1978; Davies, 1980; Brown *et al.*, 1983; Poole *et al.*, 1983). Intrapleural inoculation studies with these materials (Wagner *et al.*, 1973, 1984, 1985, 1987; Wagner *et al.*, 1972, 1980) demonstrated that erionite, chrysotile and crocidolite induce significant numbers of mesotheliomas while silica produces no such tumors. This last result is in marked contrast with the results seen following intraperitoneal inoculation where silica produced a significant number of tumors including mesotheliomas (Pott *et al.*, 1987). Since silica exposure in man is not associated with occurrence of mesotheliomas, intraperitoneal inoculation may be a less satisfactory way to test for the mesotheliomagenic nature of materials than the intrapleural approach.

This preliminary study suggests that the ultimate assay for fibrous materials is a 2 year inhalation/life-span study, although short-term inhalation and *in vitro* studies provide useful indicators of adverse biological activity.

Acknowledgements

Research sponsored by the United Kingdom Medical Research Council and the U.S. Department of Energy Office of Health and Environmental Research under Contract No. DE-AC04-76EV01013.

References

Brown RC, Poole A, Fleming GTA (1983) The influence of asbestos dust on the oncogenic transformation of C3H10T½ cells. Cancer Lett 18:221-227

Chamberlain M, Brown RC (1978) The cytotoxic effects of asbestos and other mineral dust in tissue culture cell lines. Br J Exp Path 59:183-189

Davies R (1980) The effect of dusts on enzyme release from macrophages. In: Brown RC, Gomley IP, Chamberlain M, Davies R (ed) The in vitro effects of mineral dust. Academic Press, London pp 67-71

Holland LM, Wilson JS, Tillery MI, Smith DM (1984) Lung cancer in rats exposed to fibrogenic dusts. In: Goldsmith DF, Winn DM, Shy CM (eds) Silica, silicosis and cancer. Praeger Scientific, New York, pp 267-279

Johnson NF (1987) Asbestos-induced changes in rat lung parenchyma. J Toxicol Environ Health 21:193-203

Poole A, Brown RC, Turver CJ, Skidmore JW, Griffiths DM (1983) In vitro genotoxic activities of fibrous erionite. Br J Cancer 47:697-705

Pott F, Ziem V, Reiffer FJ, Huth F, Mohr U (1987) Carcinogenicity studies on fibers, metal compounds, and some other dusts in rats. Exp Pathol 32:129-152

Simonato L, Fletcher AC, Cherrie JW, Andersen A, Bertazzi P, Charnay N, Claude J, Dodgson J, Esteve J, Frentzel-Beyme R, Gaudner MJ, Jensen O, Olsen J, Teppo L, Winkelmann R, Westerholm P, Winker PD, Zocchetti C, Saracci R (1987) The international agency for research on cancer historical cohort study of MMMF production workers in several European countries: extension of the follow up. Ann Occup Hyg 31:603-623

Wagner JC, Berry G, Timbrell V (1973) Mesotheliomata in rats after inoculation with asbestos and other materials. Br J Cancer 28:173-185

Wagner JC, Berry G, Skidmore JW, Timbrell V (1974) The effects of the inhalation of asbestos in rats. Br J Cancer 29:256-269

Wagner JC, Berry GB, Hill RJ, Munday DE, Skidmore JW (1984) Animal experiments with MMM(V)F - effects of inhalation and intrapleural inoculation in rats. In: Guthe T (ed) Biological effects of man-made mineral fibers. World Health Organization, pp 209-233

Wagner JC, Skidmore JW, Hill RJ, Griffiths, DM (1985) Erionite exposure and mesotheliomas in rats. Br J Cancer 51:727-730

Wagner JC, Griffiths DM, Munday DE (1987) Experimental studies with poly gorskite dusts. Br J Ind Med 44:749-763

Wagner MMF, Wagner JC (1972) Lymphomas in the Wistar rat after intrapleural inoculation of silica. J Natl Cancer Inst 48:81-91

Wagner MMF, Wagner JC (1980) Silica-induced malignant histiocytic lymphoma, incidence linked with strain of rat and type of silica. Br J Cancer 41:908-917

IN VITRO RESPONSES OF CANINE ALVEOLAR LYMPHOCYTES TO $BeSO_4$ AFTER INHALATION EXPOSURE TO BeO*

P. J. Haley, G. L. Finch, J. A. Mewhinney, F. F. Hahn, M. D. Hoover, and D. E. Bice

Inhalation Toxicology Research Institute
Lovelace Biomedical and Environmental Research Institute
P. O. Box 5890
Albuquerque, NM 87185

INTRODUCTION

Chronic beryllium disease (CBD) in man is a systemic immune-mediated granulomatous disease with predominantly pulmonary manifestations (Eisenbud and Lisson, 1983). CBD occurs in only 3% of individuals exposed to beryllium, and affected individuals develop beryllium-specific immune responses in blood and lung. While such beryllium-specific immune responses have been identified in lung lymphocytes from humans, similar results in laboratory animals exposed to beryllium have not been observed (Barna et al., 1984). More importantly, data from human patients with CBD indicate that peripheral blood lymphocyte reactivity alone may not accurately reflect either the predilection for, or the progression of, CBD (Rossman et al., 1988). In fact, it appears that localized lung lymphocyte reactivity to beryllium is central to the development of CBD, and that sampling of lung lymphocytes is necessary for diagnosis in humans, as well as for experimental manipulation in laboratory animals.

MATERIALS AND METHODS

Experimental Design

Twenty Beagle dogs were exposed by inhalation to aerosols of beryllium in order to achieve either a high (50 µg/kg bw)

NATO ASI Series, Vol. H30
Effects of Mineral Dusts on Cells
Edited by B. T. Mossman and R. O. Bégin

or low (17 μg/kg bw) initial lung burden (ILB) of BeO that had been calcined at either high (1000°C) or low (500°C) temperature. Lung lymphocytes were collected by periodic bronchoalveolar lavage (BAL). Lymphocyte responses to $BeSO_4$ or PHA were evaluated for BAL and peripheral blood cells at 3, 6, 7, 11, 15, 18 and 22 months after exposure.

Immunologic Assays

Lymphocyte responses to $BeSO_4$ and phytohemagglutin (PHA) were performed in triplicate, using a standard lymphocyte stimulation assay (Rossman et al., 1988). The final concentrations of PHA were 0.5 and 1.0 μg/mL; final concentrations of $BeSO_4$ were 1.0 μM, 10 μM, and 100 μM. Following incubation (37°C, 5% CO_2) of lymphocytes with PHA for 3 days or with $BeSO_4$ for 4 days, samples were pulse labeled with 1.0 μCi of tritiated thymidine (New England Nuclear, Boston, MA). The cells were harvested 6 hr later, and the radioactivity was measured by liquid scintillation spectroscopy. The stimulation index (SI) for each sample was calculated as a ratio of counts per min in experimental cultures divided by control cultures. A SI of greater than 4 was considered to be a positive response.

Lung cytology samples were prepared by cytocentrifugation, followed by staining with Diff-Quick (American Scientific Products, McGraw Park, IL). Differential cell counts were made by counting 200 cells.

RESULTS

Percentages (Fig. 1) and numbers (data not shown) of lung lymphocytes in dogs with high and low ILBs of 500°C-treated BeO only displayed a marked increase at 3 mo, which declined slowly through 22 mo.

Lymphocyte SI of both blood and BAL lymphocytes demonstrated a high degree of individual variation (Figs. 2 and 3). Positive lymphocyte stimulation in response to $BeSO_4$ incubation was seen more frequently in blood samples (14 of 16

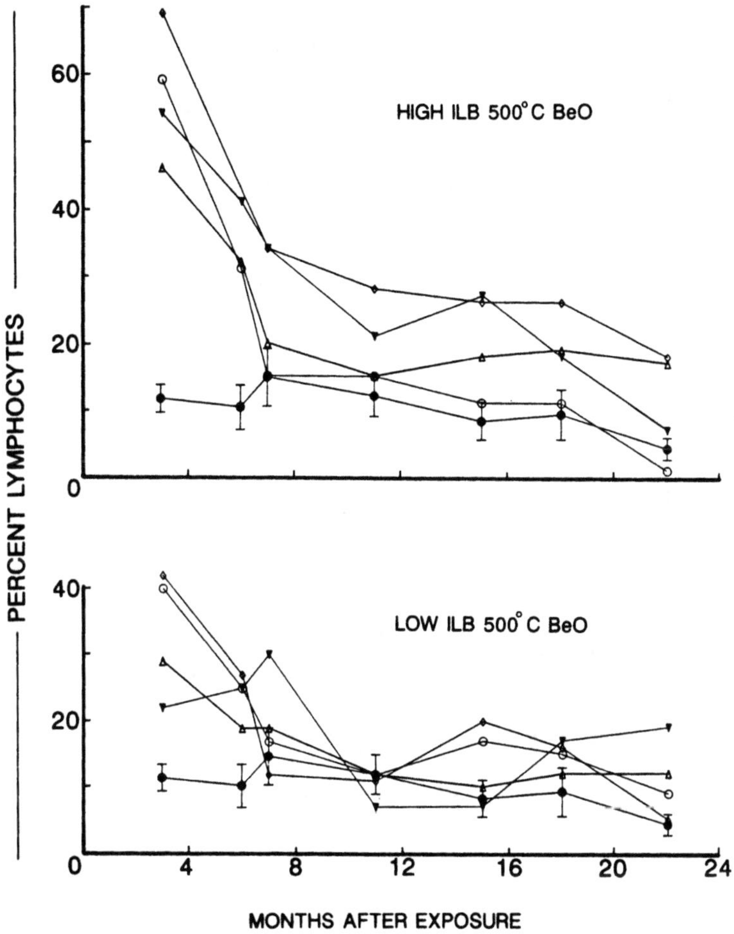

Figure 1a. Percent lymphocytes of cells recovered by bronchoalveolar lavage of dogs with high or low ILB 500°C treated BEO.

dogs) than in BAL samples (3 of 16 dogs); however, when blood and BAL samples were positive in the same animal, the BAL response was consistently greater. Dogs with positive lymphocyte stimulation indices in BAL samples usually, but not always, had positive indices in blood as well. Eleven of 16 dogs displayed a combination of negative BAL SI accompanied by positive blood lymphocyte SI. In one case, there was evidence

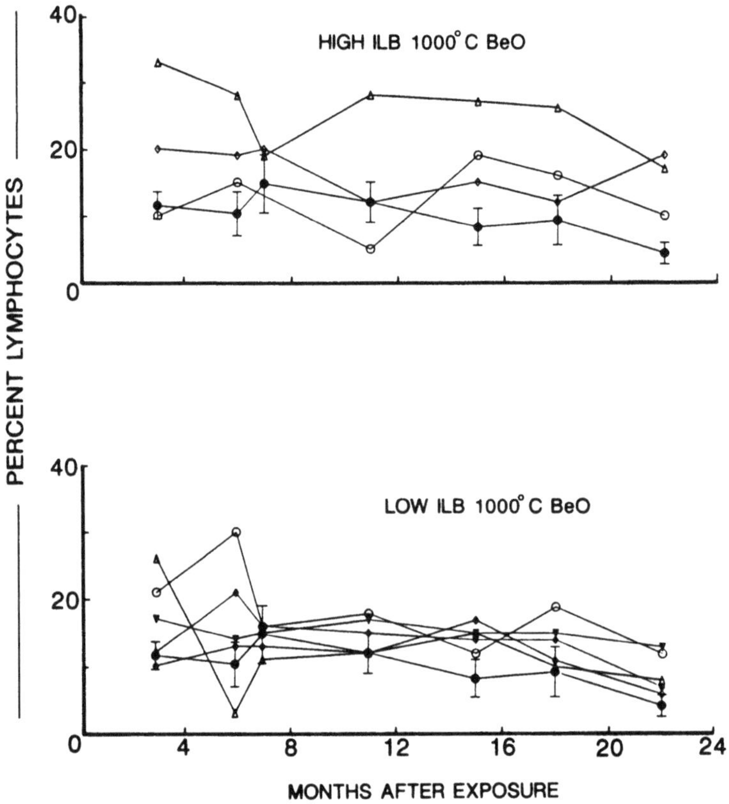

Figure 1b. Percent lymphocytes of cells recovered by bronchoalveolar lavage of dogs with high or low ILB 1000°C treated BEO.

of positive BAL lymphocyte responses in the absence of blood lymphocyte responses. Positive BAL SI were observed only in dogs with high ILB 500°C BeO. Positive blood SI were observed in some, but not all animals with high and low ILBs of BeO particles, calcined at both 1000°C and 500°C. Lymphocytes from lung and/or blood of control dogs did not respond to _in vitro_ cultivation with $BeSO_4$.

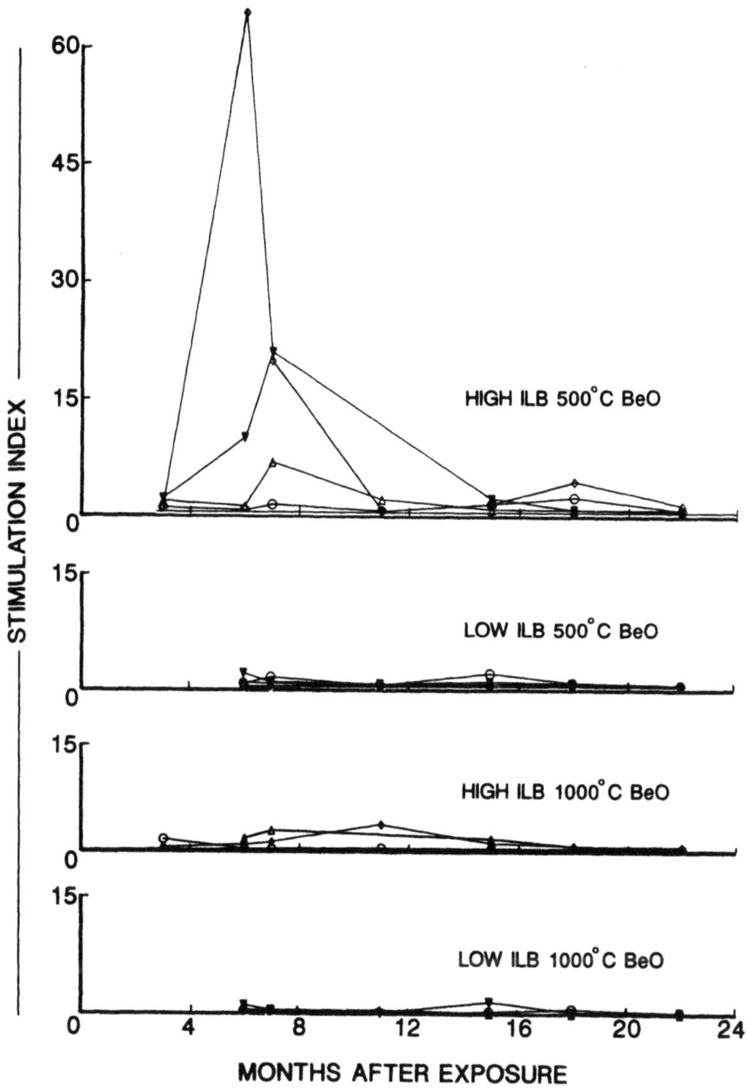

Figure 2. Stimulation indices after culture with 100 µM $BeSO_4$ of cells obtained by bronchoalveolar lavage of dogs with either high or low ILB, and 500°C- or 1000°C-treated BeO. Each symbol represents a single animal and is the same symbol for the same dog in all figures.

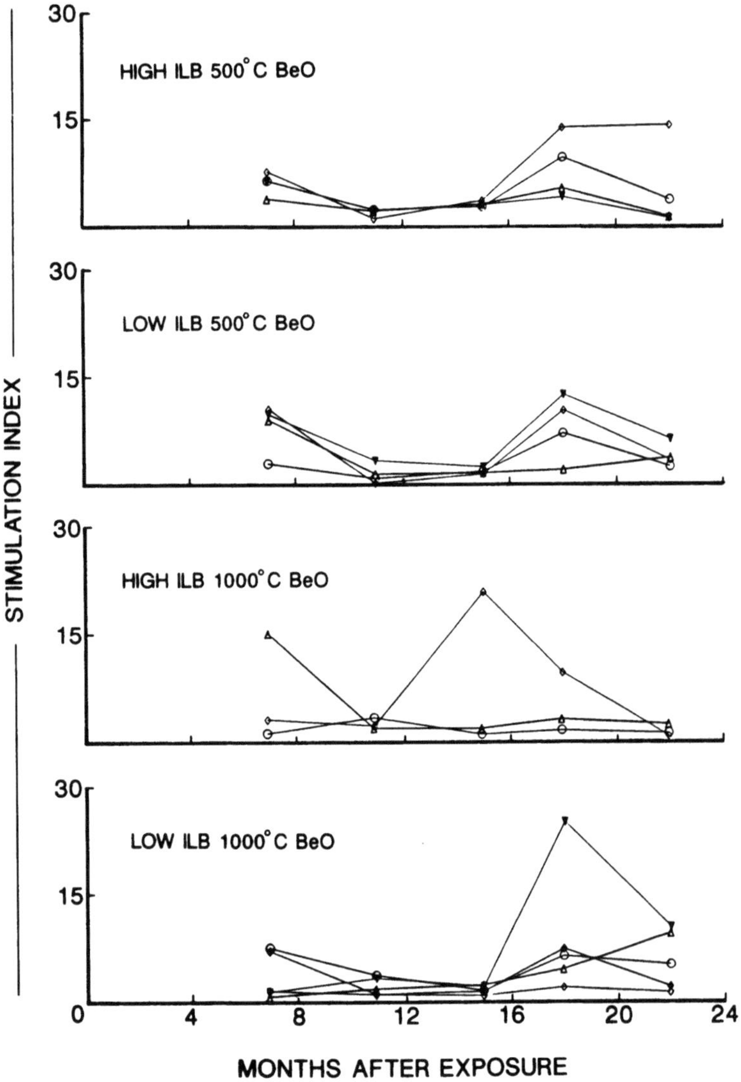

Figure 3. Stimulation indices of peripheral blood lymphocytes after culture with 100 µM $BeSO_4$ of dogs with either high or low ILB, and 500°C- or 1000°C-treated BeO. Each symbol represents a single animal and is the same symbol for the same dog in all figures.

DISCUSSION

Compartmentalized lymphocyte responses in the lung may be responsible for mediating the pathologic sequelae of CBD. The data reported here that show evidence of systemic sensitization, as measured by peripheral blood lymphocyte responses, do not necessarily mean that sensitization within the pulmonary compartment has occurred. Furthermore, these data suggest that sampling lung lymphocytes may provide a better indication of an individual's sensitivity to beryllium and possibly the propensity for developing CBD.

Despite extremely high percentages of lung lymphocytes (46 to 69%) and moderately elevated lymphocyte numbers (3.6×10^6 to 17.3×10^6) in dogs with the high ILB, 500°C-treated BeO at 3 mo after exposure, blastogenic responses of these cells following culture with $BeSO_4$ were not significantly different than the responses of cells from control dogs. Furthermore, when lymphocyte percentages and numbers were declining, stimulation indices were markedly positive. However, individual lymphocyte percentages and numbers correlated poorly with the level of stimulation indices in the same animal. In fact, at 6 mo after exposure, dogs with the high ILB, 500°C-treated BeO displayed lymphocyte numbers near those of unexposed control dogs, despite the presence of two dogs with high SI. Clearly, the presence of large lymphocyte numbers and/or percentages within the lung following exposure to beryllium is not an accurate reflection of beryllium sensitization.

Pathologic changes within the lung are probably a product of both the number of beryllium-specific lymphocytes generated as well as nonspecific inflammatory and chemotactic events that attract large numbers of lymphocytes to the lung. Subsequently, only a small portion of lymphocytes within some individuals go on to mount a beryllium-specific cell-mediated immune (CMI) response. This response may be transient or may result in CBD. Even in this context, the emergence of individuals with beryllium-specific CMI responses within the lung is not necessarily indicative of evolving CBD, but does suggest the existence of a potentially susceptible subpopulation.

In summary, a single inhalation exposure of dogs to BeO produces lung lymphocyte responses consistent with those observed in humans with chronic beryllium disease, with the caveat that such responses appear to resolve. It also appears that inhalation exposure to beryllium produces nonspecific inflammatory changes characterized by infiltration of the lung by large numbers of lymphocytes. In most cases, these lymphocytes are not beryllium-specific and rapidly leave the lung. In a few individuals, a subset of beryllium-specific lymphocytes persists within the lung and mediates beryllium-specific CMI responses. It is postulated that beryllium-induced lung lesions are also mediated by this beryllium-sensitized subpopulation of lymphocytes. Manipulation of this model may allow delineation of many of the pathogenic mechanisms responsible for the production of CBD.

REFERENCES

Eisenbud M, Lisson J (1983) Epidemiological aspects of beryllium-induced nonmalignant lung disease: A 30-year update. J Occup Med 25(3):196-202

Barna BP, Deodhar SD, Gautam S, Edinger M, Chiang T, McMahon JT (1984) Experimental beryllium-induced lung disease. II. Int Arch Allergy Appl Immunol 73:49-55

Rossman MD, Kern JA, Elias JA, Cullen MR, Epstein PE, Preuss OP, Markham TN, Daniele RP (1988) Lung proliferation to beryllium: A diagnostic test for chronic beryllium disease. Ann Internal Med, in press

OXIDANTS, ANTIOXIDANTS AND MONOKINES IN THE PATHOGENESIS OF COAL WORKERS' PNEUMOCONIOSIS. TUMOR NECROSIS FACTOR AS A MEDIATOR IN LUNG FIBROSIS.

Paul Borm, Nicole Palmen, John Engelen and Wim Buurman. Departments of Occupational Medicine, Clinical Genetics and General Surgery, University of Limburg. PO Box 616, 6200 MD Maastricht, The Netherlands.

INTRODUCTION

Historically, the lung macrophage has been allotted a central though somewhat passive role in the initiation of interstitial fibrosis. Using the adjective passive means, that its injury and death results in the release of a variety of soluble mediators playing a role in fibrogenesis. Nowadays, the central role of the macrophage is maintained and even strengthened on the basis of its role in recruitment of inflammatory cells and functional responses. Apart from releasing fibroblast growth factors (Rom, 1987) and monokines (Schmidt et al, 1984) playing a role in fibroblast proliferation, macrophages generate oxygen metabolites such as $O_2^{\cdot -}$, H_2O_2, HOCl and $OH^{\cdot}$ that are particularly toxic to surrounding cells. Moreover, recruited inflammatory cells (neutrophils, eosinophils) are also able to produce these toxic oxidant molecules. Apart from causing aspecific (membrane) damage, oxidants are known to play a role in activating or decativating proteolytic and anti-proteolytic proteins (Janoff, 1985) and causing pulmonary fibrosis (Ward et al, 1983).

What is striking is not that there are so many agents that can invoke radicals' and/or inflammatory mediators' release (both a hazard for pulmonary fibrosis), but that fibrosis can develop despite the lungs' enormous capabilities for protection and recovery. With regard to the detoxification of oxidants, the lung possesses an elaborative protection mechanism constituted by a set of anti-oxidant enzymes like SOD and radical scavenging compounds such as GSH. However, protective activity of the anti-oxidant factors is confined to the compartment in which their are located (intra/extracellular, membrane). In addition, antioxidants localized in (circulating) erythrocytes should be included in the antioxidant protection of the lung as shown by several authors (Asbeck van, 1984; Agar et al, 1986). The erythrocyte in this way should be looked upon as an (extracellular) circulating antioxidant carrier, that is able to penetrate into the smallest cappilaries.

In this paper we present some data on red blood cell and plasma anti-oxidant levels of miners suffering from Coal Workers' Pneumocionisis (CWP). We also show data on the release of tumor necrosis factor from blood monocytes during

NATO ASI Series, Vol. H30
Effects of Mineral Dusts on Cells
Edited by B. T. Mossman and R. O. Bégin

the development of CWP. The data will be used to discuss the role of oxygen radicals in lung fibrosis.

METHODS

All individuals mentioned in this paper are coal miners still active in mining, except for the patients with progressive massive fibrosis who were registered at our University's Pulmonary Clinic. The data presented in this paper are extracted from two independent case-control studies concerning a total number of 156 coal miners with (n = 60) or without (n = 96) CWP, per study matched for age and years of underground work in the coal mine. Diagnosis was obtained from panel judgement of recent (< 1 year) chest radiographs and scored according to the current UICC/ILO classification. In this study the persons with classifcation 0/1,1/0,1/1, and 1/2 were gathered in group I, subjects with 2/1,2/2 or 2/3 in group II and those with 3/2 and 3/s in group III. When evaluating the TNF data, it proved useful to distinguish the subjects with 0/1 ("suspected" cases) from the first group. During selection of cases and controls subjects were interviewed among others on their present and past smoking habits, (previous) work history, medication and family history of pneumoconiotic diseases.

Assay of Antioxidant factors

Anti-oxidant factors in red blood cells or plasma were determined as described previously (Borm et al, 1987). Catalase was measured by the method of Aebi (Aebi, 1984). Prior this measurement, the haemoglobin concentration was determined, and the lysate was diluted to 50 mg Hb per ml. The resistance of (fresh) red blood cells to imposed oxidative stress was estimated by the initiation of lipid peroxidation measured as malondialdehyde (MDA). In 2.5 ml red blood cell suspension (5% PCV) oxidative stress is induced by the addition of 2.5 ml H_2O_2 (10 mM) in the presence of sodium azide (1 mM). At different time points (0, 10, 20 and 30 min) MDA was measured using the method of Stocks & Dormandy (1971). The intercept of a graph of MDA production against time was defined as the MDA lag time, considered to be the period during which the red blood cell is able to prevent MDA production. For the determination of plasma iron we used the iron FZ tests of Roche diagnostics.

Measurement of TNF-release

Purification of blood monocytes and the ELISA for the assay of TNF released from these cells was described recently (Borm et al, to be published). Spontaneous TNF release from blood monocytes was measured and TNF release after stimulation with endotoxin (3 or 1,000 ng/ml), coal mine dust containing 6% quartz (5 mg/ml) or silica (MinUsil), 0.5 mg/ml.

RESULTS

Blood Anti-oxidant factors

When considering the CWP cases as a homogeneous group no statistical significant differences could be distinguished as compared to control miners. Glutathion levels, however, are significantly decreased in red blood cells of miners with CWP in group I (0/1-2/1). In the same group total GSH-peroxidase activity is decreased. Subjects having further developed CWP (group III) have lower plasma iron levels and less resistance to in vitro imposed oxidative stress (a shorter MDA lag time).

TABLE 1: Some anti-oxidant factors in red blood cells and plasma of miners with CWP at different stages as compared to control miners.

		GSH (µmol/g Hb)	GSH-Px (IU/g Hb)	Fe(plasma) (µmol/ml)	MDA lag time (min)
Controls	(n = 58)	3.97 [.08]	9.58 [.08]	19.0 [0.8]	3.9 [0.2]
Group I	(n = 19)	3.62 [.14]*	9.88 [.08]*	19.2 [2.2]	4.3 [0.5]
Group II	(n = 10)	3.98 [.20]	9.73 [.15]	16.8 [1.5]	4.4 [1.5]
Group III	(n = 4)	3.99 [.44]	9.79 [.09]	12.8 [1.3]*	2.2 [1.3]**

All data are mean ± standard error (between brackets).
GSH, reduced glutathione; GSH-Px, total glutathione peroxidase; Fe, total iron concent and MDA lag time, lag-time observed before onset of lipidperoxidation of red blood cells.
*) Significantly different from same value in control group, $P < 0.05$ Mann Whitney U-test.
**) Id but significance level $P < 0.08$.

Apart from above changes in <u>levels</u> of anti-oxidant factors, some factors are correlated in the CWP cases while not in the controls. This is illustrated in fig. 1 showing the correlation between red blood cell SOD and GSH-Px. The same phenomenon was observed previously in patients with progressive massive fibrosis (Borm et al, 1987). In this CWP group also red blood cell SOD and catalase were correlated well ($r = 0.68$, $P < 0.01$).

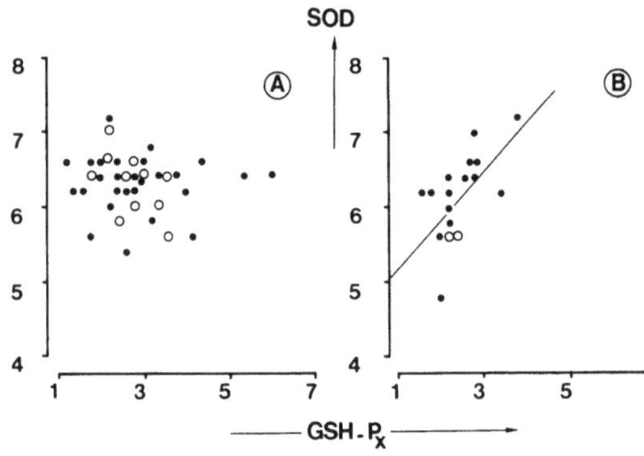

FIGURE 1: Scatterdiagrams of SOD activity (Units/g Hb) vs. GSH-Px activity (IU/g Hb) in red blood cells of control miners (panel A) and miners having CWP developed beyond stage 2/1 (panel B). Only in the latter group the two parameters are significantly correlated ($r = 0.62$, $P < 0.01$).

TNF release from blood monocytes

TNF release was measured using blood monocytes (BM) of 66 coal miners and 12 non-dust exposed individuals. Twenty seven miners had simple CWP, but were still active underground without being treated for their disease. TNF was measured with an TNF specific ELISA. Monocytes of all individuals responded to stimulants by the release of TNF (Fig. 2). First of all, control miners'monocytes release more TNF as compared to non-dust exposed controls. In contrast to dust exposure no effect of smoking, exposure-years or medication was observed on spontaneous TNF release. Smoking did, however, increase endotoxin induced TNF release. Secondly, fig. 2 shows that TNF release by BM from miners with CWP stimulated with endotoxin (3 ng/ml) or Coal mine dust (5 mg/ml) differ significantly from the same parameter in both control groups. A further analysis showes the largest difference between control miners and those cases who are seriously suspected from CWP (0/1) or having a small number of opacities (1/0,1/1,1/2). In the graphical presentation of these data (Fig. 3) we clearly observe a trend of diminishing TNF release with progression of CWP. This effect is not seen in spontaneous TNF production.

DISCUSSION

The presented data clearly show that during the development of CWP changes occur in blood anti-oxidants and monocyte TNF production. Obviously, the former changes reflect the exagerated production of oxidant species in the lung.

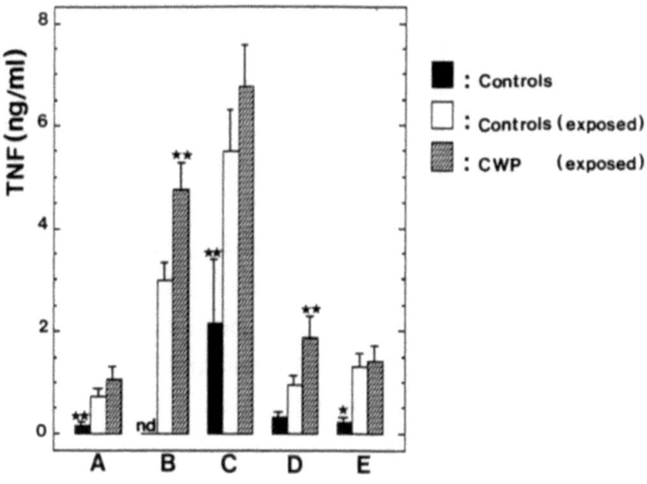

FIGURE 2: Mean TNF release plus the standard error from monocytes obtained from non-dust exposed controls (n = 12), control miners (n = 39) and miners with CWP (n = 27). The alfabetic symbols below the axis depict spontaneous TNF release (A) or TNF release in response to 3 ng/ml LPS (B), 1000 ng/ml LPS (C), 5 mg/ml coal mine dust (D) or 0.5 mg/ml silica. Testing for significance (*, P < 0.05 and **, P < 0.01) was done towards the control miners group (Mann-Whitney U-test).

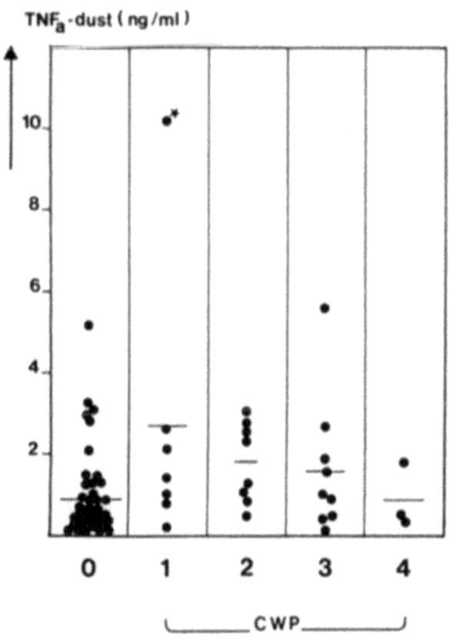

FIGURE 3: Release of TNF from peripheral BM treated with coal mine dust (TNFa-dust) in coal miners without (0, n = 39) or with CWP at various degrees of severity. A subdivision of CWP was achieved by gathering ILO classifications into the following groups: 1 (0/1), 2 (1/0-1/2), 3 (2/1 - 2/3) and 4 (3/2, 3/3).

GSH is decreased initially but is restored in later phases of CWP and is even increased in patients with PMF (Borm et al, 1987). Plasma iron is decreased in later stage CWP, probably a protective measure to reduce hydroxyradical production through the Haber-Weiss reaction. The meaning of the diminished red cell anti-oxidant capacity (MDA lag) and changes in relations between anti-oxidant enzymes are not yet explained for. In summarison, all data support the role of the eythrocyte as a circulating ant-oxidant carrier. The observed changes in blood anti-oxidant factors are most likely due to enhanced oxidant production in the lung.We do not know whether the observed differences in TNF release are acquired or genetically controlled. TNF is a pleiotropic monokine (Beutler & Cerami, 1987) capable of stimulating fibroblast growth and is very effective in enhancing production of oxidants from neutrophils adherent to vascular surfaces (Nathan, 1987). According to this, individuals showing an increased local TNF release in response to mineral dust would be more susceptible to the dusts' fibrogenic effects. Interestingly, TNF release of blood monocytes was highestin those subjects where CWP was in very early stage. During progression of the disease, the TNF release is reduced back to control levels, probably refelecting a protective feedback mechanism to prevent PMF due to the increasing load of dust particles in the lungs.

To address the final questions, i) what is the link between anti-oxidant protection and TNF? and ii) how do they fit into current concepts about lung fibrosis? the first is the easy one. TNF is the most potent inducer of PMN's oxidative burst (Nathan, 1987) and many of its in vivo effects (Pfizenmaier et al, 1987) indirectly impose an oxidative stress to the inflammatory site. Moreover, it was recently shown (White et al, 1987) that TNF and IL-1 protect against pulmonary oxygen toxity, and supposed to be result from a favored GSH/GSSG status. The second question implicitly demands for a lung fibrosis model. Present knowledge urges us to consider pulmonary fibrosis as the <u>end result</u> of a process in which the balance of normal injury/repair-protection mechanisms is disrupted. There is clearly no "single fibrogenic event" but rather a number of levels where disruption of different balances may begin. We consider the oxidant/anti-oxidant balance (Fig. 4) being just one balance in the total injury/repair clockwork-like mechanism.

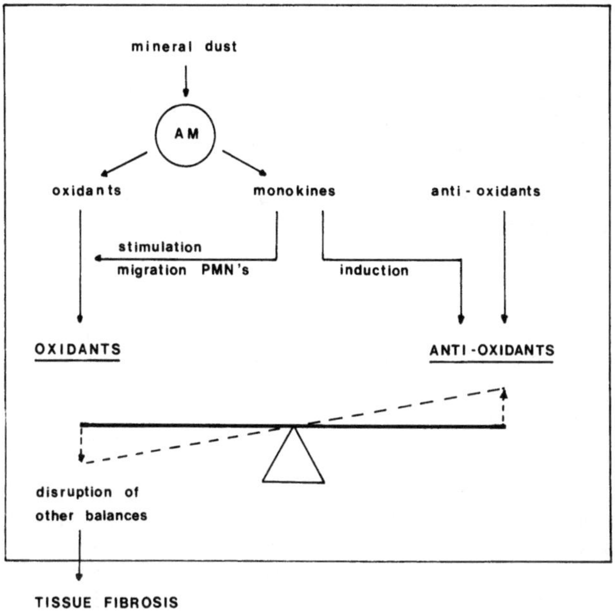

FIGURE 4: Schematical presentation of the suggested role of oxidants, anti-oxidants and monokines in mineral dust induced pulmonary fibrosis. See also text Discussion (AM: Alveolar macrophages; PMN's: Polymorphonuclear leukocytes).

Tipping over the oxidant/anti-oxidant balance in favour of the oxidants may subsequently contribute to disrupting another balance, e.g. between growth and inhibitory factors for fibroblast growth or the protease-antiprotease balance. Monokines such as TNF are considered to play different roles at the different stages "towards lung fibrosis".

ACKNOWLEDGEMENTS

The authors are seriously indebted to the Dutch Petrol Company (NAM) for its financial support of this research. We thank docters Marc van Sprundel, Luc Lenaerts and Jo Cardinaels for their support to make this study possible and for reading the X-rays, Gerard Swaen for his epidemiological advices and Marliese van Wissen for carefully preparing this manuscript.

REFERENCES

Aebi H (1984) Catalase in vitro. In: Beutler E (ed) Methods in Enzymology, vol 105. Academic Press, New York, p. 121-126.
Agar NS, Sadrsadeh SMH, Hallaway PE, Eaton JW (1986) Eythrocyte catalase a somatic oxidant defense? J Clin Invest 77: 319-321.
Asbeck van BS, Hoidal J, Vercelotti GM, Schwartz BA, Moldow CF, Jacobs HS (1984) Protection against lethal hyperoxia by tracheal insufflation of erythrocytes: role of red cell glutathione. Science 277: 756-758.
Beutler B, Cerami A (1987) Cachectin: more than a tumor necrosis factor. N Engl J Med 316: 379-385.
Borm PJA, Bast A, Wouters EFM, Slangen JJM, Swaen GMH, de Boorder Tj (1986) Red blood cell anti-oxidant parameters in silicosis. Int Arch Occup Environ Hlth 58: 235-244.
Borm PJA, Bast A, Wouters EFM, Slangen JJM, Swaen GMH, de Boorder Tj (1987) Red blood cell anti-oxidant parameters in healthy elderly control subjects versus silicosis patients. Free Rad Res Comms 3: 117-127.
Borm PJA, Palmen N, Engelen JJM, Buurman WA (to be published) Spontaneous and stimulated release of tumor necrosis factor-alpha (TNF) from blood monocytes of miners with cola workers' pneumoconiosis. Am Rev Resp Dis.
Janoff A (1985) Elastases and emphysema. Current assessment of the protease-antiprotease hypothesis. Am Rev Respir Dis 132: 417-433.
Nathan C (1987) Neutrophil activation on biological surfaces. Massive secretion of hydrogen peroxide in response to products of macrophages and lymphocytes. J Clin Invest 80: 1550-1560.
Pfizenmaier K, Kronke M, Scheurich P, Nagel GA (1987) Tumor necrosis factor (TNF) Alpha: control of TNF-sensitivity and molecular mechanisms of TNF-mediated growth inhibition. Blut 55: 1-10.
Rom WN, Bitterman PB, Rennard SI, Cantin A, Crystal RG (1987) Characterization of the lower respiratory tract inflammation of nonsmoking individuals with interstitial lung disease associated with chronic inhalation of inorganic dusts. Am Rev Resp Dis 136: 1429-1434.
Stocks J, Dormandy TL (1971) The autooxidation of human red cell lipids induced by hydrogen peroxide. Brit J Haemotol 20: 95-111.
Schmidt JA, Oliver CN, Lepe-Zuninga JL, Green I, Gery I (1984) Silica stimulated monocytes to release fibroblast proliferation factors identical to interleukin-1. A potential role for interleukin-1 in the pathogenesis of silicosis. J Clin Invest 73: 1462-72.
Ward PA, Till GO, Kunkel R, Beauchamp C (1983) Evidence for role of hydroxyl-radical in complement and neutrophil-dependent tissue injury. J Clin Invest 72: 789-801.
White CW, Ghezzi P, Dinarello CA, Caldwell SA, McMurtry IF, Repine JE (1987) Recombinant tumor necrosis factor/cachectin and interleukin 1 pretreatment decreases lung oxidized glutathione accumulation, lung injury, and mortality in rats exposed to hyperoxia. J Clin Invest 79: 1868-1873.

USE OF CHEMILUMINESCENCE ASSAYS TO MONITOR THE SURFACE CHARACTERISTICS AND BIOLOGICAL REACTIVITY OF FRESHLY FRACTURED VS AGED SILICA

V. Castranova, V. Vallyathan, K. Van Dyke, and N.S. Dalal
Division of Respiratory Disease Studies, NIOSH and
West Virginia University, Morgantown, WV 26505

Introduction

Inhalation of crystalline silica is often associated with the development of pulmonary disease (Ziskind et al., 1976). Recently, it has been proposed that freshly sheared silica may have surface properties that make it more reactive with lung tissue than aged silica, and that this unique reactivity of freshly crushed silica enhances its pathogenicity. A recent study in our laboratory strongly supports this theory (Vallyathan, et al., 1988a; 1988b). We found that grinding of silica cleaves bonds in a silica tetrahedral lattice generating reactive free radicals sites on the surface of the silica particles. The presence of surface radicals on freshly ground silica was monitored by electron spin resonance (ESR) spectroscopy. The ESR signal generated from fresh silica in air is characteristic of $Si\text{-}\dot{O}$ and $\dot{Si}$ radicals. Evidence, i.e., the ESR signal after the addition of a spin trap (DMPO) and its sensitivity to ethanol, indicates that these silicon-based surface radicals can react with water to form hydroxyl radicals ($\dot{O}H$). Results also indicate that these surface radicals make fresh silica more cytotoxic than aged silica; i.e., fresh silica causes more membrane damage as evidenced by hemolysis and lipid peroxidation. Therefore, the presence of reactive sites on the surface of freshly cleaved silica may play a significant role in its pathogenicity.

If surface free radicals are related to cytotoxicity, then screening dusts for the existence of such radicals may predict their relative pathogenicity. Since an ESR spectrometer

NATO ASI Series, Vol. H30
Effects of Mineral Dusts on Cells
Edited by B.T. Mossman and R.O. Bégin

is not available to many laboratories, the objective of this study was to develop a quick and simple technique to monitor the existence of reactive surface sites on dust. The technique chosen was the measurement of chemiluminescence using a liquid scintillation counter; i.e., an instrument which would be accessible to most investigators.

Methods

Crystalline silica (0.2 - 5.0 mm in diameter) was obtained from the Generic Respirable Dust Technology Center (Penn State University, State College, PA). Silica was ground for 30 minutes using a ball grinder with an agate ball and sieved through a 20 micron mesh filter before use. X-ray spectrometry was performed on representative samples to confirm that the ground silica was mineralogically pure and contained no detectable contaminants. All comparative studies used samples from a single stock of ground silica to assure uniformity of shearing and particle size.

Chemiluminescence generated by silica was measured as counts per minute in the tritium channel of a liquid scintillation counter (Packard Tri-carb, Downers Grove, IL) operated in the out-of-coincidence mode. The surface activity of dry quartz was monitored by adding 5 g of silica to a plastic vial and measuring the light generated. The effect of aqueous solutions on the generation of chemiluminescence by quartz was determined using a HEPES-buffered medium (145 mM NaCl, 5 mM KCl, and 10 mM HEPES; pH = 7.4). The effect of scavengers on chemiluminescence was monitored after the addition of 125 µg/ml superoxide dismutase (SOD), 125 µg/ml catalase, or 100 mM 5,5-dimethyl-1-pyrroline-1-oxide (DMPO) to quartz suspended in HEPES-buffered medium.

Alveolar macrophages were harvested from male pathogen-free Sprague-Dawley rats by pulmonary lavage with calcium and magnesium-free Hank's balanced salt solution. Macrophages from ten 8 ml lavages were sedimented by centrifugation at 500 g for 5 min at 2°C. Cells were then washed once, centrifuged, and resuspended in HEPES-buffered medium containing 1 mM $CaCl_2$ and 5 mM glucose. Cell counts and viability were determined microscopically after the addition of trypan blue dye to an aliquot of cells. Results indicate that 90-95% of the lavaged cells were viable alveolar macrophages.

Silica-induced activation of alveolar macrophages was determined by measuring lucigenin-enhanced chemiluminescence at 37°C using a Berthold 9505 Luminometer. Samples (500 µl) contained $1x10^6$ alveolar macrophages, $2.5x10^{-8}$M lucigenin, and 20 µg/ml silica. Data were expressed as counts per second per $1x10^6$ alveolar macrophages.

Results

Grinding of silica generates silicon-based radicals on the particle surface which can be monitored by ESR (Vallyathan, et al., 1988a; 1988b). Surface activity of quartz can also be monitored by measuring the generation of chemiluminescence. As is shown in Figure 1, freshly ground quartz emits substantial chemiluminescence. This chemiluminescence declines with time exhibiting a T 1/2 of approximately 37 minutes. In contrast, aged silica (stored for several days after grinding) generates little chemiluminescence.

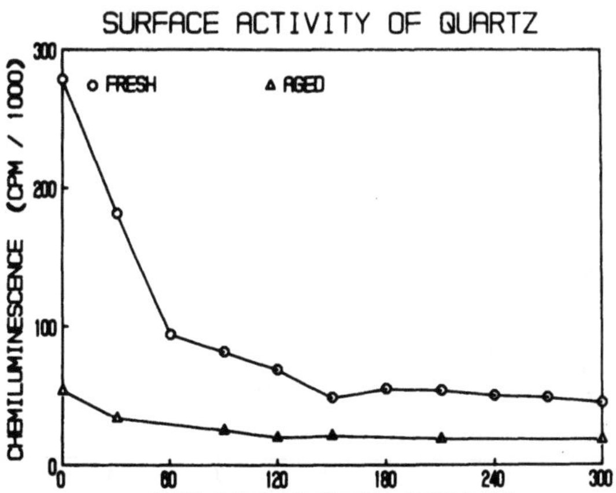

Fig. 1 Surface activity of quartz measured as chemiluminescence immediately after grinding (fresh) or several days after grinding (aged).

The decay in surface activity of freshly ground quartz is also demonstrated in Figure 2. Peak chemiluminescence emitted by ground silica decreases by 37% after 2 hours of storage in air and by 58% 3 hours after grinding.

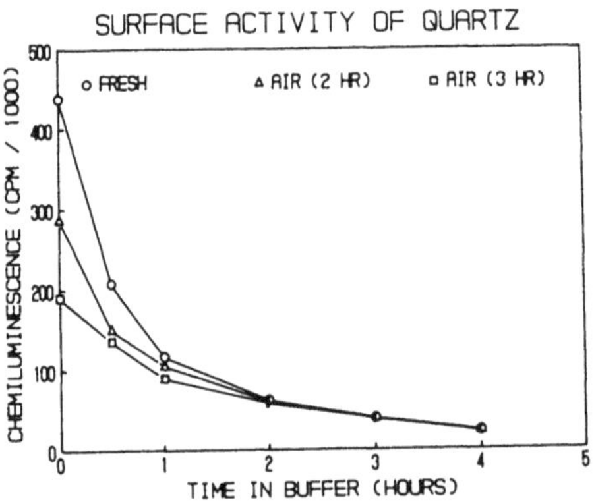

Fig. 2 Chemiluminescence generated by ground silica immediately following grinding and after 2 or 3 hours of storage in air.

Light has a substantial effect on the amount of chemiluminescence generated by silica after grinding. Figure 3 shows that silica ground in the dark generates substantial chemiluminescence (109,000 cpm over background). However, silica ground under room light generates 2 1/2 fold more chemiluminescence than silica ground in the dark.

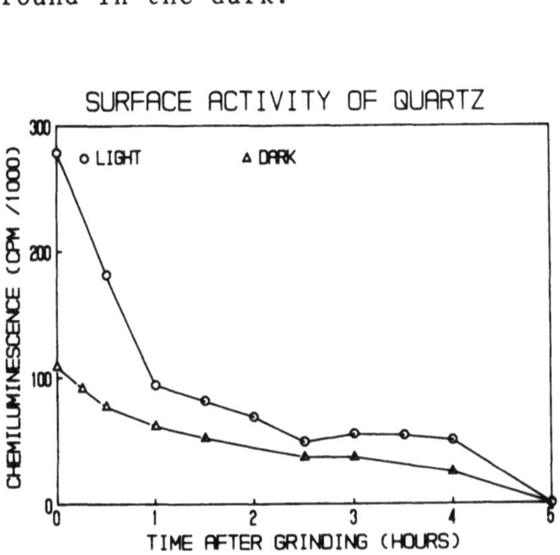

Fig. 3 The effect of light during the grinding process on chemiluminescence generated by freshly crushed quartz.

Surface activity resulting from crushing crystalline silica is measurable in buffered-medium as well as in air. The decline in activity of fresh silica suspended in medium exhibits a T 1/2 only slightly less (T 1/2 = 28 minutes) than that in air. This result suggests that an activated surface exist for fresh quartz even after it comes in contact with biological fluids.

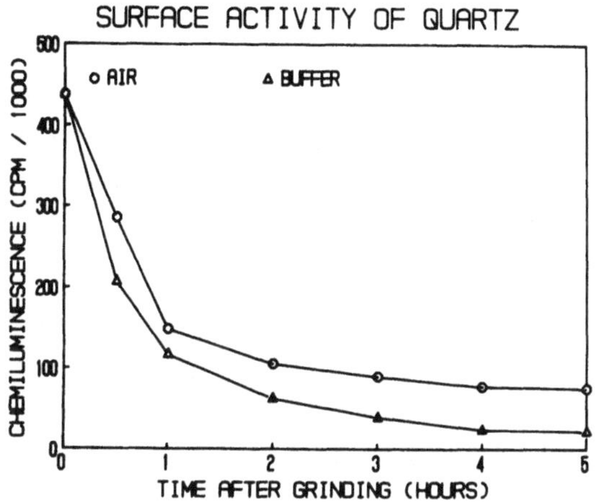

Fig. 4 Chemiluminescence generated by freshly ground silica in air and in HEPES-buffered medium.

Surface activity of freshly crushed silica suspended in medium is inhibited by scavengers of reactive oxygen species (Figure 5). SOD decreases chemiluminescence by 71% while catalase and DMPO inhibit by 88% and 97%, respectively.

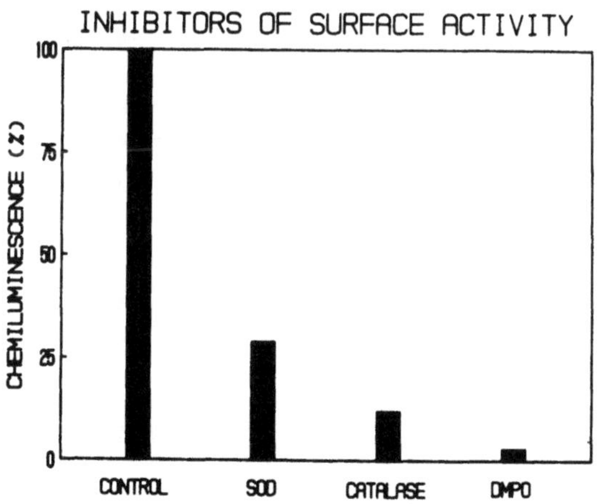

Fig. 5 Inhibition of chemiluminescence generated by freshly crushed silica with scavengers of reactive oxygen species.

SOD acts as a scavenger of superoxide anions, catalase as a hydrogen peroxide scavenger, and the spin trap DMPO as a scavenger for hydroxyl radicals (OH). Therefore, the existence of these reactive species is suggested.

Thus far, the data indicate that crushing or grinding silica forms active sites on the particle surface which generate chemiluminescence. These surface sites also make fresh silica a more potent stimulant of alveolar macrophages. Silica-induced stimulation of alveolar macrophages and the resultant release of reactive species can be monitored as chemiluminescence (Figure 6). Aged silica (2 days after grinding) does activate alveolar macrophages causing the generation of 127,135 cps (integrated counts-background) over the 40 minute measurement period. However, alveolar macrophages exposed to silica stored for only one day exhibit 8 1/2 times this activity and macrophages exposed to freshly crushed quartz are activated to 33 times this level.

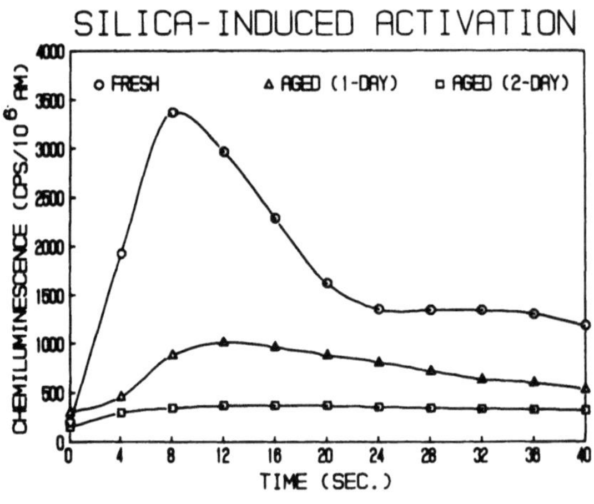

Fig. 6 The effect of freshly crushed or aged quartz on particle-stimulated chemiluminescence by alveolar macrophages.

Discussion

Previous investigations in our laboratory have shown that crushing or grinding silica cleaves bonds in the crystalline structure of quartz resulting in the formation of silicon-based radicals on the particle surface (Vallyathan et al., 1988a; 1988b). These surface radicals were identified as $Si\text{-}\dot{O}$ or $\dot{Si}$ by ESR. The present study demonstrates that the surface activity of freshly ground silica can also be monitored using a liquid scintillation counter to measure chemiluminescence. The T 1/2 for these silicon-based radicals was estimated by ESR to be approximately 30 hours. In contract, T 1/2 for the decay of chemiluminescence is much faster, i.e., ranging from approximately 37 minutes in Figure 1 to 3 hrs in Figure 2. The reason for this difference in decay rate is given in Figure 7. ESR measures the presence of radicals while chemiluminescence measures the existence of excited states. Radicals generated by crushing are initially in an excited state and rapidly decay to ground state radicals. During this process light is emitted. However, an ESR signal remains because even in the ground state the radicals have not yet reacted with H_2O to form SiOH.

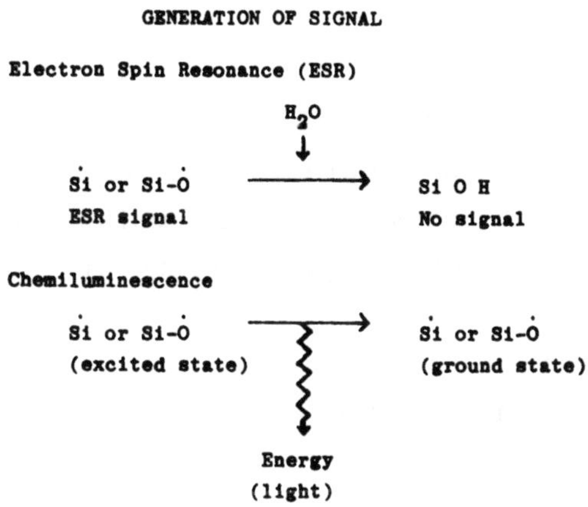

Fig. 7 Scheme for the generation of ESR signals and chemiluminescence by freshly ground quartz.

Data from the present study indicate that chemiluminescence generated from freshly crushed silica exists even in aqueous medium (Fig 4). This chemiluminescence can be inhibited by SOD, catalase, or DMPO and thus seems to result from the generation of reactive oxygen species, i.e., O_2^{-}, H_2O_2, and $\dot{O}H$ (Fig 5). This conclusion is supported by a previous report from our laboratory using ESR which demonstrated the generation of OH when freshly crushed silica reacts with aqueous medium (Vallyathan, et al., 1988b).

Data from our investigations indicate that freshly crushed silica particles possess surface radicals which can react with water to generate reactive oxygen species. The generation of these reactive species on the particle surface and in the medium can cause lipid peroxidation (Vallyathan et al., 1988b). In addition, freshly ground quartz is a potent stimulant of chemiluminescence from alveolar macrophages (Fig 6). This presumably reflects excessive secretion of reactive oxygen species from alveolar macrophages exposed to fresh silica. The combination of these oxidant stresses may overwhelm the lung's protective anti-oxidant systems and result in tissue destruction (Weiss and LaBuglio; 1982). Therefore, these oxidants may play a role in etiology of acute silicosis where inhalation of freshly cleaved silica particles occurs. In conclusion, we have proposed that surface radicals generated during the crushing of silica particles may play an important role in the cytotoxicity of this dust. We have demonstrated that measurement of chemiluminescence using a liquid scintillation counter is a simple and convenient method to screen dust for surface activity. This method may be valuable in estimating the potential cytotoxicity of various dusts.

Acknowledgement

Generic Mineral Technology Center for Respirable Dust, BOM - G1175142

References

Vallyathan V, Shi X, Dalal NS, Irr W, and Castranova V (1988) Silicon-oxygen radicals and their role in acute silicosis. Am Rev Respir Dis 137:404

Vallyathan V, Shi X, Dalal NS, Irr W, and Castranova V (1989) Generation of free radicals from freshly fractured silica dust: potential role in acute silica-induced lung injury. Am Rev Respir Dis 138:1213-1219

Weiss SJ and LaBuglio AF (1982) Biology of disease; phagocyte generated oxygen metabolites and cellular injury. Lab Invest 47:5-18

Ziskind M, Jones RN, and Weill H (1976) Silicosis. State of the art. Am Rev Respir Dis 113:643-665

Production of free radicals by non-fibrous materials in a cell-free buffer medium

D. Costa, J. Guignard, H. Pezerat
Laboratoire de Réactivité de Surface et Structure
Université Pierre et Marie Curie
75252 Paris Cédex 05, France

INTRODUCTION

Excess of lung cancers have been observed in many metallic mines : Lorraine iron mines (Pham, 1983), Swedish iron mines (Edling, 1982 ; Jorgensen, 1973 ; St Clair Renard, 1974 ; Radford et al., 1984), iron mines of Cumberland (Great Britain) (Boyd et al., 1970), the Schneeberg mines (GDR) where the historical "mountain sickness" was observed in the XVIth Century, gold mines of Gwanda (Zimbabwe) (Osburn, 1957, 1969) and of South Africa (Webster, 1973), gold mines of Ontario (Muller et al., 1984, 1986a, 1986b), and gold mine of Salsigne (France) (Perisse, 1977 , Jammes, 1985).

It appears from some of these data that radioactivity cannot be the only causal factor in these excesses of lung cancer : in Salsigne mine and in Lorraine mines, the rate of radioactivity is very low. The study of Pham and the Swedish studies excluded the role of diesel smokes and nitrate fuel explosives for the period and sites considered, and asbestos cannot be a causal factor in the sites considered here.

Thus, it is necessary to take into account the possible role of inhaled dusts in the excess of lung cancers.

Epidemiological and experimental evidences pertaining to nickel carcinogenesis have been comprehensively documented in several monographs and review articles, for example Sunderman Jr (1984a,b).

Numerous biological and physicochemical tests were realized on nickel compounds, but as yet there is no satisfactory hypothesis concerning the mechanisms of nickel genotoxicity.

The latest studies of Sunderman Jr. (1988) showed that in rats, after IM injection of several nickel compounds, Ni_3S_2 and two (Ni,Cu)O compounds induced 100 % sarcomas whereas most of the NiO compounds did not induce sar-

NATO ASI Series, Vol. H30
Effects of Mineral Dusts on Cells
Edited by B.T. Mossman and R.O. Bégin

comas. All of these compounds (NiAs, Ni_3S_2, NiO and ((Ni,Cu)O) have been investigated by us and our results are compared here with the results obtained by Sunderman using IM injection. By analogy with the results from organic carcinogenesis, we hypothezised that the ultimate carcinogenic agents are electrophilic species (oxidative species).

The compounds under study here are insoluble or very little soluble in an aqueous medium. Thus, we postulate that their activity in the biological medium is a surface activity, involving surface active sites, which after phagocytosis, are able to react with molecules present in the biological medium. O_2 can be such a target molecule, because it is present in a stationary concentration in the biological medium. Thus, we are studying the ability of the surface of the particles to produce, after interaction with dissolved oxygen, some activated species of oxygen, with a strong oxidizing capacity. This type of activity is related to the general theory of the oxidative stress, which implies active species of oxygen (Sies, 1985).

These species are created on the surface of the material after interaction between oxygen and reducing sites linked to a metallic cation in a low-valence state (Che and Tench, 1982, 1983). In our experiments we detect these species using a target molecule, here HCO_2^-, which is oxidized to $CO_2^{\cdot-}$ radicals. The quantity of $CO_2^{\cdot-}$ produced in a short time (about one hour) allows the determination of the existence and quantity of strong oxidizing oxygen species, on the surface of the particles. So our test allows to detect minerals and ores capable, after phagocytosis, to reveal an oxidizing activity, being eventually genotoxic.

The $CO_2^{\cdot-}$ has a short half-life and we use DMPO(5-5-dimethyl-1-pyrroline-N-oxide), which forms a $(DMPO,CO_2^-)^{\cdot}$ radical adduct after reaction with $CO_2^{\cdot-}$. This radical adduct has a half-life of about one hour, thus allowing its measurement by E.S.R. spectroscopy. As a preliminary approximation, one can consider the intensity of the signal $(DMPO,CO_2^-)^{\cdot}$ as a measure of the radical oxidizing activity of the material under study, thereby allowing a classification of the samples. The quantity of radicals produced is not reported as such here. We just classify the material in seven categories ranging from "non active" to "out of category". Between the categories "little active" and "out of category" there is a multiplicative factor, for the number of radicals, of about one to two hundreds.

The typical experimental procedure is as follows. Forty-five mg of the solid under study was hand-ground for 1 minute and then immediately introduced into the open reactor. 1 ml of 1M sodium formate (dissolved in a potassium phosphate buffer 1 $mol.l^{-1}$, pH = 7.4) was added. The open reactor had a water jacket thermostated at 37 °C and protected from light exposure. Then 1 ml of an aqueous solution of DMPO 100 mM was added to the mixture. Time was counted from this moment. The mixture was stirred with the aid of an oscillating table.

Aliquots were withdrawn at 25 min. and 60 min. without affecting the relative concentrations of the various reactants. These aliquots were then filtered through a filter 0.65 μm porosity and immediately analyzed by ESR spectroscopy (Varian CSF instrument, field set 3380 G, scan range 100 G, microwave power level 10 mW, time constant 1 sec.). (For more details, see Costa et al., 1988a, 1988b, Zalma et al., 1987).

CHARACTERIZATION

The iron-containing compounds are principally from French iron mines : iron ores from the Lorraine mines (the tree essential phases are goethite, ($\alpha FeOOH$), siderite ($FeCO_3$) and berthierine ($(Si,Al)_2(Fe,Mg,Al)_3O_5(OH)_4$), and phyllosilicates from French iron mines (including biotite $K(Si,Al)_4(Fe,Mg)_3O_{10}(OH)_2$, chlorite $(Si,Al)_4(Fe,Mg)_6O_{10}(OH)_8$, and berthierine). Pyrite and Fe^{3+} oxides and oxyhydroxides are samples from mineralogical collections. Pyrrhotites come from the Salau and Salsigne French mines.

The Ni-Cu containing materials originate from mineralogical collections (minerals from the Schneeberg mine or of unknown origin) or have been synthesized, or are commercial products. The synthetic NiAs and NiS are the same as the ones studied by Sunderman Jr. (Alfa products), and the NiO and (Ni,Cu)O compounds were provided by Mr Bozec from NIPERA (all of these were also tested by Sunderman Jr. in IM injection).

The nickel mattes rich in Ni_3S_2 and the one rich in NiO were given to us by Mr. Bozec. They come from nickel refineries, where the mattes are treated in order to obtain nickel.

Table 1. Classification of the samples in regard to their ability to oxidize formate to $CO_2^{\overline{\cdot}}$ radicals in the buffer medium

Iron containing samples	
Air-aged (more than 4 months) powder of pyrite (FeS_2)	extremely active
Fresh powders of French iron ores Fresh powders of Fe^{2+} containing phyllosilicates (biotite, chlorite, berthierine)	very active
Some pyrrhotites ($Fe_{1-x}S$) Fresh powder of siderite ($FeCO_3$)	active
Fe^{3+} oxides and oxyhydroxides (goethite FeOOH, haematite ...) Fresh powder of pyrite (FeS_2), magnetite (Fe_3O_4)	non active (blank level)
Ni,Cu containing samples	
Mineral (Ni,Cu)As	out of category
Synthetic Cu_3As	extremely active
Nickel mattes (rich in Ni_3S_2) Fresh powder of synthetic NiAs Synthetic Ni_3S_2	very active
Air-aged powder of synthetic NiAs	moderately active
NiS (hexagonal phase) NiS (millerite) NiO(A-F)(NIPERA) (Ni,Cu)O(G-J)(NIPERA)	(blanck level) non active

RESULTS

Table 1 summarizes the results of the test of activity for two categories of samples : iron containing samples and nickel and copper containing samples.

<u>Iron samples</u> : Among the most active iron minerals, one find the air-aged powder of pyrite (FeS_2), which is an exceptional case where the air-aged powder is active whereas freshly ground powder is inactive. This is due to the slow formation of a coating of Fe^{2+}-sulfate and thiosulfate on the surface of the particle.

In the second category, are found freshly formed (by grinding) powders of some iron ores and of iron-rich phyllosilicates as chlorite and biotite. In this case, the air-aging is a very important factor of passivation, because the oxidized surface sheet is very difficult to dissolve in the buffer medium used and probably also in a biological medium. Some pyrrhotites ($Fe_{1-x}S$) and fresh powders of siderite ($FeCO_3$) (i.e. ground just before test) are active.

Finally, iron oxides and oxyhydroxides (goethite, haematite,magnetite), are inactive.

Nickel compounds :

- mineral (Ni,Cu)As is "out of category"
- synthetic Cu_3As is extremely active
- nickel mattes, rich in Ni_3S_2, are very active ; these mattes are air-aged powders. Freshly ground powder of synthetic NiAs is very active, and so is synthetic Ni_3S_2.
- air-aged powder of synthetic NiAs is moderately active. Air-aging is a strongly passivating factor.
- NiO, copper sulphides, (Ni,Cu)O are non active. Both phases of NiS (hexagonal and millerite) are non-active.

DISCUSSION

In all the cases, the activity is determined by the presence and the availability to oxygen, of metal cations in an electron donor state (Fe^{2+}, Ni^+, Cu^+) near or on the surface of the particles.

The main parameters playing a role in the activity are :

- The composition of the sample. For two samples of the same mineral, the rate of metal cation in a low valence state will influence the activity ; this is apparent for example with the chlorites : the activity increases with the Fe^{2+} content.

- The nature of the cation. The reducing power of the cations under study can be classified as $Ni^+ \sim Cu^+ > Fe^{2+}$.

- The coordination of the low valence state cation and the nature of its neighbours are of determining influence on the activity : in the magnetite Fe_3O_4 for example, the divalent irons of the magnetite are surrounded by trivalent irons, which have a strong electron acceptor character through the oxygens shared by the coordination polyedra of Fe^{2+} and Fe^{3+}.

In the case of nickel compounds, Ni^{+} can be stabilized in the Ni_3S_2, because of its tetraedral coordination and of Ni-Ni bonds (Nag, 1980), whereas it is not the case in NiS.

The arsenic ligands have a greater donor character than sulfur ligands, thus Ni^{+} may be stabilized on the surface of NiAs and not in the surface of NiS.

- The nature of the surface sheet. The surface sheet has in general a different composition than the bulk of the mineral : in the case of magnetite for example, the surface reorganizes to maghemite (γ-Fe_2O_3), which has the same oxygen sublattice as magnetite, but where all the iron is in the trivalent state. To the contrary, the surface sheet appearing by slow air-aging of the pyrite has stabilized Fe^{2+}, and thus is active. However, incubation in water during a few hours quickly lixiviates this sheet.

Generally, air-aging is rather a passivating factor : synthetic NiAs and phyllosilicates are good examples.

- The solubility of the surface sheet in the medium used is another important parameter of activity. An insoluble surface sheet inhibits the access of O_2 to reducing sites under the surface. Thus, the composition of the liquid medium used is of primary importance. In pure water for example, the quantity of radicals produced is very weak, even for very active materials. The presence of phosphate ions allows the scouring of the surface sheet.

Our results are compatible with those of Sunderman in the case of Ni_3S_2. The divergences for the other compounds could be explained by the fact that Sunderman perhaps tested the NiAs as "air-aged powder" (one month of contact with air is enough to render the surface passivated) ; in the case of NiS compounds the long latent times observed (40 weeks, for 30 weeks in the case of Ni_3S_2)(Sunderman, 1984b) in this case suggests that surface transformations are necessary before any genotoxic activity is observed. Such transformations have not yet been modelled.

CONCLUSION

The work presented here allows a broad screening of a great number of materials and permits us to point out a certain number of compounds as active in the model proposed. This is especially interesting in the case of the investigation of the toxic properties of such complex atmospheres as those inhaled in metallic mines. Thus, our work on Lorraine iron ores shows that the Fe^{2+}-containing phyllosilicates, as "fresh powders", may have a toxic activity.

It is also possible in some cases to determine the most favorable treatment, for a given compound, so that the material exhibits an activity in the biological tests.

Biological studies are still necessary to validate the model of carcinogenesis that we have been investigating. Tests realized on air-aged pyrite and one French iron ore in the lipid peroxidation (Fontecave M.) and in the hydroxylation of guanosine (Cadet J.) (personal communications) confirm that active oxygen species are created on the surface of these compounds.

REFERENCES

Boyd J.T., Doll R., Faulds J.R. anf Leiper J; (1970) Cancer of the lung in iron ores (haematite) miners. Brit. J. Industr. Med., 27, 97-105.

Che M. and Tench A.J. (1982) Characterization and reactivity of molecular oxygen species on oxide surfaces. Adv. Catalysis, 31, 77-133.

Che M. and Tench A.J. (1983) Characterization and reactivity of molecular oxygen species on oxide surfaces. Adv. Catalysis, 32, 1-48.

Costa D., Guignard J., Zalma R. and Pezerat H. (1988) Production of free radicals arising from the surface activity of minerals and oxygen. Part I : Iron mine ores. J. Tox. Ind. Health, in press.

Costa D., Guignard J. and Pezerat H. (1988) Production of free radicals arising from the surface activity of minerals and oxygen. Part II : sulfides and arsenosulfides of iron, nickel and copper. J. Tox. Ind. Health, in press.

Edling Ch. (1982) Lung cancer and smoking in a group of iron ore miners. Am. J. of Indus. Med., 3, 191-199.

Jammes Ph. (1985) Le cancer bronchopulmonaire des travailleurs des mines d'or et usines de Salsigne est-il professionnel ? Medical Thesis, University of Toulouse, France.

Jorgensen H.S. (1973) A study of mortality from lung cancer among miners of Kiruma 1950-1970. Work Environn. Health, 10, 126-133.

Muller J., Wheeler C., Gentleman J.F., Suranyi G. and Rusiak R. (1984) Study of the mortality of Ontario miners. Proc. Int. Conf. "Occupa-

tional Radiation Safety in Mining". October 14-18, Toronto, Canada, p. 335-343.
Muller J., Kusiak R.A., Suranyi G. and Ritchie A.C. (1986a) Study of mortality of Ontario gold miners. Rapport ministry of Labour, Ontario.
Muller T. Kusiak R.A., Suranyi G. and Ritchie A.C. (1986b) Study of mortality of Ontario gold miners. 1955-1977. Addendum to the Report. Ministry of Labor, Ontario.
Nag K. and Chakravorty A. (1980) Monovalent, trivalent and tetravalent nickel. Coordination Chem. Rev., 33, 87-147.
Osburn H.S. (1957) Cancer of the lung in Gwanda. Centr. Afr. J. Med., 3, 215.
Osburn H.S. (1969) Lung cancer in a mining district of Rhodesia. S.A. Medical Journal. October , 1307-1312.
Perisse R. (1977) Neoplasies bronchopulmonaires chez les travailleurs des mines et usines de Salsigne. Medical Thesis, University of Toulouse, France.
Pham Q.T., Gaertner M., Mur J.M., Braun P., Gabiano M., Sadoul P. (1983) Incidence of lung cancer among iron miners. Eur. J. Respir. Dis. (64), 534-540.
Radford E., St Clair Renard K. (1984) Lung cancer in Swedish iron miners exposed to low doses of radon daughters. New eng. J. Med., 310, 1485-1494.
Saint Clair Renard K.G. (1974) Respiratory cancer mortality in an iron ore mine in Northen Sweden. Ambio, 3, (2), 67-69.
Sies H. (1985) Oxidative stress. Acad. Press London.
Sunderman F.W. Jr. (1984a) Recent progress in nickel carcinogenesis. Tox. and Env. Chem., 8, 235-252.
Sunderman F.W. Jr. (1984b) Pouvoir carcinogène des dérivés du nickel chez l'animal, Nickel et Environnement humain, Ed. INSERM, Paris 1984, pp. 129-144.
Sunderman F.W. Jr. (1988) Oral communication at Helsinki.
Webster I. (1973) Malignancy in relation to crocidolite and amosite. Biological effects of asbestos, Ed. IARC, Lyon; 195-198.
Zalma R., Bonneau L., Jaurand M.C., Guignard J., Pezerat H. (1987) Formation of oxy-radicals by oxygen reduction arising from the surface activity of asbestos. Can. J. Chem., 65, 2338-2341.

DETOXIFIED CROCIDOLITE EXHIBITS REDUCED RADICAL GENERATION WHICH COULD EXPLAIN ITS LOWER TOXICITY: ESR AND MOSSBAUER STUDIES

M. Gulumian, J.A. van Wyk[1], and B. Kolk[2]
National Centre for Occupational Health
P. O. Box 4788
Johannesburg 2000, South Africa

Asbestos fibres are widely used commercially because of their insulating and friction properties. Exposure to these fibres however can produce asbestosis and cancer of the lung and mesothelium. Attempts have therefore been made to detoxify these fibres to reduce their biological toxicity but retain their desirable characteristics. One such attempt was reported by Flowers (1982). The procedure involved exposure of fibres to ferric ammonium sulfate for 30 min and then the addition of ammonium hydroxide. This modified the surface of the fibres by metal oxides to a metal micelle form of asbestos. This treatment preserved the commercial properties of the fibres but reduced their toxicity to human lung macrophages and decreased the hemolytic activity to red blood cells (Hahon et al. 1986).

The mechanism by which asbestos fibres exert their toxicity is a subject of much speculation. Reports in the literature and results obtained in this laboratory have indicated that asbestos fibres cause the peroxidation of lipids in rat microsomes (Gulumian et al. 1983), phospholipids (Weitzman and Weitburg 1985), and in the lung tissues of rats after prolonged deposition of fibres (Jajte et al. 1987). They initiate lipid peroxidation due to their ability to generate hydroxyl radical in the presence of an oxidizing agent such as superoxide anion or hydrogen peroxide (Weitzman and Graceffa 1984; Eberhardt et al. 1985).

---

1. Department of Physics, University of the Witwatersrand, Johannesburg 2001, South Africa

2. Deceased 22nd November, 1987.

NATO ASI Series, Vol. H30
Effects of Mineral Dusts on Cells
Edited by B.T. Mossman and R.O. Bégin

Metal ions such as iron present in the structure of the fibres or as contaminants on their surfaces can react with these oxidizing agents and generate the hydroxyl radical (Weitzman and Weitberg 1985; Gulumian and Kilroe-Smith 1982).

$$\text{Asbestos-M}^{2+} + H_2O_2 \longrightarrow \text{Asbestos-M}^{3+} + OH^- + {}^{\bullet}OH$$

Thus, metals are important factors in the production of the above reaction. It therefore seemed paradoxical to detoxify asbestos fibres by coating them with extra metal.

Materials and Methods

Crocidolite fibres were standard UICC samples. Detoxified UICC crocidolite was obtained from Dr E.S. Flowers of Flow General Incorporated McLean, Virginia. The spin trap 5,5-dimethyl-1-pyrroline N-oxide (DMPO) was purchased from Sigma Chemical Campany, St Louis, U.S.A. All reagents used were AR grade. The glassware was washed in 30% nitric acid and thoroughly rinsed with distilled, deionized water. The hydroxyl radical spin trap-adduct $(DMPO\text{-}OH)^{\bullet}$ was detected as reported by Gulumian and van Wyk 1987. The procedure of Zalma et al. 1987, was employed to detect the $CO_2^{\bullet -}$ radical in the presence of sodium formate. Electron spin resonance (ESR) spectra of these spin-adducts were recorded in quartz aqueous cells with a Varian E-line Spectrometer equipped with an online Hewlett Packard 9816 microcomputer.

Mössbauer effect (ME) measurements were carried out at room temperature in the conventional transmission geometry. The ME spectrometer consisted of an Austin Science Associates K3 linear motor combined with a laser interferometer for calibrating the source velocity. The electronics for driving the electron-mechanical motor in the constant-acceleration mode as well as the laser-interferometer electronics were developed in our ME laboratory. The source, prepared by Amersham, consisted of ^{57}Co embedded in a rhodium matrix. The X-rays were detected by a proportional counter.

Metals present in the two crocidolite samples were quantitated as reported by Gulumian and Kilroe-Smith (1987).

Results

ESR experiments - When untreated UICC crocidolite fibres were incubated with hydrogen peroxide and the spin-trap DMPO, hydroxyl radical spin-adduct hyperfine splitting spectrum was obtained (Fig. 1b). A similar spectrum was observed with detoxified crocidolite (Fig. 1a). However, the same concentration of the untreated fibres (0.4mg/reaction mixture) produced a 4.5 times stronger signal than that of the detoxified crocidolite.

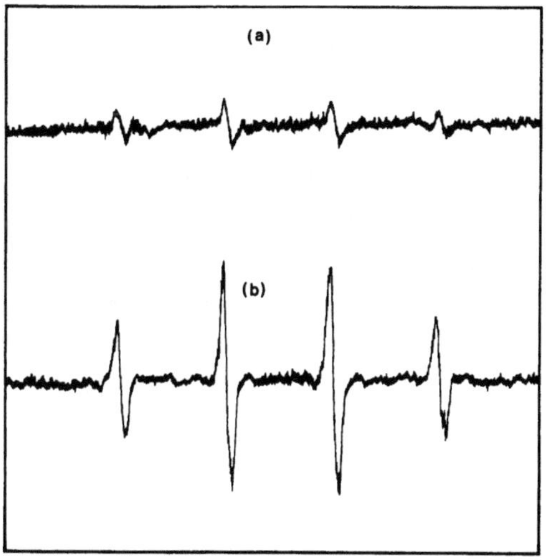

Fig. 1. Detection of $(DMPO\text{-}OH)^{\cdot}$ spin-adduct
a. Detoxified UICC Crocidolite
b. UICC Crocidolite

the same was true when the fibres were incubated in the presence of the formate anion (Fig. 2 b and c).

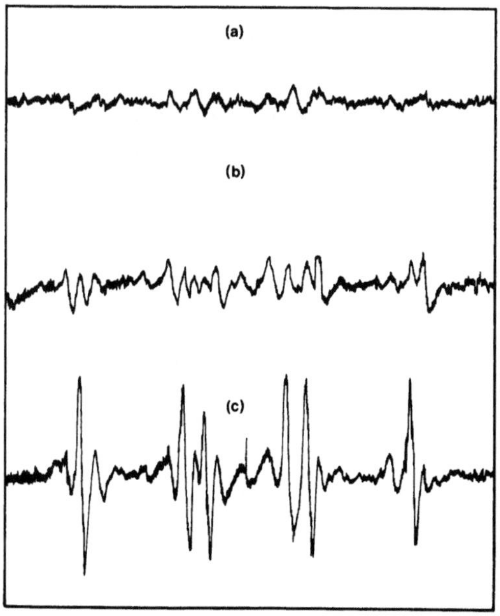

Fig. 2. Detection of $(DMPO\text{-}CO_2)^{\cdot -}$ spin-adduct
a. Control containing no fibres
b. 45mg/ml detoxified UICC crocidolite
c. 45mg/ml UICC crocidolite

Mössbauer - effect studies - ME spectra of the untreated and of the treated crocidolite are displayed in Figs. 3a and 3b. In these preliminary measurements, both spectra were fitted with two sets of two Lorentzian absorption lines. Each set of absorption line (doublet) arises from a quadrupole splitting of the nuclear energy levels, and is characteristic for ^{57}Fe atoms at specific lattice sites with non-cubic symmetry.

The positions of the two absorption lines forming a doublet are characterized by two parameters: the quadrupole interaction constant, ΔE, which represents the distance between the two lines, and the isomer shift, δ which represents the position of the center of the doublet. It is common to adjust the isomer shift for the source used, and to report the value of the δ relative to pure iron. The isomer of the large quadrupole doublet is typical of that of Fe^{2+} ions, while that of the smaller quarupole doublet is typical for Fe^{3+} ions. The line widths of the individual absorption lines are about twice the natural line width. This broadening indicates that the Fe^{2+} ions as well as the Fe^{3+} ions in the unit cell of crocidolite do occupy slightly different sites. In addition, the line width of the ferric component in the treated samples is 15% broader than the line width of the ferric component in the untreated sample. Since both samples were measured under identical experimental conditions, this broadening can possibly be attributed to the presence of additional Fe^{3+} sites in the treated samples.

Hence, in the ME spectra, the contribution arising from the Fe^{2+} and Fe^{3+} ions can be readily distinguished. The areas of the 2+ and 3+ doublets represents the number of Fe ions in the 2+ and 3+ states, respectively, assuming that the recoilless fraction of the Fe^{2+} and Fe^{3+} ions are approximately the same.

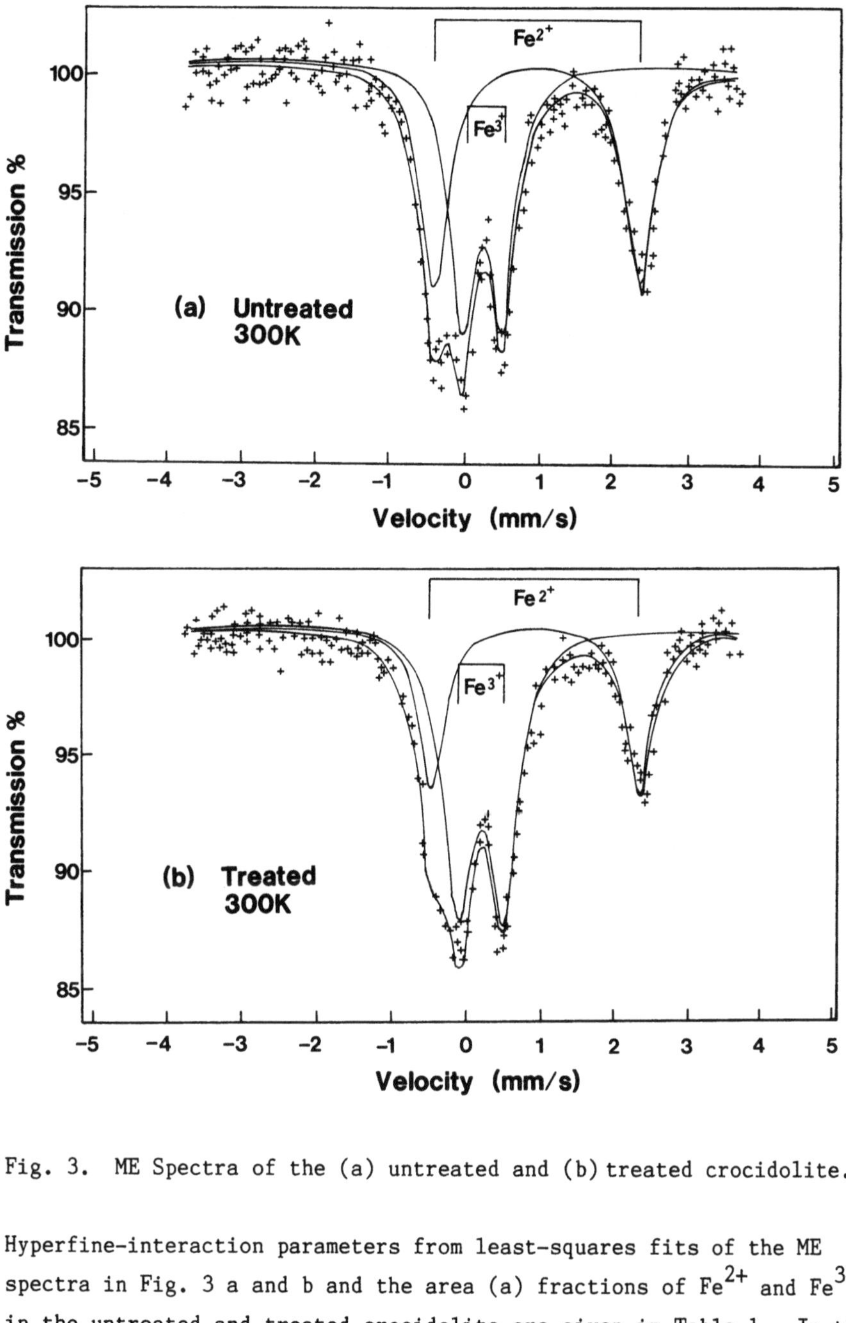

Fig. 3. ME Spectra of the (a) untreated and (b) treated crocidolite.

Hyperfine-interaction parameters from least-squares fits of the ME spectra in Fig. 3 a and b and the area (a) fractions of Fe^{2+} and Fe^{3+} in the untreated and treated crocidolite are given in Table 1. In the untreated crocidolite sample the fraction of Fe^{2+} ions is slightly larger than that of the Fe^{3+}. In the treated sample the situation is reversed: the fraction of Fe^{3+} ions is substantially larger than that of the Fe^{2+} ions.

Table 1: Hyperfine-interaction parameters from least-squares fits of the ME spectra in Fig. 3: ΔE, the quadrupole splitting, is the distance between the two lines forming the quarupole doublet; δ is the isomer shift relative to iron; a is the ratio of the area of the doublet to that of the total absorption area.

	Doublet 1			Doublet 2		
	ΔE (mm/s)	δ (mm/s)	a (%)	ΔE (mm/s)	δ (mms)	a (%)
untreated	1.28(1)	1.13(1)	53(3)	0.26(1)	0.40(1)	47(2)
treated	1.39(1)	1.14(1)	40(3)	0.29(1)	0.39(1)	60(3)

Results of quantitative measurements of different metals are shown in Table 2. Iron is the major constituent of both crocidolite samples. The concentration of this metal in the detoxified crocidolite is substantially higher (Fe^{2+}14%, Fe^{3+}21%) than the concentration measured in the native untreated crocidolite (Fe^{2+}15.5%, Fe^{3+}13%). This confirms the findings with ME reported earlier.

Table 2: Quantitative measurements of different elements present in native and detoxified crocidolite fibres

Sample	Ni	Si	Fe^{2+}	Fe^{3+}	Mg	Al	Ca	Cu	Pb	Ti	Cr	Mn	Mo	Sn	V
	ppm	%	%	%	%	ppm	ppm	ppm	ppm	ppm	ppm	ppm	ppm	ppm	ppm
UICC Crocidolite	100	22.2	15.5	13	1.14	349	6820	100	200	500	25	910	50	100	200
UICC detoxified Crocidolite	100	16.5	14	21	0.69	304	6690	100	200	500	25	680	50	100	200

Discussion

Ferrous is the oxidation state of iron which is involved in a Fenton-type reaction to produce $^{\bullet}OH$ from hydrogen peroxide.

$$Fe^{2+} + H_2O_2 + {}^{+}H \longrightarrow Fe^{3+} + H_2O + {}^{\bullet}OH$$

In the presence of an organic molecule such as formate anion, the $^{\bullet}OH$ radical can then react and produce the $CO_2^{\bullet -}$ radical. The decreased $^{\bullet}OH$ and $CO_2^{\bullet -}$ radical generation by detoxified crocidolite (Figs 1 and 2) may be due to either the reconstruction of the fibre surface to an oxidized sheet, impermeable to air even in aqueous medium (Zalma et al. 1987), or to the conversion of all iron by the process of detoxification into the ferric state. Elemental analysis and Mössbauer effect (ME) spectra however did not support the latter possibility as results obtained from these two studies have indicated that iron in the detoxified crocidolite existed in its two oxidation states. Moreover, ME spectra have further indicated the broadening of Fe^{3+} line width (15%) compared to the untreated fibres. This broadening can be attributed to the presence of additional Fe^{3+} sites in the treated sample. Therefore, the added ferric ion in the detoxified crocidolite has not only coated the surface of the fibres but has also occupied sites which were not occupied in the untreated fibres. This produced a reduction of electron flow and therefore made the exhisting iron in the ferrous state less available for reaction to generate the $^{\bullet}OH$ and $CO_2^{\bullet -}$ radicals.

Hydroxyl radicals are known to initiate lipid peroxidation (Piette et al. 1984), cause DNA single strand breaks (Kasai and Nishimura 1984), and act as tumour promotors (Troll and Wiesner 1985). Thus, the ability of asbestos fibres to generate these radicals may be one of the mechanisms by which fibres exert their harmful effects. The detoxification process was able to reduce this ability to generate radicals leading to a reduced toxicity of the treated crocidolite.

References

Eberhardt MK, Román-Franco AA, Quiles MR (1985) Asbestos-induced decomposition of hydrogen peroxide. Environ Res 37: 287-292

Flowers ES (1982) Metal micelle asbestos and treatment of asbestos and other silicate minerals to reduce their harmful properties. U.S. Patent 4, 328, 927, Assigned to Flow General Inc, McLean, Virginia.

Gulumian M, Kilroe-Smith TA (1987) Crocidolite-induced lipid peroxidation in rat lung microsomes I. Role of different ions. Environ Res 43: 247-273

Gulumian M, Sardianos F, Kilroe-Smith TA, Ockerse G (1983) Lipid peroxidation in microsomes induced by crocidolite fibres. Chem-Biol Interact 44: 111-118

Gulumian M, van Wyk J A (1987) Hydroxyl radical production in the presence of fibres by a Fenton-type reaction. Chem-Biol Interact 62: 89-97

Hahon N, Vallyathan V, Booth JA, Sepulveda MJ (1986) In vitro biologic responses to native and surface-modified asbestos. Environ Res 39: 345-355

Jajte J, Lao I, Wiśniewska-Knypl J M (1987) Enhanced lipid peroxidation and lysosomal enzyme activity in the lungs of rats with prolonged pulmonary deposition of crocidolite asbestos. Br J Ind Med 44: 180-186

Kasai H, Nishimura S (1984) DNA damage by asbestos in the presence of hydrogen peroxide. Gann 75: 841-844

Piette LH, Baxley LH, Grover TA, Harwood PJ (1984) A comparative kinetic study of the initiation of lipid perxidation with OH radicals and ferrous iron. In: Bors W, Saran M, Tait D (eds) Oxygen radicals in chemistry and biology. Walter de Gryter, Berlin, 134-145

Troll W, Wiesner R (1985) The role of oxygen radicals as a possible mechanism of tumor promotion. Ann Rev Pharmacol Toxicol 25: 509-529

Weitzman SA, Graceffa P (1984) Asbestos catalyzes hydroxyl and superoxide radical generation from hydrogen peroxide. Arch Biochem Biophys 228: 373-376

Weitzman SA, Weitberg AB (1985) Asbestos-catalysed lipid peroxidation and its inhibition by desferroxamine. Biochem J 225: 259-262

Zalma R, Bonneau L, Guignard J, Pezerat H (1987) Formation of oxy radicals by oxygen reduction arising from the surface activity of asbestos. Can J Chem 65: 2338-2341

THE FORMATION OF OXYGEN REACTIVE RADICALS AT THE SURFACE OF THE CRUSHED QUARTZ DUSTS AS A POSSIBLE CAUSE OF SILICA PATHOGENICITY

B.Fubini, V.Bolis, E.Giamello, L.Pugliese and M.Volante
Dipartimento di Chimica Inorganica, Chimica Fisica e Chimica dei Materiali - Università di Torino
Via Pietro Giuria 9
10125 Torino
Italy

Introduction

In spite of the massive work done on the various pathological effects of crystalline silica dusts their mechanism of action at the molecular level is not yet fully clarified. It is generally accepted that it starts up as a failed phagocytosis by alveolar macrophages, whereby silica exhibits its membranolytic action and in the meanwhile induces the release of the abnormal growth factor (MFF) (Heppleston 1984) yielding fibrogenicity. Because of the complexity of the silica-cell interaction it is likely that several surface functionalities may act simultaneously or at different stages of the process. Much emphasis has been given so far to the role of silanols (Hobza and Hurich 1978; Langer and Nolan 1986), which indeed may play a crucial role in membrane damage. However their role in the MFF sinthesis is not straightforward and the inertness of chemically prepared silicas, all rich in silanols (Iler 1979), remains unexplained. Besides hydroxyls, when the surface is created by grinding, the cleavage of Si-O bonds originates surface radicals (Ratdzig and Bystrikov 1978). Some of us have shown that these radicals are present and stable in samples ground in air (Bolis et al. 1983; Fubini et al. 1987) and that several radical species are originated by contact of Si and SiO with atmospheric components (Fubini et al. 1988), which may be involved in the pathogenic process. The present paper concerns the stability and the distribution of these radicals on various

NATO ASI Series, Vol. H30
Effects of Mineral Dusts on Cells
Edited by B. T. Mossman and R. O. Bégin

SiO_2 dusts and their reactivity on oxidizing media, such as H_2O_2, mimicking the environment of the inhaled particle during phagocytosis.

Experimental

Materials. Very pure quartz crystals (purified by melting, 99.999% Atomergic) were ground in air, water and aqueous hydrogen peroxide solution in an agatha ball mill. The concentration of the H_2O_2 solution employed was 35% in weight. Two standard quartz dusts, usually employed in biological tests, were investigated: Min-U-Sil (BET specific surface 2.1 m^2g^{-1}) from Pennsylvania Glass Sand Co., and DQ-12 (BET specific surface 7.4 m^2g^{-1}), kindly supplied by Dr. K. Robock (K. Robock 1973).

Methods. EPR spectra were taken in vacuo at 298K or at 77K on a Varian E 109 spectrometer operating in the X band mode (9.4 GHz). Varian pitch (g = 2.0029) was used for g values calibrations.

Results and discussion

a) Surface radicals obtained by grinding pure quartz crystals.

A pure quartz crystal does not bear detectable surface radicals or other paramagnetic centers, as indicated by the absence of any EPR spectrum. Upon grinding in air several EPR signals appear due to the formation of surface radicals, originated by the interaction of the cleaved Si-O-Si bonds with atmospheric components. Among these we have identified (Fubini et al. 1988) partially reduced oxygen forms, such as $SiO^{\bullet}$, $SiO_2^{\bullet}$, $SiO_3^{\bullet}$ and $O_2^{\bullet -}$, which may be relevant to the toxicity of ground quartz dusts. Immediately after grinding the surface radicals undergo slow modifications, likely related to a partial surface reconstruction, which can be monitored by the decrease in the overall intensity of the spectrum. Figure 1 reports the variation of the intensity of the EPR spectrum with time, measured as a peak to peak height (see inset) and reported as a

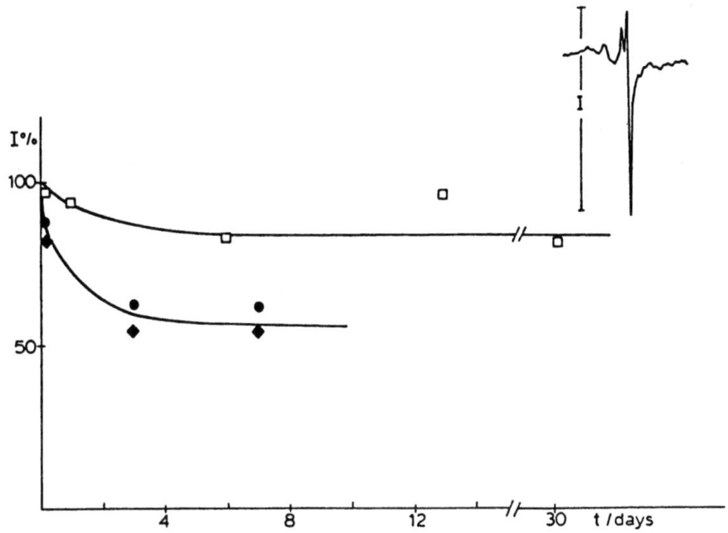

Fig. 1: Evolution with time of the intensity of the EPR spectrum of a freshly ground pure quartz crystal expressed as a percentage of the spectrum recorded immediately after grinding. Spectra recorded at r.t. (inset). Empty points: sample kept in air; full points: sample kept in a moist atmosphere at 310K (37°C).

percentage intensity of the spectrum recorded just after grinding. Experiments have been performed either by keeping the sample in air or by keeping it under the vapour pressure of water at 310K (37°C), in order to simulate the conditions experienced by the newly fractured particle in the lung. Three main results can be drawn from the figure: i) the number of free surface radicals visible by EPR decreases within the first ≈ 48 hours whereas in the subsequent 30 days is practically unchanged; ii) a more marked decrease is found in the presence of water vapour, which likely reacts with the radicals and assists surface reconstruction; iii) the decrease in radical population in any case is not dramatic and a consistent part of the free radicals still remains at the surface over a long period of time. The short-live radicals, probably consisting in particularly reactive dangling bonds, may be relatable to acute toxicity, i.e., diseases occurring in a short period of time when workers are exposed to freshly ground quartz, as also reported by Dalal et al. (1987).

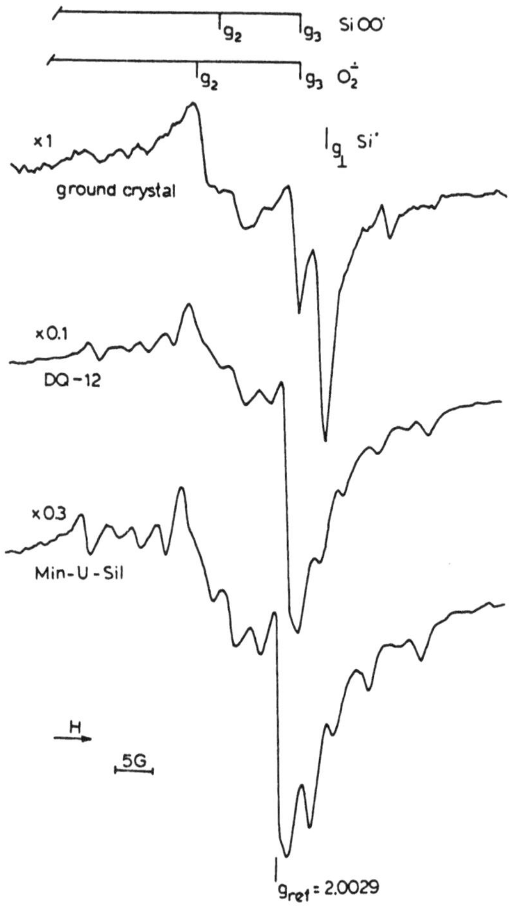

Fig. 2: Spectra of a pure quartz crystal freshly ground in air, of Min-U-Sil and DQ12. All spectra recorded in vacuo at 77K. Spectra are reported on an expanded scale to show all components present near the free electron value. The g values of $Si^{\bullet}$, $SiO_2^{\bullet}$ and $O_2^{\dot{-}}$ are indicated.

On the other hand, some among the radicals stable at the surface, such as $SiO_2^{\bullet}$ or $O_2^{\dot{-}}$ may play a role in long term toxicity, e.g., releasing free oxygen radicals in aqueous media. The reactivity of these radicals towards mild reducing agents such as CO has already been reported by our group (Fubini et al. 1988). In aqueous solution a production of a $OH^{\bullet}$ radical has been in fact found by one of us (Costa, Fubini, Pezerat unpublished results) and other authors (Gulumian and Van Wyk 1987; Dalal et al. 1987). The total amount of $OH^{\bullet}$ released is too low in our opinion to justify per se silica toxicity: its presence, however, confirms the reactivity potential of the surface radicals and their ability to trigger redox reactions in biological media.

b) <u>comparison with dusts used as standard in biological tests</u>

The stability of the various surface radicals over a long period of time is confirmed by the fact that the same kind of radicals are present on quartz dusts kept standing in air for long periods of time. Figure 2 compares the spectrum obtained on a freshly ground sample with those from DQ12 and Min-U-Sil dusts. The spectra have been recorded at 77K because only at this temperature a good resolution between signals arising from the various oxygen species can be achieved (Fubini et al. 1988). In order to examine in detail the central part of the spectra near the free electron value, an expanded scale has been used. Consequently some spectral components, namely the g_1 of the $SiO_2^{\bullet}$ (g = 2.06) and of the $O_2^{\dot{-}}$ (g = 2.04) signals are not visible as they fall out of the examined field range. With the only exception of the two sets of three lines at both ends of the spectra (only present on DQ12 and Min-U-Sil) the three samples exhibit spectral features at the same g values which indicate that the same paramagnetic centers are present on the three samples. The remarkable differences visible from one spectrum to the other are due to a different abundance of each radical form, complicated by the fact that many lines are partially superimposed one with the other. Obviously differences in the biological response from one to the other standard sample can be expected if these radicals do play some role in pathogenicity. Some discrepancies found between the results of various authors with different quartz dusts (e.g. Hemenway et al. 1986) may arise from differences in radical availability. The fact, however, that in all samples $SiO_2^{\bullet}$ and $O_2^{\dot{-}}$ are present, indicates that any quartz dusts is a carrier of potentially toxic oxygen forms. The absence of any radical species in chemically prepared amorphous silicas is in agreement with the established low toxicity of this silica form.

c) Effect of grinding in oxidizing aqueous media

The grinding atmosphere strongly influences the formation and evolution of the various spectral components. During the grinding process and in the immediately following period in which the originated dangling bonds are rearranged, radicals can evolve in different ways, depending on the molecules present at the surface. The spectrum obtained by grinding in argon is much different from what obtained in oxygen atmosphere as in the former case no oxygen was available except few traces to react with the cleaved bonds. Upon subsequent contact with oxygen however the spectrum was markedly different from the one obtained by directly grinding in O_2 (Fubini et al. 1988), the presence of freshly cleaved dangling bonds being necessary for the reduction of O_2 into superoxide anion or a peroxyradical.

In order to mimic the environment encountered in the lung by the newly fractured particle experiments were carried out in aqueous solutions. In pure water the development of surface

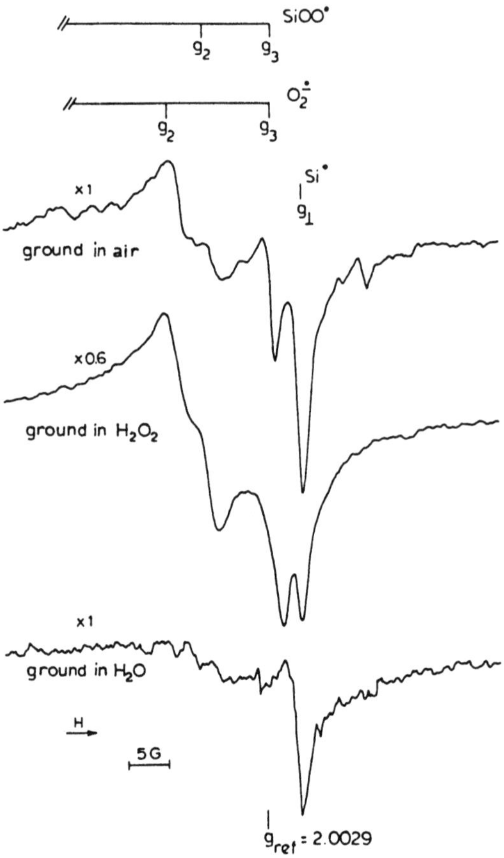

Fig. 3: Comparison between the spectra obtained by grinding in air, in water and in aqueous H_2O_2 solution, a pure quartz crystal. Spectra recorded in vacuo at 77K, same scale as in figure 2.

radicals and particularly of the two main oxygen forms is largely depressed by comparison to air (fig.3): this because on the one hand water assists surface reconstruction, e.g., yielding surface hydroxylation (in agreement with results in fig.1); on the other one the partial pressure of oxygen is much lower in water than in air, therefore active sites are saturated by water before contact with O_2. Quite a different result is obtained when grinding in aqueous hydrogen peroxide solution: the spectrum obtained, reported in fig. 3 shows the same components as the one ground in air but the overall spectrum intensity and particularly both components arising from SiO_2 and $O_2^{\cdot-}$ are markedly enhanced. Taking into account that also in this case the fractured surface is mainly contacted by simple water the intense spectrum formed on the solid has to be regarded as a result of a specific reaction between H_2O_2 and the fractured surface.

Various mechanism can be hypothized for the reaction between hydrogen peroxide and the cleaved silicon-oxygen bonds. As, besides peroxyradicals, also the superoxide anion signal is enhanced in comparison to grinding in air, a series of redox reaction will necessarily occur. A possible mechanism is the following:

$$Si^{\cdot} + H_2O_2 \longrightarrow SiOH + OH^{\cdot}$$
$$SiO^{\cdot} + H_2O_2 \longrightarrow SiOOH + OH^{\cdot}$$

$$OH^{\cdot} + H_2O_2 \longrightarrow H_2O + H^+ + O_2^{\cdot-}$$

the last reaction being the well known one of production of $O_2^{\cdot-}$ (which can be stabilized on the Si^+ sites) proposed for asbestos toxicity by Weitzmann and Graceffa (1984).

If instead of a pure quartz crystal, dusts such as DQ12 or Min-U-Sil are reground in hydrogen peroxide the spectra obtained (to be published) show modifications which are in agreement with the above hypothesis: the spectral components corresponding to active oxygen forms are markedly enhanced by comparison with the starting material.

It is noteworthy that a treatment of the samples ground in H_2O_2 in a buffered SOD aqueous solution modifies the spectrum by reducing the intensity of the superoxide anion components. These modifications are not very large because the interaction between the active site of the enzyme and the superoxide anion at the surface are inhibited by steric factors, their occurrence however confirms the biological "availability" of the radical sites.

The concentration of the oxygen radicals at the surface ($\approx 10^{18}$ sites per m^2) do not justify per se a toxicity related directly to these forms, taking into account the abundance of oxidizing agents in the macrophage during phagocytosis. However the presence of oxygen radicals, their relative stability with time and their capability of reaction in solution indicate that certainly they are able to induce various reactions in biological medium, the results of which can only be seen on the solid through small modifications in radical population.

It has moreover to be pointed out that the hypothesis that only the surface originated by grinding bears sites active in silica pathogenicity is in agreement with what found with the different SiO_2 polymorphs, namely the non toxicity of chemically prepared silicas (amorphous) opposed to the toxicity of silica glass (amorphous but obtained by grinding) and of the various crystalline polimorphs (all obtained by grinding) (IARC Monographs 1987). As to the latter ones, the toxicity decreasing in the series trydimite > cristobalite > quartz > coesite and the inertness of stishovite may be explained by a different stability of the radicals on the various crystal lattices. It is noteworthy that the series follows a scale of increasing density: the larger the density, the higher is the possibility for the dangling bonds to recombine, i.e., the less stable are the free surface radicals. On trydimite in fact the oxygen radicals are more stable than on quartz (Fubini et al., 1988). As to stishovite, it is possible that the octahedral coordination may eliminate even the formation of stable radicals, surface reconstruction into siloxane bridges being facilitated by the compactness of the structure.

Conclusions

All quartz dusts obtained by mechanical grinding in air exhibit at the surface a series of radical species among which those obtained from a partial reduction of the atmospheric oxygen to $SiO_2^{\bullet}$, $SiO_3^{\bullet}$, $O_2^{\dot{-}}$. The amount of each single form depends on the history of the sample and varies from one dust to the other: this can be relevant when comparing data obtained from different quartz dusts sources.

The study of the radicals obtained by directly grinding a very pure quartz crystal indicates that only a fraction of the radicals is unstable and undergoes modifications in the first 48 hours whereas the mayor part is fairly stable with time, if not submitted to redox reactions.

If the freshly ground particle is contacted with an oxidizing agent such as those present during phagocytosis, e.g., H_2O_2, a specific reaction occurs yielding OH and superoxide anions. Although a direct correlation between surface radicals and quartz pathogenicity has not yet been found we suggest that these radicals may interfere with biological substrates at different stages, e.g.:

- via membranolytic action caused by a lipid peroxidation as hypothized some years ago (Gabor and Anca 1974; Chvapil 1977)
- inducing acute silicosis in workers exposed to very freshly ground samples, in agreement with what reported by Dalal et al. (1987)
- by acting as a triggering agent and/or a catalyst for the series of reactions occurring within the macrophage yielding the abnormal production of the MFF (macrophage fibrogenic factor).

References

Bolis V, Fubini B, Venturello G (1983) Surface characterization of various silicas. A tentative correlation between the energy of adsorption sites and the different biological activities. J. Thermal Anal. 28: 249-258

Chvapil M (1977) Old new problems of silicotic fibrosis. In Bendz G, Lindqvist I (eds) Biochemistry of silicon and related problems. Plenum Press, New York London, p 381
Dalal NS, Suryan MM, Jafari B, Shi X, Vallyathan V, Green FHY (1987) Electron Spin Resonance detection of reactive free radicals in fresh coal dust and quartz dust and its implications to pneumoconiosis and silicosis. Proc. Int. Symp. on Respir. Dusts in the Mineral Ind.: 24-29
Fubini B, Bolis V, Giamello E (1987) On the possible role of the silica surface chemistry in silicosis. In Le Bouffant (ed) Silicosis and mixed-dusts pneumoconiosis. Colloque de l'INSERM 155: 115-122
Fubini B, Bolis V, Giamello E (1987) The surface chemistry of crushed quartz dusts in relation to its pathogenicity. Inorg. Chim. Acta, Bioinorg. Chem. 138: 193-197
Fubini B, Giamello E, Pugliese L, Volante M (1988) Mechanically induced defects in quartz and their impact on pathogenicity. Solid State Ionics, in press
Gabor S, Anca Z (1974) Effect of silica on lipid peroxidation in the red cells. Int. Arch. Arbeitsmed 32: 327-332
Gulumian M, Van Wyk A (1987) Free radical scavenging properties of polyvinylpyridine N-oxide: a possible mechanism for its action in pneumoconiosis. Med. Lav. 78: 124-128
Hemenway DR, Absher M, Landesman M, Trombley L, Emerson RJ (1986) Differential lung response following silicon dioxide polymorph aerosol exposure. In Goldsmith DF, Winn DM, Shy CM (eds) Silica, Silicosis and Cancer. Praegher Publisher, New York. p 105
Heppleston AG (1984) Pulmonary toxicology of silica, coal and asbestos. Environ. Health. Persp. 55: 111-127
Hobza P, Hurich J (1978) Quantum chemical study of properties and reactivity of quartz dust. Environ. Res. 15: 432-442
IARC Monographs (1987) Evaluation of the carcinogenic risk of chemicals to humans. In Silica and some silcates, vol. 42, IARC Lyon
Iler RK (1979) The chemistry of silica, John Wiley and Sons, New York
Langer AM, Nolan RP (1986) Physicochemical properties of quartz controlling biological activity. In Goldsmith DF, Winn DM, Shy CM (eds) Silica, silicosis and cancer. Praeger Publisher New York. p 125
Ratdzig VA, Bystrikov AV (1978) ESR study of chemically active centers on the surface of quartz. Kinetika i Kataliz 19: 713-719
Robock K (1973) Standard quartz DQ 12 < 5 μm for experimental pneumoconiosis research project in the Federal Republic of Germany. Ann. Occup. Hyg. 16: 63-66
Weitzmann S, Graceffa P (1984) Asbestos catalyzes hydroxyl and superoxide radical generation from hydrogen peroxide. Arch. Biochem. Biophys. 228: 373-376

ARACHIDONIC ACID METABOLITE PRODUCTION BY ALVEOLAR MACROPHAGES CULTURED IN GAZEOUS PHASE. EFFECTS OF NO_2 AND DIESEL EXHAUST.

S. KOUZAN, T. FOURNIER, C. VOISIN*, M.C. JAURAND, J. BIGNON.
*Institut Pasteur, Lille
Unité INSERM 139, Hôpital Henri Mondor
Créteil 94010 cedex
France

Introduction

Exposure to Diesel exhaust can occur either in the general population or in occupationnal groups. Since a variety of mutagenic and carcinogenic substances have been identified from the gazeous and the particulate phases of Diesel exhaust, there is concern that such exposure might induce an excess of respiratory tract tumors.This concern has been substantiated by the recent finding of an increase in lung cancer risk in US (Garschik, 1987) or Canadian (Howe, 1983) railroad workers.Since pulmonary alveolar macrophage is endowed with an armamentarium with putative carcinogenic potencies, we have initiated a study assessing the effects of Diesel exhaust on alveolar macrophages. Cell reactivity was evaluated through assessment of release of arachidonic acid metabolites. In order to mimic physiological conditions, cells were cultured in gazeous phase. This enables the cells to be directly in contact with gazes. In addition, the respective role of the gazeous and the particulate phases of the Diesel exhaust were evaluated.

MATERIAL AND METHODS

1) Preparation of radioactive non-adherent alveolar macrophages

Procedure is schematically described on figure 1. Male CD rats (Charles Rivers) were killed by IP pentobarbital, the lungs were removed, and lavaged with phosphate bufferered saline, as previously described (Kouzan,1985).The cells, more than 95% of them macrophages were then washed, spun down and resuspended in RPMI 1640 medium (GIBCO) containing 5% fetal bovine serum (FBS). All the cells obtained by the procedure (approximately 50 millions) were plated on a single 35mm plastic dish (Falcon) and incubated at 37°C, 5% CO_2. This limited adherence to a minimal number of cells, leaving the vast majority floating above the monolayer of adherent macrophages. After a 1-hour period, the non-adherent cells were washed, spun down and resuspended in RPMI containing tritiated arachidonic acid (35µCi per dish) (CEA). Incorporation of arachidonic acid was allowed to procede for 3 hours, after wich the radioactive

NATO ASI Series, Vol. H30
Effects of Mineral Dusts on Cells
Edited by B. T. Mossman and R. O. Bégin

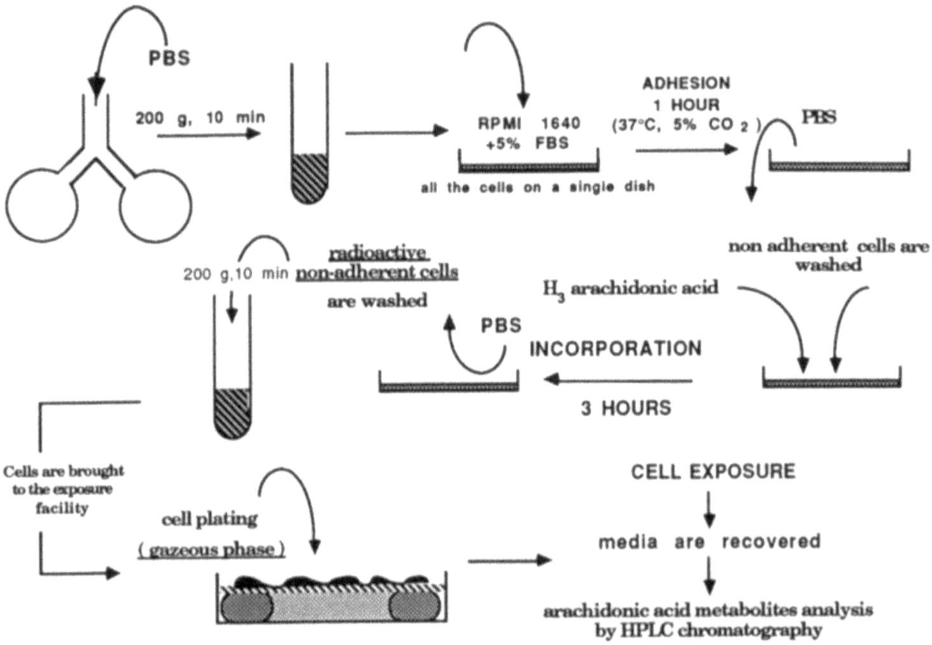

figure 1:*experimental design. Cell culture in gazeous phase.*

non-adherent cells were washed, spun down and brought to the exposure facility.

2) Macrophage culture in gazeous phase.

Cells were layered on to a filtration membrane (Gelman®) that was placed on a plastic ring, in a 50 mm culture dish (3 millions cells/dish), with RPMI medium underneath. This procedure allows the cells to be in close contact with ambient gazes. The cell nutritionnal requirement are delt with by the pseudopodes emitted through the membrane pores. Following cell setting in gazeous phase, the culture dishes were placed into an exposure chamber laying in a 37°C water bath. Parallel sets were either exposed to control air or to Diesel exhaust.

3) Cell exposure

a) exposure setting (see fig. 2)

Exhaust from a Diesel vehicle was collected, diluted twice in tunnels with fanned purified air (control air). This resulted in approximately a 1/20 dilution . A dilution flap located in the 1st tunnel allowed for a coarse adjustment of the dilution factor, resulting in either a "low" or a "high" concentration . Diluted exhaust was then flown to one of the exposure chambers, the other being vented with control air, 5% CO_2 being added in either case for purpose of pH stability.

b) exposure modality

Diesel vehicle front wheels were placed on rolls. Various engine speed were imposed, together with various gear positions, throughout a cycle, termed "European", enabling a mimicry of real traffic conditions. Cycle duration was 3 minutes 45 seconds, and cells were exposed to 12

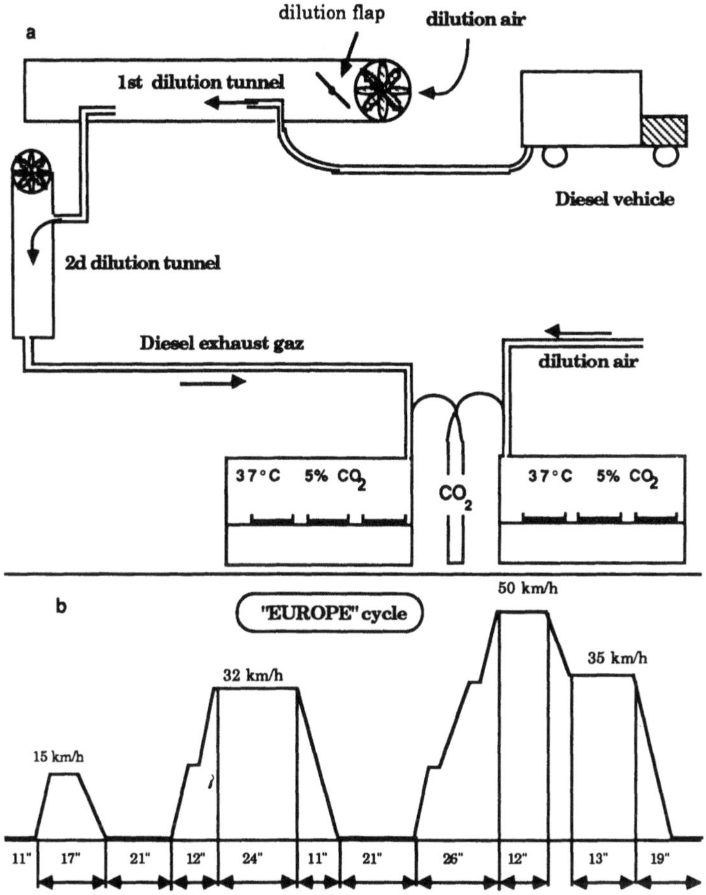

figure 2: *(a) experimental setting for cell exposure to Diesel exhaust (b) Chart of vehicle speed throughout a "european" cycle. X axis represents duration of cycle segments.*

such cycles, i.e. 45 minutes total exposure.

c) exposure conditions. Conditions were designed for evaluation of the effect of Diesel exposure on cell metabolic and phagocytic function. Due to the length of the path followed prior to macrophage exposure, the Diesel exhaust was devoid of particles at the entry of the exposure chamber. Thus, in order to evaluate the role of the particulate phase, Diesel particles were recovered on filtration membranes placed for 5 minutes inside the first dilution tunnel, while the vehicle was running at a stabilized speed of 50 km/hr, and macrophages were plated on top. Approximately 0.25 mg of particles were layered on a 10 cm^2 membrane, giving a theoretical dose of 25 $\mu g/cm^2$. However, due to the high hydrophobicity of Diesel particles, contact with macrophages was probably uneven, and the actual challenging dose was probably lower.

Therefore exposure conditions were the following:

a) Macrophages were exposed, for 45 minutes, either to Diesel exhaust gazeous phase or to control air. Within each group, cells were either unstimulated, or submitted to a phagocytic challenge (i.e. carbonyl iron beads particles), or treated with Diesel particles.

b) In separate experiments, macrophages cultured in gazeous phase were exposed to NO_2 at increasing concentrations, for 30 minutes. Cells were exposed either without any additional stimulus, or with a concomitant iron beads phagocytic challenge.

At the end of the exposure period, the culture media were recovered, the membranes rinsed, and the media stored until processing.

4) Assessment of arachidonic acid metabolites release.

The media were treated as published previously (Henke, 1984) . Briefly, 4 volumes of absolute ethanol were added, the precipitated protein spun down, and after evaporation the residue was dissolved in 600 µl 30% Methanol/water.

Arachidonic acid metabolites were separated by reverse phase HPLC chromatography, using 4 successive isocratic steps with increasing methanol proportions in water (45/55%, 55/45%, 65/35%, 100%) . This enables successive elution of prostaglandins, leukotrienes, monohydroxylated fatty acids, then unmetabolized arachidonic acid . Quantification of eluting radioactive peaks was performed with an online radioactivity counter.

RESULTS

1) Stability of Diesel exhaust composition

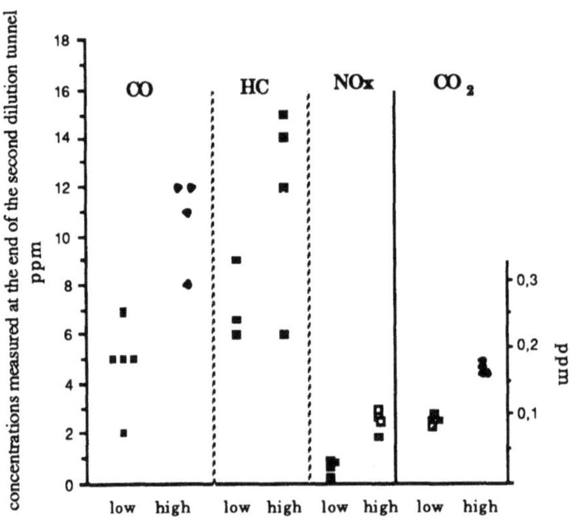

Figure 3 shows the concentration of carbon monoxide (CO), polycyclic aromatic hydrocarbons (HC), Nitrogen oxides (NOx), and carbon dioxide (CO_2), at the entry of the exposure chambers, according to the flap position in the 1st dilution tunnel (giving a "low" or a "high" gaz concentrations). There was a narrow range of variation of the NOx family and CO_2, from one experiment to the other. This range was wider for carbon monoxide and hydrocarbons, the concentrations of which seemed more depending on atmospheric conditions such as temperature or humidity.

Arachidonic acid metabolite secretion by alveolar macrophages cultured in gazeous phase.

Diesel exhaust-exposed macrophages released an array of metabolites that was qualitatively identical to the one seen in control conditions.

Macrophages exposed to control air without any additional stimulus exhibited a spontaneous secretion of metabolites, mainly from the cyclooxygenase pathway (Thromboxane B_2 (TXB_2) and 12-hydroxy-5,8,10-heptadecatrienoic acid (HHT)). Macrophages stimulated by carbonyl iron beads released an array of metabolites from both pathways (cyclo as well as lipoxygenase),

<u>figure 4</u>: *HPLC chromatography of metabolites secreted by alveolar macrophages cultured in gazeous phase, in various conditions.*

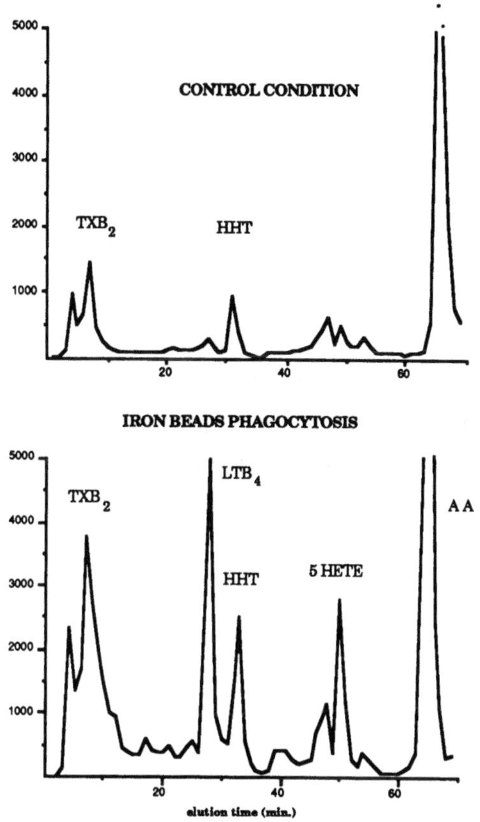

identical to the spectrum seen after stimulation of cells cultured in liquid phase. The main metabolites identified were TXB_2, HHT, Leukotriene B_4 (LTB_4) and 5-hydroxy-eicosatetraenoic acid (5 HETE) (see fig 3). Thus cell culture in gazeous phase did not obviate cell capability to secrete arachidonic acid metabolites following appropriate stimulation. Since LTB_4 and HHT represent the main metabolites from their respective pathway, they were used for quantification purposes thereafter.

<u>Secretory response following macrophage exposure to Diesel exhaust</u>

<u>1) exposure to"low" concentration</u>

Macrophages exposed to Diesel exhaust, with or without Diesel particles, released the same qualitative array of metabolites than unstimulated cells, that is mainly cyclooxygenase metabolites (TXB_2 and HHT) (fig. 4).
Cell exposure to the "low" gaz concentration, either to the sole gazeous phase or to the particulate phase alone, or to both simultaneously, had no effect on the lipoxygenase pathway metabolite secretion (fig. 5).
On the other hand, exposure to low concentration Diesel exhaust had a stimulatory effect on the cyclooxygenase pathway. Indeed, when the cells were simultaneously exposed to the gazeous and to the particulate phases, there was a 40 to 50% increase in release of HHT and TXB_2. This contrasted with the lack of any consistent stimulation when the cells were challenged with either stimulus alone (fig. 5).

<u>2) exposure to "high" concentration</u>

When macrophages were exposed to a "high" concentration of Diesel exhaust gazeous phase, a decrease in secretion of metabolites was observed, as compared to cells exposed to control air. Indeed the LTB_4 and HHT amounts released by macrophages following stimulation with carbonyl iron beads were lower than those observed when iron-phagocytizing cells were cultured

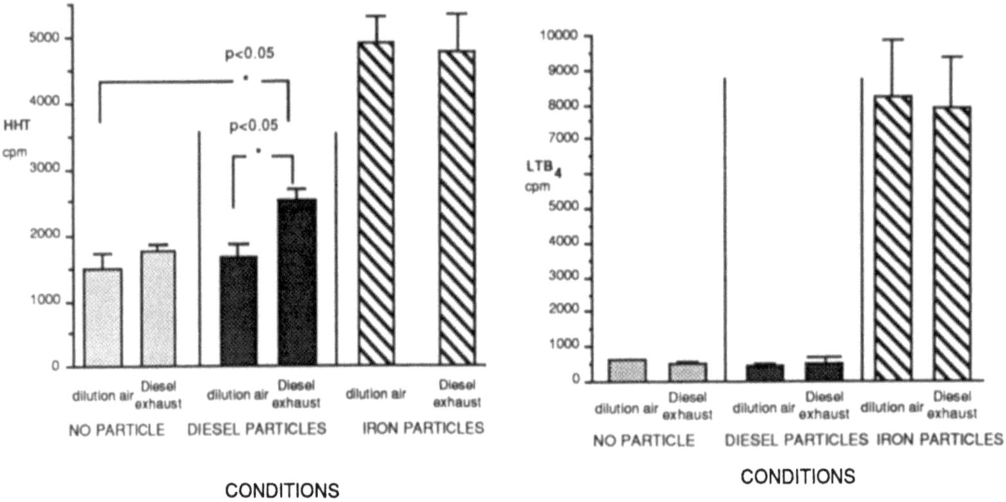

Figure 5: *quantitative comparison of the release of LTB_4 (right) and HHT (left) following various conditions. Comparison is made between cells exposed either to Diesel gazeous phase in low concentration or to control air. Each subset is in addition either submitted or not to a phagocytic challenge. Bars represent the mean ± SEM of 3 measurements. Data are representative of 4 experiments.*

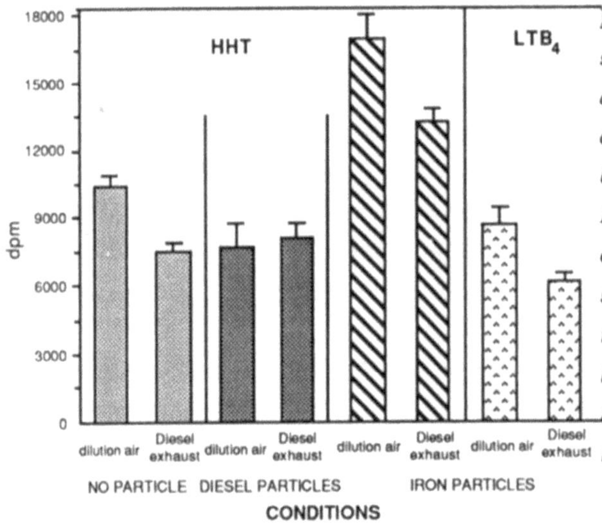

Figure 6: *quantitative comparison of the release of LTB_4 (right) and HHT (left) following various conditions. Comparison is made between cells exposed either to Diesel gazeous phase in high concentration or to control air. Each subset is in addition either submitted or not to a phagocytic challenge. Bars represent the mean ± SEM of 3 measurements. Data are representative of 4 experiments.*

in control air (fig. 6). Similarly, this purported cytotoxicity phenomenon blunted the stimulatory effect seen previously on HHT secretion when the cells were exposed simultaneoulsy to the gazeous and the particulate phases of Diesel exhaust (fig. 6).

Macrophage response following exposure to NO_2.

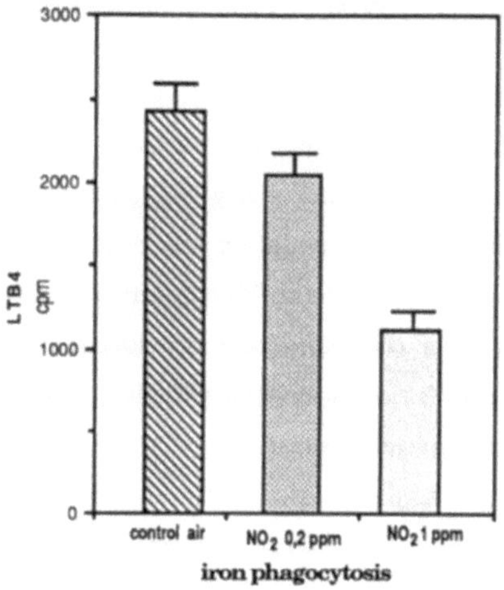

figure 7: *amounts of LTB_4 released following cell exposure to NO_2.*

Cell exposure to 0.1ppm NO_2 did not induce any stimulation on either AA metabolism pathway, since the metabolic profile observed was identical to the one seen when cells are exposed to control air. This was true for the phagocytosis-triggered secretion as well. Similarly to Diesel gazeous phase, a further increase in NO_2 concentration (to 0.2 and to 1 ppm) did bring about cytotoxicity, as shown by a decrease in LTB_4 production by iron-phagocytizing cells (fig.7).

DISCUSSION

Assessement of pulmonary cell exposure to gazes has been hampered so far by the fact that models of cell culture in liquid media are not ideally suited for an assessement of interactions between effector and target, and do not reflect the reality of phenomenons taking place in vivo at the alveolar level. A number of methods have been implemented to address this problem, with either systems exposing cells alternatively to air and to nutritionnal media, or cell culture in gazeous phase. We adapted a methodology previously devised to evaluate cell reactivity to toxic gazes . In a first step, it was shown that alveolar macrophage culture on membranes, with cells in contact with ambiant gazes, did not prevent cell secretion of arachidonic acid metabolites following cell stimulation. However, as opposed to the methodology previously used with cells cultured in liquid phase (Kouzan, 1985) , unstimulated cells exhibited a spontaneous secretion of metabolites from the cyclooxygenase pathway (namely TXB2 and HHT). This can putatively be attributed to cell interaction with the underlying substratum. Indeed we have previously shown that when macrophages are allowed to adhere to plastic, a notable stimulation of arachidonic metabolism is induced, with the subsequent inactivation of the self-destructive cyclooxygenase enzyme (Kouzan 1988) . Here, prior to cell plating on membranes, non-adhering macrophages incorporated tritiated AA for 3 hours. In other experiments (unpublished results), we showed that when such cells are subsequently allowed to adhere to plastic in a serum-free environment, they exhibit the same "spontaneous" cyclooxygenase metabolites secretion that the one observed here . Thus cell-substratum interaction is probably instrumental in this basal secretion. Further studies are needed to render the susbtratum used non-reactive towards alveolar macrophage membrane.

In spite of this activation process, macrophage exposure to Diesel exhaust had a further stimulatory effect on the cyclooxygenase enzyme, when both components (gazeous and particulate) were employed simultaneously. This was not the consequence of a cytotoxicity phenomenon, inusmuch as metabolite production following carbonyl iron beads stimulation was not modified when cells were exposed to "low" concentration Diesel gazeous phase. This is further substantiated by the finding that increasing Diesel gazeous phase concentration did induce a decrease in metabolites production, in all the situations tested (i.e. spontaneous secretion, iron beads phagocytosis, and exposure to particulate phase from Diesel exhaust).

Macrophage exposure to an oxidant burden (namely NO_2) was devoid of any stimulatory effect on AA metabolism. Increasing concentration led only to cytotoxicity. Thus, the experiments reported here seem to suggest that exposure to the sole gazeous component of a toxic emission (either Diesel or NO_2) does not induce release of inflammatory mediators. In keeping is the previously reported finding that only the particulate phase from cigarette smoke was stimulating secretion, by alveolar macrophage, of chemotactic factors .

It has been recently demonstrated that chronic exposure to Diesel exhaust increases lung cancer risk. One can wonder whether the cyclooxygenase enzyme stimulation found after macrophage exposure to the complete Diesel exhaust could have a bearing in this context. Indeed, the cyclooxygenase enzymatic system is able, by means of cooxidation,to metabolize some carcinogens to their active species . The enzymatic stimulation seen after Diesel exposure could increase, at the lung level, biodisponibility of carcinogens prone to cooxidative activation.

REFERENCES

Garschik E, Schenker MB, Munoz A et al.(1987) A case-control study of lung cancer and Diesel exhaust exposure in railroad workers. Amer Rev Respir Dis 135:1242-1248

Henke DC, Kouzan S, Eling TE (1984) Analysis of leukotrienes, prostaglandins, and other oxygenated metabolites of arachidonic acid by high-performance liquid chromatography. Anal Biochem 140:87-94

Hunninghake GW, Crystal RG(1983) Cigarette smoking and lung destruction Amer Rev Respir Dis 128:833-838

Howe GR et al.(1983) Cancer mortality (1965-1977) in relation to Diesel fume and coal exposure in a cohort of retired railway workers. J Natl Cancer Inst 70:1015-1019

Kouzan S, Brody AR, Nettesheim P, Eling TE (1985) Production of arachidonic acid metabolites by macrophages exposed in vitro to asbestos, carbonyl iron particles, or calcium ionophore Amer Rev Respir Dis 131:624-632

Voisin C, Aerts C, Fournier E, Firlik M (1985) Acute effects of tobacco smoke on alveolar macrophages cultured in gas phase. Eur J Respir Dis 66, Suppl, 139:76-81

SUPEROXIDE ANION PRODUCTION INDUCED BY CHRYSOTILE ASBESTOS IN THE GUINEA PIG ALVEOLAR MACROPHAGE

Andrij Holian and Paul L. Roney*
Pulmonary Division/Department of Internal Medicine
University of Texas Medical School
6431 Fannin
Houston, TX 77030

INTRODUCTION

It is well known that environmental exposure to asbestos fibers can lead to the development of various pathological conditions in the lung which include fibrosis (asbestosis), bronchogenic carcinoma and mesothelioma. Although the etiology of these conditions has not been established, there is good evidence that the interaction of asbestos fibers with the alveolar macrophage and the subsequent release of macrophage derived factors are central to the disease process (Mossman et al., 1987). One of the factors that could account for some of the tissue damage and carcinogenic potential of the fibers is superoxide anion. The destructive capability of superoxide anion and other oxygen radicals have been documented (Mossman et al., 1987; Weitzman and Stossel, 1981) as has the ability of the alveolar macrophage to release superoxide anion by a number of soluble stimulants (Holian and Daniele, 1979; Holian, 1986). Consistent with this hypothesis is evidence that asbestos can directly stimulate and/or enhance the release of superoxide anion from rodent alveolar macrophages (Hansen and Mossman, 1987). Furthermore, macrophages from patients with asbestosis have been reported to spontaneously produce elevated levels of superoxide anion (Rom et al., 1987).

If the above hypothesis is valid, it remains to be established by what mechanism the fibers stimulate superoxide anion production and to what sites on the macrophage surface fibers bind, resulting in macrophage stimulation. The model for this study comes from previous work (Daniele et al., 1982; Holian and Stickle, 1986; Holian, 1986; Holian et al., 1988) which has shown that soluble stimulants such as N-formyl-nle-leu-phe (FNLP) and leukotriene B4 bind to specific surface receptors which transduce receptor occupancy through a guanine nucleotide coupling protein (G protein) to activate phospholipase C. Phospholipase C hydrolyzes phosphatidyl inositols (PI) (eg. phosphatidyl inositol-4,5-bisphosphate (PIP2)) yielding 1,2-diacylglycerol (which can activate protein kinase C) and, in the case of PIP2 hydrolysis, inositol-1,4,5-trisphosphate (which can release intracellular calcium). These events in turn activate a membrane bound oxidase responsible for reducing molecular oxygen to superoxide anion. The purposes of this study were to: 1) to test the hypothesis that asbestos fibers may stimulate this transduction pathway in the guinea pig alveolar macrophage and 2) obtain preliminary evidence as to what surface sites asbestos may bind, which stimulates this pathway.

*Present address: RiskFocus/Versar, Inc.
6850 Versar Center
Springfield, Virginia 22151

NATO ASI Series, Vol. H30
Effects of Mineral Dusts on Cells
Edited by B. T. Mossman and R. O. Bégin

METHODS

Alveolar macrophages were isolated from guinea pigs as described (Dauber and Daniele, 1978). Cells were suspended in calcium- and magnesium-free Hank's balanced salt solution (MHBSS). Macrophage purity (ranged from 85-95%, PMN were always < 5%), viability to trypan blue (>93%) and total cell counts were assayed.

Superoxide anion production was monitored continuously at 37°C in a stirred cuvette as described in Holian and Daniele (1979) using a Perkin-Elmer Lambda 3B recording spectrophotometer by measuring cytochrome c reduction (100 uM, Type VI; Sigma, St. Louis, MO) at 550 nm. Pertussis toxin (List Biological Laboratories, Cambell, CA) treatment of macrophages was carried out as described (Holian, 1986).

The flux of 32Pi-labelled metabolites through the PI cycle was measured as described (Holian and Stickle, 1985). Analysis of phosphatidic acid (PA), phosphatidylinositol-4-phosphate (PIP) and PIP2 was carried out by as described (Imai et al., 1983).

Intracellular calcium mobilization was monitored continuously in a stirred cuvette by fura-2 fluorescence using a Shimadzu RF-540 fluorimeter in a similar manner as described (Stickle et al., 1984). Macrophages were loaded for 30 min with 5 uM fura-2, ester derivative (Molecular Probes, Eugene, OR) then washed and resuspended in MHBSS. Fluorescence excitation was at 335 nm and emission at 505 nm. Using a Kd of 224 nM, the free calcium concentration was calculated from $[Ca]i = Kd(F-Fmin)/(Fmax-F)$, where F, Fmin and Fmax represent the fluorescence values of the sample, lysed cells in 2 mM excess EDTA and 2 mM excess calcium, respectively.

Chrysotile (RG-144) sized 1-5 um was a kind gift from P. Thorne (Thorne et al., 1985). Crocidolite, amosite and anthophyllite were UICC standards. FNLP, phorbol-12,13-dibutyrate (PDB) and lectins were purchased from Sigma. Staurosporine was purchased from Calbiochem (San Diego, CA), fluphenazine was donated by E.R. Squibb and Sons, Inc. (Princeton, NJ), and endoglycopeptidase F (PNGase) was purchased from Boehringer Mannheim Biochemicals (Indianapolis, IN).

RESULTS

All four forms of asbestos were tested for their ability to stimulate superoxide anion production by guinea pig alveolar macrophages _in vitro_ using the stirred assay over a ten min period as described above. Using this model system _none_ of the amphiboles stimulated superoxide anion production or were cytotoxic towards macrophages in concentrations ranging from 5 - 100 ug/ml. Furthermore, in additional studies, they were not stimulatory over a one hour time period. In contrast, as shown in Table 1, chrysotile was a potent stimulant of superoxide anion production and was dose dependent. Chrysotile was partially cytotoxic only at the highest dose tested (23% decrease in viability after 15 min using 50 ug/ml chrysotile).

In order to evaluate whether the same biochemical pathway was stimulated by chrysotile, as was described for FNLP, the potential involvement of protein kinase C was tested by using protein kinase C

Table 1. Chrysotile Stimulation of Superoxide Anion Production by Alveolar Macrophages (n = 3)

Chrysotile (ug/ml)	Max Rate O_2^- nmoles/10^6 cells/5 min	O_2^- Production in 10 min nmoles/10^6 cells
2.5	5.5 ± 0.5	3.8 ± 0.7
5	9.3 ± 1.3	6.1 ± 1.0
10	12.4 ± 1.8	7.4 ± 1.3
25	21.8 ± 3.5	14.2 ± 2.5
50	27.5 ± 3.5	17.7 ± 2.4

inhibitors (25 uM fluphenazine and 2 uM staurosporine) on chrysotile, FNLP as well as direct protein kinase C stimulation by PDB. As shown in Table 2, both fluphenazine and staurosporine were effective in inhibiting superoxide anion production by all three stimulators.

Table 2. Effect of Protein Kinase C Inhibitors on Superoxide Anion Production

Stimulant	Control	25 uM Fluphenazine		2 uM Staurosporine	
	Rate	Rate	Per Cent Inhibition	Rate	Per Cent Inhibition
50 ug/ml Chrysotile	24.8 ± 6.0	5.0 ± 1.5	80	0.5 ± 0.3	98
0.5 uM PDB	31.7 ± 5.2	3.2 ± 1.0	90	0.0 ± 0	100
50 nM FNLP	41.0 ± 4.2	9.0 ± 3.3	78	15.4 ± 6.0	62

10^6 macrophages/ml were incubated with or without inhibitor for 5 min at 37°C then assayed for superoxide anion production. Rates are expressed as nmoles/10^6 cells/5 min of the mean (± SE) of three experiments.

In order to determine whether chrysotile stimulated phospholipase C, measurements were made of the flux of 32Pi intermediates of the PI cycle. Previous experience has shown that the intermediate which is the most sensitive determinant of stimulated PI cycle activity is an increase in labelled PA from the 1,2-diacylglycerol formed during phospholipase C activation (Holian and Stickle, 1985). Shown in Figure 1 is the percent increase over prestimulated levels of 32-PA following addition of 25 ug/ml chrysotile at both 22°C (for comparison with studies done at 22°C) and 37°C. There were also transient, but slight, decreases in both PIP and PIP2 (not shown).

Stimulation of the PI cycle should result in mobilization of intracellular calcium which is most easily detected using an intracellular fluorescence probe such as fura-2. Shown in Figure 2 is the result from a series of such measurements in which the dose of chrysotile was varied from 10-50 ug/ml. As can be seen, following the addition of chrysotile there was an increase in fluorescence indicating a transient increase in intracellular calcium. Furthermore, the increase in intracellular calcium was dose dependent.

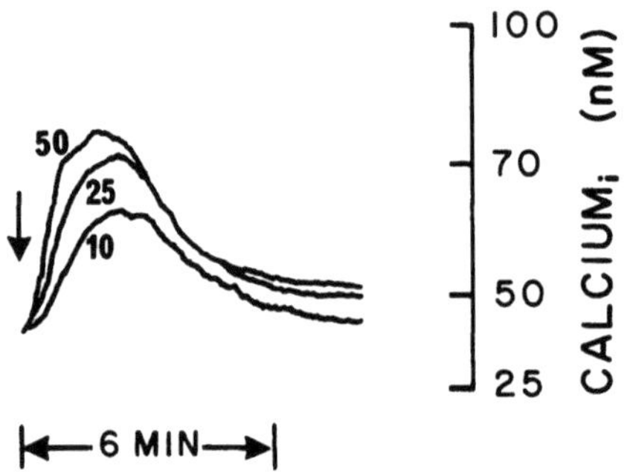

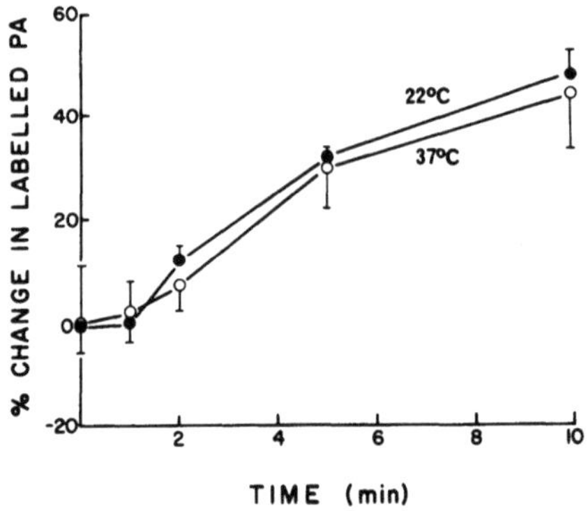

Figure 1. Percent increase in 32P-labelled PA following 25 ug/ml chrysotile at 22°C and 37°C. Chrysotile was added at zero time. The data represents the mean ($\pm$ SE) of three experiments.

Figure 2. Increase in cytosolic calcium, measured using fura-2, following addition of chrysotile. Each dose of chrysotile (ug/ml) was added as indicated by the arrow.

Soluble stimulants such FNLP and leukotriene B4 appear to activate phospholipase C through a G protein which can be inactivated using pertussis toxin (Holian, 1986). In order to determine whether chrysotile stimulation also involved a G protein, macrophages were preincubated with 2 ug/ml pertussis toxin for one hour at 37°C and compared to sham-treated cells. Similar to previous our results, FNLP stimulation was inhibited 93% while PDB stimulation was not affected. Stimulation by 50 ug/ml chrysotile decreased from 18.4 $\pm$ 2.6 to 11.8 $\pm$ 1.4 (36% decrease) nmoles superoxide anion/10^6 cells/5 min (mean $\pm$ SE for n=3, $P < 0.05$).

Previous studies have suggested that chrysotile binds to sialic acid residues of rat alveolar macrophages (Gallagher et al., 1987). Therefore, if the binding of chrysotile to stimulatory sites on the macrophage surface requires N-linked glycoproteins then treatment of macrophages with PNGase, which hydrolyzes the oligosaccaride from the protein at the N-linked asparagine, should decrease the stimulatory activity of chrysotile. Macrophages were treated with 3 units/ml of PNGase for one hour at 37°C and compared to sham-treated cells. Both PDB and FNLP stimulation were inhibited by approximately 25%, while 25 ug/ml chrysotile stimulation was inhibited by 53%.

In order to further test the role of glycoproteins in chrysotile stimulation of macrophage superoxide anion production and to determine whether there may be any specificity to the various sugar residues macrophages were pretreated with a number of lectins. Lectins specific for sialic acid, N-acetylgalactosamine and fucose (horseshoe crab, winged bean and asparagus pea agglutinins, respectively) did not stimulate superoxide anion production nor did they affect the ability of chrysotile to stimulate superoxide anion. Lectins specific for mannose (concanavalin A and garden pea agglutinin) stimulated superoxide anion production and partially blocked chrysotile stimulation (76 and 33%, respectively). Lectins specfic for N-acetylglucosamine (e.g., wheat germ and jimson weed agglutinins) also stimulated superoxide anion production, however, they completely blocked superoxide anion production by 50 ug/ml chrysotile.

DISCUSSION

The results from this study indicate that chrysotile asbestos is different from the amphibole forms of asbestos in that it can stimulate superoxide anion production by guinea pig alveolar macrophages. The mechanism by which chrysotile stimulates macrophages appears to be similar to that reported for soluble stimulants (Holian and Stickle, 1986; Holian, 1986; Holian et al., 1988). A feature that distinguishes chrysotile from the amphibole fibers is the surface charge which are positive and negative, respectively. Consistent with these observations are preliminary results indicating that positively charged latex beads can stimulate macrophages while neutral and negatively charged latex beads are inactive. Therefore, particle surface charge may be an important factor in determining the mechanism of particle interaction with the alveolar macrophage.

The mechanism by which chrysotile stimulates superoxide anion production appears to involve activation of phospholipase C causing the hydrolysis of polyphosphatidyl inositols. This hydrolysis would result in the formation of inositol-1,4,5-trisphosphate to cause calcium mobilization, which was observed as an increase in fura-2 fluorescence. 1,2-Diacylglycerol would also be formed, which along with the elevated calcium activate protein kinase C. In the presence of 32P-labelled ATP the 1,2-diacylglycerol would be converted 32P-labelled PA by 1,2-diacylglycerol kinase which was also observed. The potential role of protein kinase C was confirmed using two inhibitors of protein kinase C, fluphenazine (Holian et al., 1988) and staurosporine (Watson et al., 1988), which were able to block stimulation by chrysotile, FNLP and PDB. Therefore, this biochemical chain of events appears to be involved in chrysotile stimulation. Of less certainty is the involvement of a G protein. In this case the results indicate that a G protein may be involved since pertussis toxin partially inhibited chrysotile stimulation, however the mechanism is probably not one of a classical receptor-G protein interaction.

The interaction of chrysotile with the macrophage surface is not well understood. Some binding studies have been done with rat alveolar macrophages suggesting that sialic acid residues are important binding sites for chrysotile and other positively charged particles (Gallagher et al., 1987). However, this does not preclude the possibility that chrysotile may bind to an additional site resulting in macrophage stimulation. In the present studies no information was obtained on nonstimulatory binding sites, but rather the results suggest that N-acetylglucosamine and possibly mannose residues on surface N-linked glycoproteins are involved in chrysotile stimulatory binding sites. In fact Mossman et al. (1982) proposed that mannose residues may be involved in the stimulation of rodent tracheal explants. These sites may be independent of other binding sites such as those described by Gallagher et al. (1987). It could be speculated that the positive charge distribution on the surface of particulates such as chrysotile may function to aggregate surface glycoproteins which are receptors for physiological agonists.

ACKNOWLEDGEMENTS

The authors would like to thank Mark Jordan for expert technical assistance. This study was supported by grant BC-614 from the American Cancer Society.

REFERENCES

Daniele RP, Diamond MS, Holian A (1982) Demonstration of a formyl peptide receptor on lung macrophages. Am Rev Respir Dis 126:274-280
Dauber JH, Daniele RP (1978) Chemotactic activity of guinea pig alveolar macrophages. Am Rev Respir Dis 117:673-684
Gallagher JE, George G, Brody AR (1987) Sialic acid mediates the initial binding of positively charged inorganic particles to alveolar macrophages. Am Rev Respir Dis 135:1345-1352
Hansen K, Mossman BT (1987) Generation of superoxide from alveolar macrophages exposed to asbestiform and nonfibrous particles. Cancer Res 47:1681-1686
Holian A (1986) Leukotriene B4 stimulation of phosphatidylinositol turnover in macrophages and inhibition by pertussis toxin. FEBS Lett 201:15-19
Holian A, Daniele RP (1979) Stimulation of oxygen consumption and superoxide anion production in pulmonary macrophages by N-formyl methionyl peptides. FEBS Lett 108:47-51
Holian A, Stickle DF (1985) Calcium regulation of phosphatidyl inositol turnover in macrophage activation by formyl peptides. J Cell Physiol 123:39-45
Holian A, Jordan MK, Nguyen HV, Devenyi ZJ (1988) Inhibition of macrophage activation by isoquinolinesulfonamides, phenothiazines, and a napthalenesulfonamide. J Cell Physiol 137:45-54
Imai A, Nakashima S, Nozawa Y (1983) The rapid polyphosphoinositide metabolism may be a trigger event for thrombin mediated stimulation of human platelets. Biochem Biophys Commun 110:108-115
Mossman BT, Adler KB, Jean L, Craighead JE (1982) Mechanisms of hypersecretion in rodent tracheal explants after exposure to chrysotile asbestos. Chest 81:23s-24s
Mossman BT, Marsh JP, Shatos MA, Doherty J, Gilbert R, Hill S (1987) Implication of active oxygen species as second messengers of asbestos toxicity. Drug Chem Toxicol 10:157-180
Rom WN, Bitterman PB, Rennard SI, Cantin A, Crystal RG (1987) Characterization of the lower respiratory tract inflammation of nonsmoking

individuals with interstitial lung disease associated with chronic inhalation of inorganinc dusts. Am Rev Respir Dis 136:1429-1434
Stickle DF, Daniele RP, Holian A (1984) Cytosolic calcium, calcium fluxes, and regulation of alveolar macrophage superoxide anion production. J Cell Physiol 121:458-466
Tamaoki T, Nomoto H, Takahashi I, Kato Y, Morimoto M, Tomita F (1986) Staurosporine, a potent inhibitor of phospholipid/Ca^{++} dependent protein kinase. Biochem Biophys Res Commun 135:397-402
Thorne PS, Lightfoot EN, Albrecht RM (1985) Physicochemical characterization of cryogenically ground, size separated, fibrogenic particles. Environ Res 36:89-110
Watson SP, McNally J, Shipman LJ, Godfrey PP (1988) The action of the protein kinase C inhibitor, staurosporine, on human platelets. Biochem J 249:345-350
Weitzman SA, Stossel TP (1981) Mutation caused by human phagocytes. Science 212:546-547

ALTERED CALCIUM HOMEOSTASIS AND MINERAL DUST TOXICITY

Agnes B. Kane, George F. Gleva, and Lee A. Goodglick
Department of Pathology and Laboratory Medicine
Brown University
Providence, RI 02912
USA

INTRODUCTION

Asbestos fibers are acutely toxic to a variety of cells *in vitro* including macrophages, tracheal epithelial cells, and lung fibroblasts (Goodglick and Kane, 1986; Shatos *et al*., 1987) In these model systems, acute asbestos toxicity is mediated by oxidants. Macrophages exposed to asbestos fibers produce $O_2^{\cdot-}$ and H_2O_2. Asbestos toxicity is decreased under hypoxic conditions or by addition of exogenous superoxide dismutase or catalase. Asbestos fibers have been shown to catalyze the formation of $^{\cdot}OH$ in a cell-free system. This catalysis is prevented by coating the fibers with the iron chelator, deferoxamine (Weitzman and Graceffa, 1984). In cell cultures, deferoxamine-coated fibers are not toxic and toxicity is decreased by the $^{\cdot}OH$ scavenger, dimethylthiourea (Goodglick and Kane, 1986; Shatos *et al*., 1987). The subsequent metabolic consequences of oxidant-mediated injury responsible for cell death are unknown. Oxidant-induced injury is associated with depletion of glutathione, oxidation of membrane lipids and proteins, damage to DNA, and altered calcium homeostasis (reviewed by Farber, 1982). In this paper, we test the hypothesis that altered calcium homeostasis is responsible for acute asbestos cytotoxicity *in vitro*.

Intracellular calcium levels are tightly regulated. Free or ionized cytosolic calcium is maintained at 100-200 nm by the relative impermeability of the plasma membrane to extracellular calcium, one or more extrusion pumps in the plasma membrane, and sequestration of calcium within cytoplasmic organelles (Carafoli, 1987). Altered calcium homeostasis has

NATO ASI Series, Vol. H30
Effects of Mineral Dusts on Cells
Edited by B.T. Mossman and R.O. Bégin

been hypothesized to be responsible for irreversible injury produced by ischemia or chemical toxication (Farber, 1982). It is unknown whether influx of extracellular calcium ions or redistribution of calcium from intracellular stores or both contribute to toxic or ischemic cell death. We have developed techniques to monitor cell viability and calcium content directly in single cells in monolayer cultures using an inverted fluorescence microscope equipped with a microspectrophotometer. Fluorescent probes were used to monitor viability, the mitochondrial membrane potential, and membrane-bound and free cytosolic calcium in macrophages exposed to crocidolite asbestos fibers in vitro.

MATERIALS AND METHODS

Cell Injury Protocol

Thioglycollate-elicited mouse peritoneal macrophages were collected and cultured as described previously (Goodglick and Kane, 1986). Cells were plated at a density of 2.5 - 5.0 X 10^4 onto glass coverslips 12 mm in diameter. For toxicity experiments, the coverslips were rinsed and placed in 1 ml of Dulbecco's modified Eagles' medium in the absence of serum. Cells were exposed to 25-50 ug/cm^2 of UICC crocidolite asbestos fibers or 100 ug of latex beads for up to six hours as described in the text. Where indicated, cultures were exposed to 100 mM 2-deoxyglucose (2-dg), 5 mM sodium azide (NaN_3), 10 ug/ml antimycin A, or 1 uM p-trifluoromethoxyphenylhydrazone (FCCP) as described in Kane et al. (1985).

Fluorescent Probes

Cell viability was assessed by exclusion of erythrosin B or hydrolysis of fluorescein diacetate with retention of fluorescein as described previously (Goodglick and Kane, 1986). Membrane-bound and free cytosolic calcium were monitored by chlorotetracycline (CTC)-induced fluorescence and fura-2, respectively, as described by Schneider et al. (1983) and Tsien et al. (1985). The fluorescence intensity of single cells was quantified with a microspectrophotometer attached to a Leitx inverted, phase-contrast, fluorescence microscope. Free cytosolic calcium $[Ca^{++}]_i$ was calculated as described by

Scanlon *et al*. (1987) using the fluorescence intensity ratio at 350 and 380 nm measured during a 0.25 second interval. The final concentration of fura-2/AM was 12 uM; cells were incubated for 1.5-2 hours at 37^{0}C in the dark. The final concentration of CTC was 25 uM; cells were incubated for 20 minutes at 37^{0}C in the dark before the end of the experiment. The fluorescent probe rhodamine 123 was used to monitor the mitochondrial membrane potential in individual cells (Johnson *et al*., 1981). Cells were preloaded with 0.5 - 2 uM rhodamine 123 for 10 minutes at 37^{0}C prior to exposure to asbestos fibers or metabolic inhibitors. These probes, alone or in combination with metabolic inhibitors, were not toxic to macrophages under these conditions.

ATP and NAD^+ Assays

ATP content was quantitated by high performance liquid chromatography ; cellular protein was measured as described previously (Kane *et al.*, 1985). NAD^+ content was determined as described by Schraufstatter *et al*., 1986).

All data are expressed as the mean ± standard deviation of triplicate cultures.

TABLE 1

Time Course of Mitochondrial Membrane Depolarization, CTC or Fura-2 Flourescence, and Viability of Macrophages Exposed to Crocidolite Asbestos Fibers

Sample	% with depolarized mitochondria	$[Ca^{++}]_i$ nM	CTC fluorescence intensity	% of control viability
control	0 ± 1.2	116 ± 3.4	1.4 ± 0.6	100. ± 1.6
asbestos:				
1 hr	11.2 ± 13.9	315 ± 93	7.5 ± 1.7	103 ± 4.7
3 hr	46.8 ± 7.3	313 ± 43	15.5 ± 2.7	76.1 ± 5.7
6 hr	63.9 ± 1.1	n.d.	20.4 ± 3.1	57.4 ± 3.8

RESULTS

Macrophages show a dose and time-dependent loss of viability when exposed to crocidolite asbestos fibers in vitro (Goodglick and Kane, 1986). At a dose between 25-50 ug/cm^2, 40-50% of the cells are killed after six hours. Loss of viability is preceded by increased membrane-bound calcium as monitored by CTC-induced fluorescence. After six hours, most of the cells show bright green flourescence of the plasma membrane and cytoplasmic organelles. There is a 20-fold increase in CTC flourescence intensity after exposure to asbestos fibers in comparison to untreated cultures. This corresponds to 1.8 - 3.6 mM calcium bound to cellular membranes after six hours of exposure to asbestos fibers. Free cytosolic calcium, is increased nearly 3-fold after only one hour of exposure to asbestos fibers. This increased level of free cytosolic calcium is sustained in this population of cells exposed to asbestos fibers throughout the experiment. These data are summarized in Table 1. These changes in membrane-bound and free cytosolic calcium are not seen after phagocytosis of nontoxic latex beads.

Oxidant stress in hepatocytes is accompanied by oxidation of mitochondrial pyridine nucleotides and release of mitochondrial calcium stores (Richter and Frei, 1988). In order to monitor the integrity of mitochondria in macrophages exposed to asbestos fibers, the fluorescent probe rhodamine 123 was used. This probe is actively sequestered within mitochondria and produces discrete, granular cytoplasmic fluorescence. If the mitochondrial membrane is depolarized, the dye leaks into the cytoplasm causing dim, diffuse fluorescence (Johnson et al., 1981). As summarized in Table 1, loss of viability in macrophages exposed to asbestos fibers is preceded by depolarization of the mitochondrial membrane.

We tested whether disruption of the mitochondrial membrane potential is sufficient to cause release of mitochondrial calcium stores, increased cytosolic and membrane-bound calcium, and death of macrophages. Depolarization of the mitochondrial membrane is produced directly by the proton ionophore, FCCP, or the inhibitor of electron transport, antimycin A. These inhibitors also release mitochondrial calcium stores (Bellomo et al., 1982). As shown in Table 2, after six hours of exposure to FCCP or antimycin A, 99-100% of macrophages show loss of the mitochondrial membrane potential as monitored

TABLE 2

Mitochondrial Membrane Depolarization, CTC or Fura-2 Fluorescence, and Viability of Macrophages Exposed to Asbestos or Metabolic Inhibitors for Six Hours

Sample	% with depolarized mitochondria	$[Ca^{++}]_i$ nM	% with CTC fluorescence	% of control viability
control	7.3 ± 1.4	116 ± 3.4	3.3 ± 0.9	98.3 ± 4.9
asbestos	89.9 ± 0.9	315 ± 93	73.3 ± 5.0	49.1 ± 8.4
FCCP	99.7 ± 0.5	146 ± 33	2.8 ± 0.4	94.9 ± 2.4
anti-mycin A	100. ± 0	n.d.	3.5 ± 0.6	96.3 ± 6.7

TABLE 3

ATP content, CTC Fluorescence, and Viability of Macrophages Exposed to Asbestos or Metabolic Inhibitors for Six Hours

Sample	nmoles ATP/ mg protein	% with CTC fluorescence	% of control viability
control	18.6	3.0 ± 1.1	100. ± 3.5
asbestos	6.5	73.3 ± 5.0	49.1 ± 8.4
2-dg + NaN_3	0.6	2.5 ± 1.0	98.3 ± 4.9
2-dg + Na_3+ FCCP	0.6	5.5 ± 1.0	100. ± 3.2

by rhodamine 123. FCCP also increases free cytosolic calcium, although not to the same extent as asbestos. In contrast to asbestos fibers, neither of these inhibitors increases membrane-bound calcium nor kills macrophages even after six hours of exposure.

Macrophages exposed to toxic mineral particles also have decreased ATP content (Kane et al., 1985). Loss of ATP may indirectly contribute to calcium accumulation by compromising membrane calcium pumps. As shown in Table 3, ATP levels in macrophages exposed to asbestos for six hours are 30% of control cultures. We tested whether a decline in ATP content alone is sufficient to alter calcium homeostasis and kill macrophages. Glycolysis was inhibited by 2-deoxyglucose combined with NaN_3 to inhibit mitochondrial ATP synthesis. These inhibitors reduced ATP levels even further; however, this inhibition is reversible after removal of the inhibitors and no cell death occurs. Even energy-depleted macrophages exposed to FCCP for up to six hours do not accumulate calcium nor lose viability. We conclude that depolarization of the mitochondrial membrane and release of sequestered calcium, even in combination with reduced ATP content, are not sufficient to kill macrophages.

Macrophages exposed to asbestos fibers also have altered pyridine nucleotide metabolism as monitored by NAD^+ content. The decrease in NAD^+ can be prevented by treatment with nicotinamide (2 mM) or 3-aminobenzamide (10 mM; Goodglick et al., in press). These agents also prevent depolarization of the mitochondrial membrane in macrophages exposed to asbestos fibers. Under these conditions, 3-aminobenzamide does not interfere with phagocytosis, inhibit respiratory burst activity, nor scavenge $O_2^{\cdot-}$ and H_2O_2. In other models of oxidant injury, including H_2O_2 toxicity to the P338D$_1$ macrophage cell line (Schraufstatter et al., 1986) and ionizing radiation (Berger, 1985), NAD^+ depletion is thought to be the consequence of activation of the nuclear enzyme, poly (ADP-ribose) polymerase, during repair of DNA strand breaks. This enzyme is inhibited by 3-aminobenzamide or nicotinamide. Oxidant-induced DNA damage is not repaired in the presence of these agents; however, NAD^+ levels are maintained and the cells remain viable. Alternatively, these inhibitors also interfere with ADP-ribosylation of an inner mitochondrial membrane protein involved in calcium release. Under oxidative stress, there is hypothesized to be continual cycling of calcium across the mitochondrial membrane leading to mitochondrial damage, increased cytosolic calcium and cell death (Richter and Frei, 1988). Additional experiments are in progress to determine whether either of these mechansims is responsible for acute asbestos cytotoxicity.

DISCUSSION

Asbestos cytotoxicity is an oxidant-mediated process which is accompanied by lipid peroxidation, depletion of ATP and NAD^+, and loss of the mitochondrial membrane potential (Goodglick et al., in press; Nadeau and Lane, 1988). In the face of such widespread biochemical changes in macrophages injured by asbestos fibers, we must define those changes which are causally related to irreversible injury. These experiments show a temporal relationship between increased free cytosolic and membrane-bound calcium and loss of viability of macrophages exposed to asbestos fibers in vitro. Experiments described in this paper also explore whether release of mitochondrial calcium stores is sufficient to irreversibly injure macrophages. Direct release of mitochondrial calcium by uncouplers or proton ionophores was not sufficient to kill macrophages. In contrast to these metabolic inhibitors, asbestos fibers also impose an oxidant stress on macrophages. Richter and Frei (1988) have proposed that oxidative stress causes oxidation of mitochondrial pyridine nucleotides, hydrolysis of NAD^+, and repeated cycles of calcium release and uptake by mitochondria. We hypothesize that similar mitochondrial injury occurs in macrophages exposed to asbestos fibers. This mitochondrial damage, combined with reduced ATP content and oxidative damage to membrane calcium pumps, is responsible for calcium accumulation and cell death.

Acknowledgement. This work was supported by a research grant from the National Institues of Health (R01 ES 03189).

Bellomo G, Jewell SA, Thor H, Orrenius S (1982) Regulation of intracellular calcium compartmentation. Proc Natl Acad Sci USA 79 : 6842-6846
Berger NA (1985) Symposium : cellular response to DNA damge :

the role of poly(ADP-ribose). Rad Res 101: 4-15
Carafoli E (1987) Intracellular calcium homeostasis. Ann Rev Biochem 56 : 395-433
Farber JL (1982) Membrane injury and calcium homeostasis in the pathogenesis of coagulative necrosis. Lab Invest 47 : 114-123
Goodglick LA, Kane AB (1986) Role of reactive oxygen metabolites in crocidolite asbestos toxicity to mouse macrophages. Cancer Res 46 : 5558-5566
Johnson LV, Walsh ML, Bockus BJ, Chen LB (1981) Monitoring of relative mitochondrial membrane potential in living cells by fluorescence microscopy. J Cell Biol 88 : 526-535
Kane AB, Petrovich DR, Stern RO, Farber JL (1985) ATP depletion and loss of cell integrity in anoxic hepatocytes and silica-treated P388D, macrophages. Am J Physiol 249 : C256-C266
Nadeau D, Lane DA (1988) The cytotoxicity of chrysotile asbestos fibers to pulmonary alveolar macrophages. I. Effects of inhibitors of ADP-ribosyl transferase. Cell Biol Toxicol 4 : 13-30
Richter C, Frei B (1988) Ca^{2+} release from mitochondria induced by prooxidants. Free Rad Biol Med 4 : 365-375
Scanlon M, Williams DA, Fay FS (1987) A Ca^{2+} -insensitive form of fura-2 associated with polymorphonuclear leukocytes. J Biol Chem 262 : 6308-6312
Schneider AS, Herz R, Sonenberg M (1983) Chlorotetracycline as a probe of membrane -associated calcium and magnesium. Biochemistry 24 : 6309-6312
Schraufstatter IU, Hinshaw DB,Hyslop PA, Spragg RG, Cochrane CG (1986) Oxidant injury of cells. J Clin Invest 77: 1312-1320
Shatos MA, Doherty JM, Marsh JP, Mossman BT (1987) Prevention of asbestos-induced cell death in rat lung fibroblasts and alveolar macrophages by scavengers of active oxygen species. Environ Res 44 : 103-116
Tsien RY, Rink TJ, Poenie M (1985) Measurement of cytosolic free Ca^{2+} in individual small cells using fluorescence microscopy with dual excitation wavelengths. Cell Calc 6 : 145-157
Weitzman SA, Graceffa P (1984) Asbestos catalyzes hydroxyl and superoxide radical generation from hydrogen peroxide. Arch Biochem Biophys 238 : 373-376

FORMATION OF 8-HYDROXYDEOXYGUANOSINE IN DNA BY MMMF.

P. Leanderson[1,2], P. Söderkvist[1,3], M. Gillner[3] and C. Tagesson[1,2]

Department of Occupational Medicine
Faculty of Health Sciences
S-581 85 Linköping
Sweden

Footnote; [1]Department of Occupational Medicine and [2]Clinical Chemistry, Faculty of Health Sciences, S-581 85 Linköping, Sweden.

[3]Department of Medical Nutrion, F 69, Huddinge University Hospital, S-141 86 Huddinge, Sweden.

NATO ASI Series, Vol. H30
Effects of Mineral Dusts on Cells
Edited by B. T. Mossman and R. O. Bégin

Introduction

Several epidemiological and experimental studies have shown that asbestos- and man-made mineral fibres (MMMFs) are carcinogenic, but the underlying molecular mechanisms have not yet been resolved. Recent studies indicate that asbestos and MMMFs can induce the formation of reactive oxygen species. Furthermore, it is considered probable that oxygen radicals are invonlved in mutagenesis and carcinogenesis (Cerutti, 1985), since many exogenous chemicals, that are mutagens and carcinogens, produce vast amounts of oxygen radicals. Although the molecular mechamisms for hydroxyl radical-induced carcinogenesis is not known, reactive oxygen metabolites attack the DNA-molecule producing many different types of damage, including DNA strand breaks and base hydroxylations. Therefore, it is not unlikely that hydroxyl radical-induced DNA damage, e.g. formation of 8-hydroxydeoxyguanosine (8-OH-dG) represents a critical event in mutagenesis and carcinogenesis.

Kuchino et al. have shown that insertion of 8-OH-dG in a synthetic oligodeoxynucleotide affects the fidelity of DNA replication in vitro using E. coli DNA polymerase I (Klenow fragment) (Kuchino et al., 1987). The modified template was misread at the position of 8-OH-dG residue itself and at adjacent bases (Kuchino et al., 1987). An ab initio molecular orbital study of the caracteristics of 8-OH-dG indicated that the most stable configuration is probably the 6,8-diketo tautomer (Aida et al., 1987). In Fig.1, a model of 8-OH-dG in a DNA-molecule is shown. If a normal Watson-Crick base pairing is assumed, the oxygen of 8-OH-dG would project into the major groove of the DNA-helix. If the local conformation of DNA is changed by formation of 8-OH-dG, this may be assessed by energy minimization and NMR studies. However, addition of an oxygen atom in the 8 position of guanine remarkably influences the intire electrostatic potential, giving a negative character of the whole molecule (Aida et al., 1987). Electrostatic forces play an important role in the interactions between enzymes and substrates and between proteins and DNA. This large change in electrostatic potential of one base may therefore interfere in protein-DNA recongnition, especially when located in the major groove were most proteins (i.e. DNA polymerase I) is postulated to interact with DNA (Hayes et al., 1976). This means that induction of misreadings of 8-OH-dG and adjacent bases in an oligonucleotide (Kuchino et al., 1987) may be due to a defect protein-DNA recognition caused by 8-OH-dG.

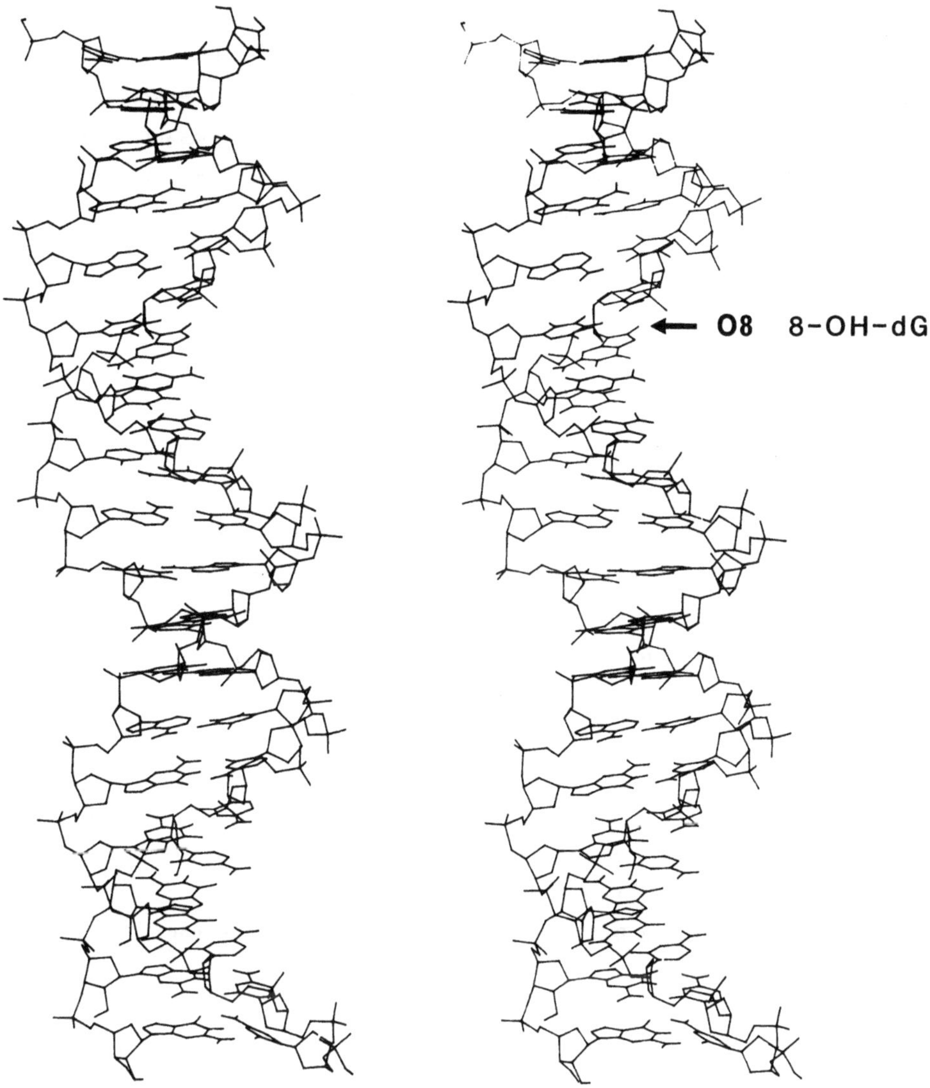

Fig.1 Stereo plot of a model of 8-OH-dG in DNA. The model was built and plotted using Silicon Graphics 3130 work station running the package QUANTA from Polygen. The oligonucleotide sequence containing 8-OH-dG corresponds to fragment D in Kuchino et al., (1987). When a normal Watson-Crick base-pairing is assumed, the oxygen atom O8 of 8-OH-dG (residue 7) projects into the major groove of the DNA-helix.

It is possible, therefore, that the formation of 8-OH-dG in the DNA by oxygen radicals, generated by chemicals or physical agents such as radiation or fibres, is an important cause of mutations and carcinogenesis.

Recent studies have shown that both rockwool and glasswool can induce hydroxyl radicals (Zalma, 1988) and that these fibres mediate the formation of 8-hydroxydeoxyguanosine (8-OH-dG) in isolated DNA (Leanderson et al., 1988). In this study we show that two important MMMFs, rockwool and glasswool fibre, both induce the formation of 8-OH-dG in DNA and that this hydroxylation can be influenced by different pretreatments of the fibres.

Materials and Methods

Rockwool and glasswool fibres were investigated with regard to their ability to hydroxylate deoxyguanosine (dG)-residues in isolated calf-thymus DNA. The reaction mixtures contained 10 mg fibre and 0.5 mg DNA dissolved in 1.0 ml PBS (phosphate buffered saline). To compare the hydroxylating capacity of two fibre samples the fibre-DNA mixtures were incubated up to 180 min. in 37°C, before analysis of 8-OH-dG. In other experiments, we used rockwool fibres that had been crushed in a glass mortlar prior to fibre-DNA incubations for 1 hour in 37°C. In one series of experiments fibres were preheated (200 or 400°C, 1 hour) in air or oxygen atmosphere (crushed fibres was placed on a watch glass at the bottom of an owen), or pretreated with deferoxamine (Desferal®),(Weitzman et al., 1985). Following incubation, the fibres were pelleted and supernatants were extracted with 3.0 ml chloroform:isoamylalcohol (24:1), and centrifuged (2000 g, 6 min.). The upper aqueous phase containing DNA was collected and organic solvents remaining in the aqueous layer were evaporated under a stream of nitrogen. After addition of 100 µl 3 M sodium acetate pH 5.2 , the DNA was precipitated with 2 vol. 70 % ice-cold ethanol, centrifuged, washed once with 70 % ethanol, dried, and redissolved in 50 µl 0.10 M sodium acetate buffer, pH 4.8. The DNA was digested at 37°C in a water bath with nuclease P_1 for 30 min. and alkaline phosphatase for 1 hour (pH was raised from 4.8 to 7.5 by addition of 1.0 M Tris-HCl, pH 8.0, before treatment with alkaline phosphatase). The amounts of 8-OH-dG and dG were then determined with the HPLC-technique originally described by Floyd et al.,(1986). The ratio between 8-OH-dG and dG was used to measure the DNA-damage. A typical HPLC-chromatogram is shown in figure 2.

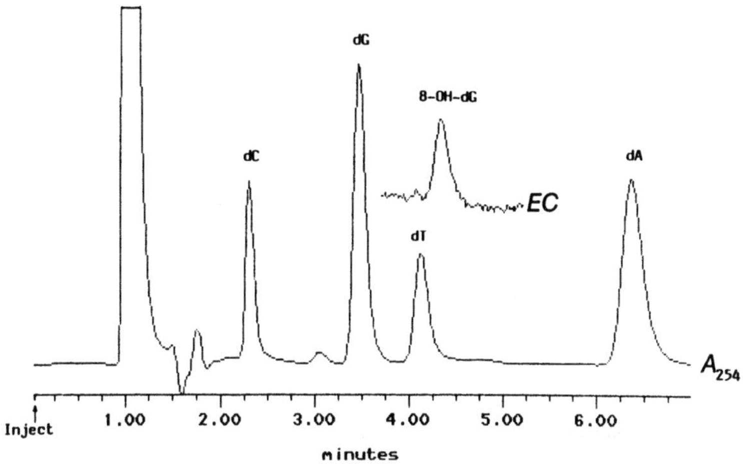

Fig.2 HPLC-chromatogram of digested DNA. A 10 µl sample was injected and the nucleosides detected on line with an UV-detector (254 nm) and an electrochemical detector (8-OH-dG). dC= deoxycytosine, dG= deoxyguanosine, dT= deoxythymidine, 8-OH-dG (inserted) = 8-hydroxydeoxyguanosine and dA= deoxyadenine. The amounts of dG and 8-OH-dG, correspond to 4.8 nmol and 0.15 pmol, respectively.

Results and discussion

As illustrated in Figure 3, both fibres showed a time dependent increase in the formation of 8-OH-dG. The surface area of the rockwool and glasswool fibres were 0.62 m^2/g and 0.91 m^2/g, respectively. When rockwool fibres were crushed prior to the incubation, the 8-OH-dG formation was enhanced. The hydroxylating capacity of rockwool fibres increased with the surface area as shown in figure 4. Thus the surface area seems to be one factor that affects the hydroxylation of DNA.

Results of experiments with pretreated rockwool fibres are shown in figure 5. When fibres were preheated prior to incubation with DNA the formation of 8-OH-dG was decreased. This may be due to an oxidation of the fibre surface. When heating was performed in an oxygen atmosphere the inhibition of the 8-OH-dG formation was even further accentuated which supportes the contention that an oxidation of the fibre surface attenuates the hydroxylating capacity.

Pretreatment with the iron chelator deferoxamine also resulted in a deactivation of the fibres. Iron is important for generation of hydroxyl radicals (acting as an electron donator) in the so called Fenton reaction. By chelating iron at the fibre surface, deferoxamine has earlier been shown to inhibit asbestos mediated lipid peroxidation (Weitzman et al., 1985). Other studies with different asbestos fibres have shown that fibres rich in iron

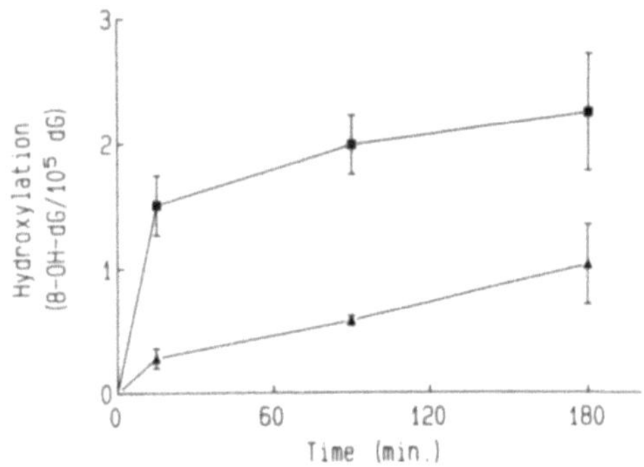

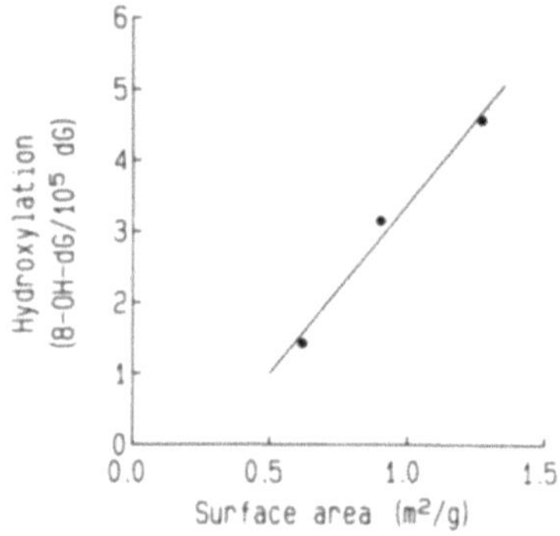

Fig.3 (left) Hydroxylation of DNA caused by rockwool (■), and glasswool (▲). Means ± SD of four experiments.

Fig.4 (right) Relation between the surface area of a mineral fibre and its ability to hydroxylate dG residues in DNA. Rockwool were crushed for different lengths of time (0, 30 and 60 sec.) incubated with calf-thymus DNA for 1h, and the amount of 8-OH-dG detected.

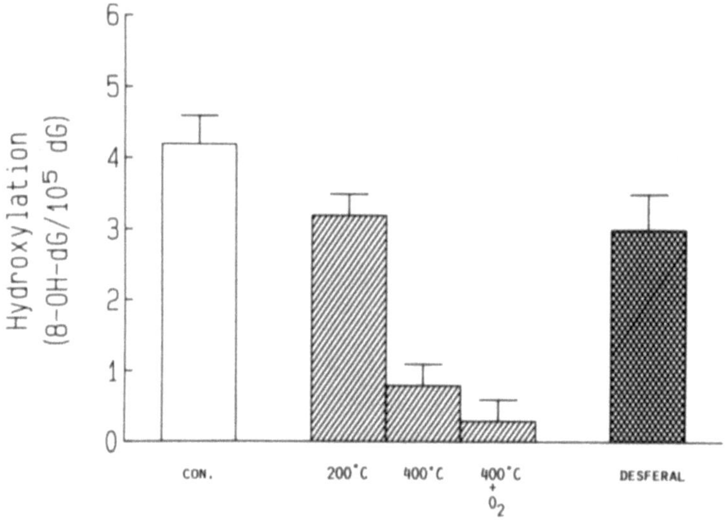

Fig.5 Inhibition of MMMF-mediated DNA-hydroxylation by various treatments of rockwool fibres. Fibres were treated as described in "Materials and Methods". Means ± SD of six experiments.

possessed an increased capacity to hydroxylate DNA compared to fibres with less iron (Kasai et al., 1984). Figure 6 shows Energy Dispersive X-Ray Fluorecence spectra of the two fibres. Rockwool fibre contained a substantial amount of iron compared to glasswool fibres and this may be responsible at least partly to the differences in the hydroxylation capacity between the fibres.

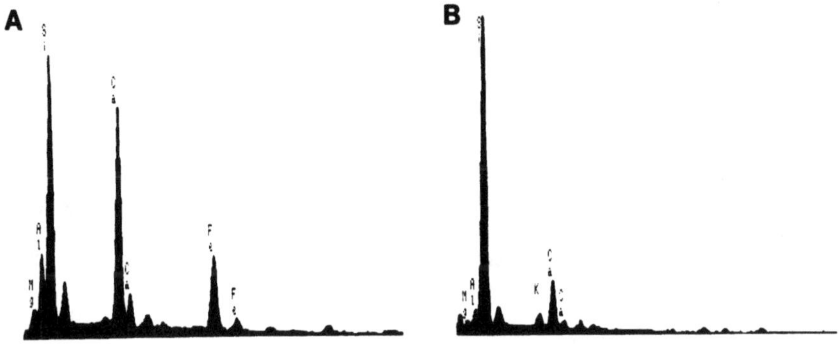

Fig.6 Energy Dispersive X-Ray Fluorescence spectrum of (A) rockwool and (B) glasswool. Crushed fibres were placed on colloidal graphite and coated with gold. Spectra was collected and processed with a Link AN 10 000 X-Ray Analytical System coupled to a Jeol JSM-840 scanning microscope (accelerating voltage of 20 kV, 600 sec.).

In analogy to asbestos fibres one possible mechanism for MMMF induced hydroxyl radical production is that iron at the fibre surface acts as an electron donator and reduce oxygen and hydrogen peroxide according to the following reactions (Zalma et al., 1987);

$$O_2 \xrightarrow{e^-} O_2^{\cdot -}$$

$$O_2^{\cdot -} \xrightarrow{e^-, 2H^+} H_2O_2$$

$$H_2O_2 + \text{Fibre-Fe}^{2+} \longrightarrow OH^{\cdot} + OH^- + \text{Fibre-Fe}^{3+}$$

Our results show that rockwool and glasswool fibres can catalyze the formation of hydroxyl radicals that can hydroxylate DNA in vitro, and it is tempting to speculate that generation of hydroxyl radicals could represent an important step in the carcinogenicity of MMMFs.

Acknowledgement; We thank Olav Axelson for valuble support and acknowledge gratefully the skilled technical assistance of Anna-Lena Sääf. P S is a recipient of a research fellowship from the Swedish Natural Science Research Council. This work was supported by the Swedish Work Environment Fund (87/0898).

References

Aida M, Nishimura S (1987) An ab initio orbital study on the characteristics of 8-hydroxydeoxyguanine. Mutation Res 192:83-89

Cerutti PA (1985) Pro-oxidant states and tumor promotion. Science 227:375-380

Floyd R, Watson JJ, Wong PK, Altmiller DH, Rickard RC (1986) Hydroxyl free radical adduct of deoxyguanosine: Sensitive detection and mechanism of formation. Free Radical Research Communication 1:163-172

Hayes DM, Kollman PA (1976) Electrostatic potential of proteins. I; Carboxy-peroxidase A. J Am Chem Soc 98:3335-3345

Kasai H, Nishimura S (1984) DNA damage induced by asbestos in the presence of hydrogen peroxide. Gann 75:841-844

Kuchino Y, Mori F, Kasai H, Inoue H, Iwai S, Miura K, Ohtsaka E, Nishimura S (1987) Misreading of DNA templates containing 8-hydroxydeoxyguanosine at the modified base and at adjacent residues. Nature 327:77-79

Leanderson P, Söderkvist P, Tagesson C, Axelson O (1988) Formation of the DNA adduct 8-hydroxydeoxyguanosine induced by man-made mineral fibres. In: Bartsch H, Hemminki K and O'Neill I K (ed) Methods for Detecting DNA Damaging Agens in Humans; Applications in Cancer Epidemiology and Prevention. IARC Scientific Publications 89. pp 422-424, Lyon

Weitzman SA, Weitberg AB (1985) Asbest-catalyzed lipid peroxidation and its inhibition by desferroxamine Biochem J 225:259-262

Zalma R, Bonneau L, Guignard J, Pezerat H, Jaurand M-C (1987) Formation of oxy radicals by oxygen reduction arising from the surface activity of asbestos. Can J Chem 65:2338-41

Zalma R (1988) Contribution a l'etude de la reactivite de surface des fibres minerales. Relations possibles avec leurs proprietes cancerogenes., Thesis, Universite Pierre et Marie Curie, March 25, Paris

Effect Of Chrysotile, Amosite and Crocidolite on the Oxidative Burst of Rabbit Alveolar Macrophages

Q. N. Myrvik, P. Wood and H. Hayakawa
Department of Microbiology and Immunology
Bowman Gray School of Medicine
Winston-Salem, NC 27103

Introduction.

Numerous publications have documented the relative toxicities of asbestos preparations (Miller and Harrington, 1972; Kaw, et al., 1975; Kang, et al, 1979; McLemore, et al., 1979; Sirois, et al., 1980; Donaldson, et al., 1985; Fisher, et al., 1985; Mossman and Marsh, 1985; Myrvik, et al., 1985; Goodglick and Kane, 1986; Shatos, et al., 1987). Models have included the effect of asbestos on the viability of alveolar macrophages (AM) and other cell types, effect on random migration, release of hydrolases as well as production of reactive oxygen intermediates. Collectively, these data indicate that essentially all asbestos preparations are toxic for AM.

The data to be presented in this study indicate that at moderately large doses asbestos can function as an eliciting agonist to trigger substantial oxidative bursts in immunologically primed AM. Furthermore, preincubation of AM with asbestos impairs their subsequent PMA-induced oxidative response.

Materials and Methods.

Animals. New Zealand white rabbits of either sex ranging in age from 5 to 6 months old used in these experiments were obtained from Franklin's Rabbitry (Wake Forest, NC).

NATO ASI Series, Vol. H30
Effects of Mineral Dusts on Cells
Edited by B.T. Mossman and R.O. Bégin

Culture Medium.

RPMI 1640 (Sigma) medium without phenol red was prepared from stock powder and supplemented with 100 U of penicillin per ml, 100 ug of streptomycin per ml, sodium bicarbonate (1 g/L), and 0.1 M L-glutamine (final concentrations). This medium did not contain serum unless specified. Fetal bovine serum came from Hyclone Laboratories.

Procurement of Alveolar Macrophages (AM).

The AM were obtained aseptically by lavage with cold (4°C) saline using a previously described technique (McGee and Myrvik, 1981). The AM were washed 2 times at 55 x g for 10 min in cold RPMI 1640. The viability of the AM detected by the trypan blue staining technique always exceeded 90%.

Sensitization with Heat-killed BCG.

Groups of 5-6 month-old rabbits were vaccinated i.v. with 200 ug of heat-killed Mycobacterium bovis BCG (HK-BCG) suspended in 0.2 ml light mineral oil. The AM were harvested 21 days after vaccination by lavage.

Chemiluminescent (CL) Assay.

We have described this assay in a previous publication (Chida, et al., 1987). CL was monitored in a Beckman CPM liquid scintillation counter; the CPM were recorded every 1-2 min for a 45 min period. Luminol and phorbol myristate acetate (PMA) were obtained from Sigma.

Assays for Measuring the Capacities of Asbestos Preparations to Elicit Chemiluminescent Responses of BCG-immune AM.

In these experiments 1-3 x 10^6 AM obtained from BCG-vaccinated rabbits were suspended in HBSS (37°C). After resting values were established, the asbestos preparations were added to the cell suspensions in various concentrations and their CL responses were monitored.

Asbestos Preparations.

The asbestos preparations (chrysotile A, crocidolite and amosite) were kindly provided by Dr. A. Brody, National Institute of Environmental Health Sciences, Research Triangle Park, NC. The asbestos preparations (1 mg/ml) were sonicated in a probe Fisher Sonic Dismembrator-Model 300 until greater than 80% of the particles were in the phagocytosable range.

Statistical Analysis.

Differences between two means were assessed by student's t test.

Results.

Asbestos Preparations.

The stock preparations of chrysotile A, crocidolite and amosite were sonicated as previously described and stored in aliquots at -20°C until use. The respective samples were thawed, diluted and examined by scanning electron microscopy. Typical representations of the 3 preparations are presented in Figure 1. In all instances greater than 80% of the fibers were in the phagocytosable range.

Effects of Chrysotile, Crocidolite and Amosite on the Viability of Alveolar Macrophages.

Viabilities were determined using the trypan blue exclusion technique and expressed in percent. It can be noted that 200 ug/ml chrysotile or crocidolite produced a 78% or 96% loss in viability at 3 hr, respectively, in serum-free RPMI 1640 (Fig 2.) The addition of 5% fetal bovine serum (FBS) provided substantial protection against the toxicity of chrysotile, and crocidolite at the highest concentrations. Control preparations which included 200 ug/ml of latex particles exhibited no detectable toxicity even in serum-free RPMI 1640 medium (data not shown).

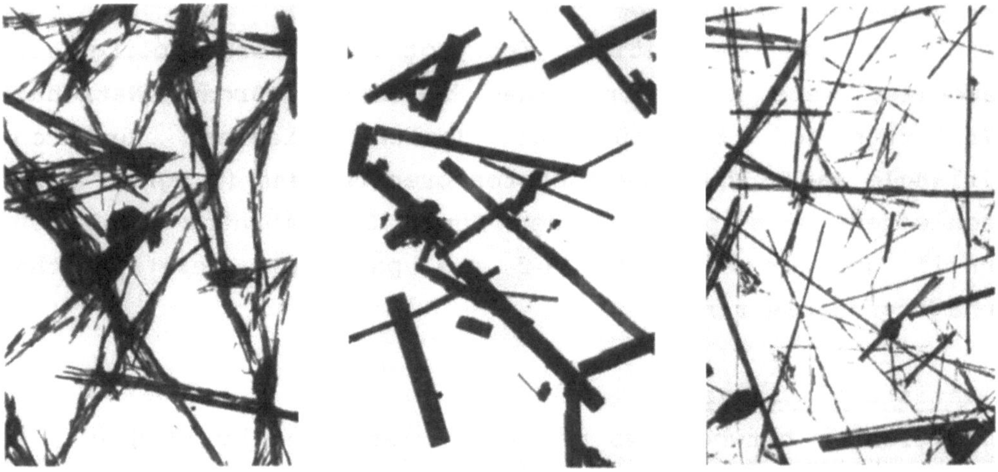

Figure 1. Sonicated preparations of asbestos used in these studies. Greater than 80% of the fibers were in the phagocytosable range. ⊢—⊣ = 5 uM

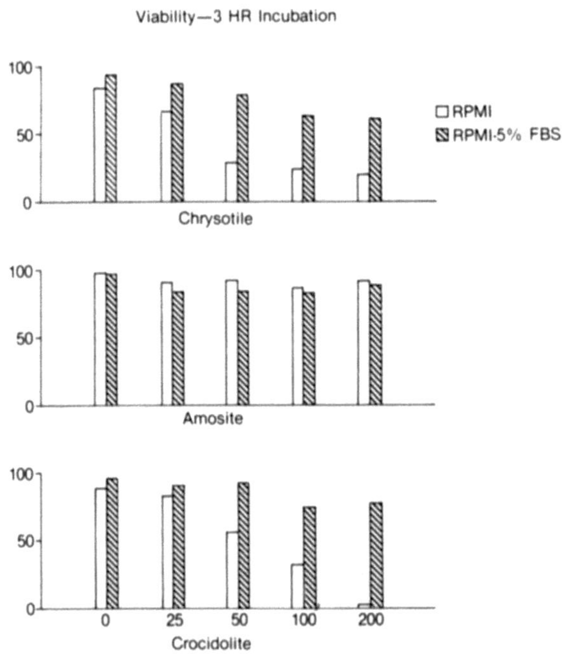

Figure 2. Effect of asbestos on viability of normal AM.

Effect of a 3 hr Incubation of AM with Chrysotile, Crocidolite or Amosite on their Response to PMA.

Normal AM in RPMI 1640 medium and 5% FBS plus a final concentration of 0, 25, 50, 100 or 200 ug/ml of chrysotile, crocidolite or amosite were incubated for 3 hr at 37°C. The cells were washed 2 times with HBSS in the cold and resuspended in HBSS (37°C) and assayed for their CL response to 0.5 ug/ml PMA. The results presented in Figure 3 indicate that all 3 asbestos preparations reduced the CL responses of AM to PMA by about 50% at a concentration of 200 ug/ml of asbestos. This impairment was reduced to about 20% when the asbestos concentrations were 100 ug/ml.

Capacity of Asbestos to Elicit an Oxidative Burst in AM from BCG-Vaccinated Rabbits.

In this series of experiments we challenged the BCG-immune AM with chrysotile, crocidolite and amosite in the standard CL assay and compared their responses with PMA. The data are presented in Figures 4, 5 and 6. It can be noted that chrysotile produced a low CL burst compared to PMA.

However, crocidolite and amosite at 200 ug/ml produced CL responses significantly higher than PMA.

Discussion.

The data derived from this study clearly indicate that preincubations of normal AM for 3 hr with the asbestos preparations (100 or 200 ug/ml) impairs their ability to give a normal CL burst when subsequently challenged with PMA. Although this impairment may reflect toxicity of the asbestos, it is also possible that this incubation period caused the AM to produce an initial burst which impaired their abilitiy to respond to a second agonist. This principle can be demonstrated by using a double challenge with PMA spaced 3 hr apart.

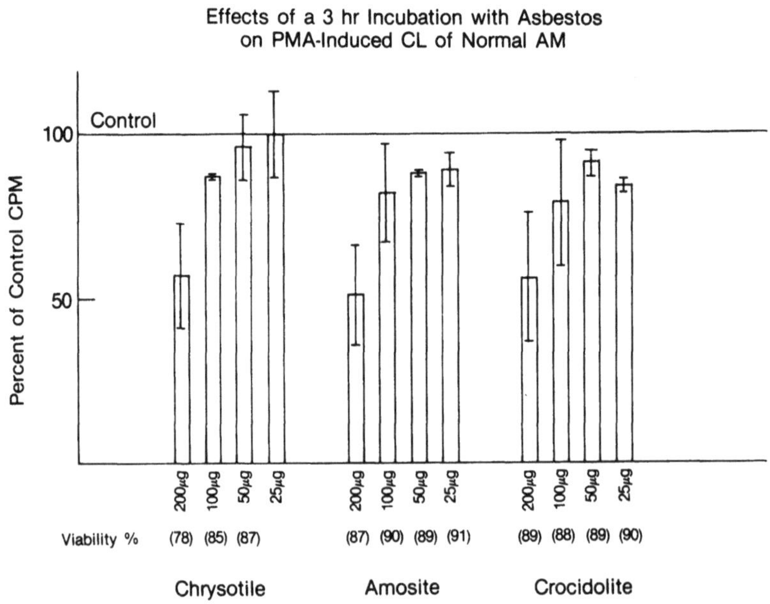

Figure 3. Effects of asbestos on a subsequent response of AM to PMA.

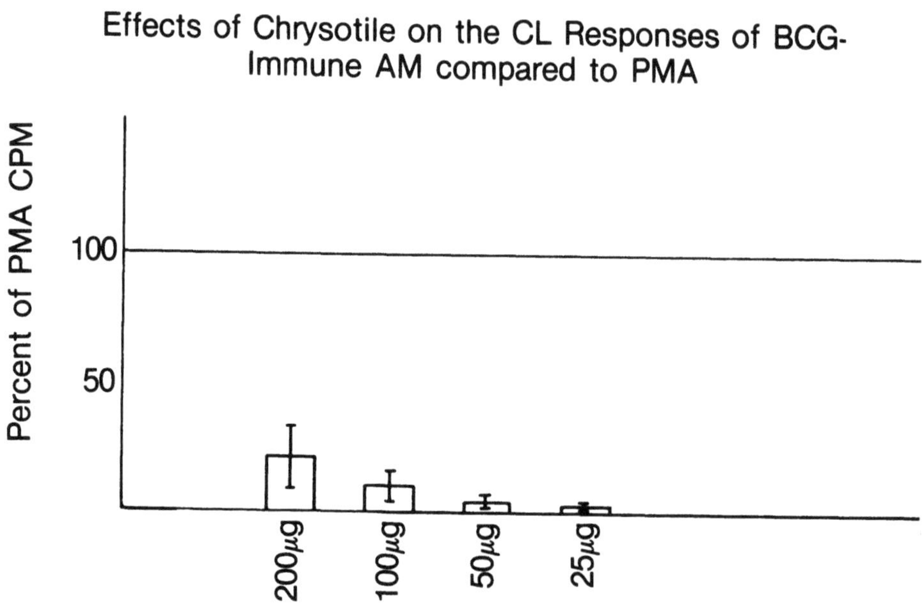

Figure 4. Comparative responses of chrysotile and PMA as eliciting agents for BCG-immune AM.

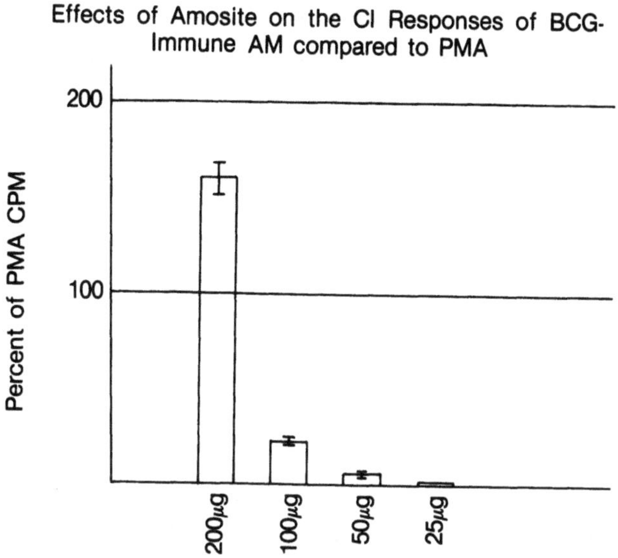

Figure 5. Comparative responses of amosite and PMA as eliciting agents for BCG-immune AM.

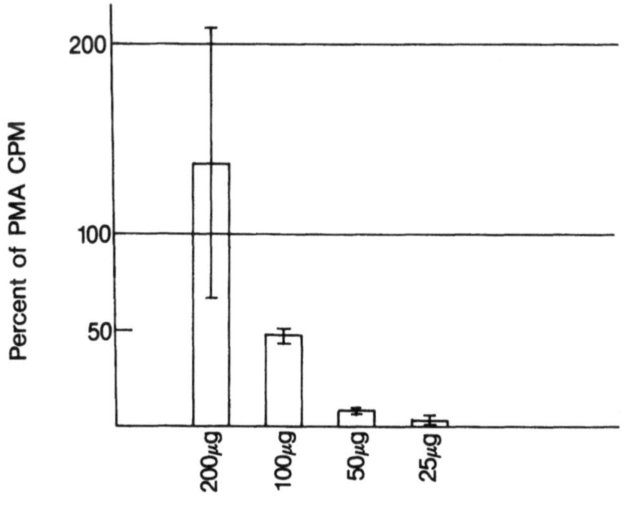

Figure 6. Comparative responses of crocidolite and PMA as eliciting agents for BCG-immune AM.

The viability studies demonstrated that serum protects AM from the primary toxicity of chrysotile and crocidolite. It is noteworthy that amosite produced no detectable loss in viability in these experiments. It can also be noted that crocidolite and amosite elicited a CL burst in BCG-immune AM which was substantially higher than that produced by PMA. In contrast, chrysotile produced a CL response which was only about 20% of the PMA response. We interpret this to be a reflection of the toxicity of chrysotile. The high responses produced by crocidolite and amosite could involve an increased production of superoxide anion because of the iron present in these amphiboles (Hatch, et al., 1980). If this is the case, it must be a synergistic interaction with the BCG-immune AM because when normal AM were challenged with asbestos the responses to crocidolite and amosite were about 10% of the responses elicited with BCG-immune AM even though their responses were equal or greater than PMA.

A remaining unanswered question pertains to the effect of asbestos on the priming of AM incubated with macrophage activation factor (MAF) or interferon gamma. Our preliminary experiments suggest that phagocytosed asbestos appears to prevent their priming by MAF (data not shown). However, this does not preclude the priming of newly arrived AM in the lung that are exposed to appropriate lymphokines but free of asbestos.

References

Chida K, Myrvik QN, Leake ES, Gordon MR, Wood PH and Ricardo, Jr. MJ (1987) Chemiluminescent responses of alveolar macrophages from normal and BCG-vaccinated rabbits as a function of age. Infect Immun 55:1476-1483

Donaldson K, Slight J and Bolton RE (1985) Release of superoxide anion and hydrogen peroxide by macrophages in response to asbestos. In Beck EB, Bignon J (eds) In vitro effects of mineral dusts. Springer-Verlag, Berlin Heidelberg New York, p 75-81

Fisher GL, McNeill KL, Mossman BT, Marsh J, McFarland AR and Hart RW (1985) Investigations into the mechansisms of asbestos toxicity. In Beck EB, Bignon J (eds) In vitro effects of mineral dusts. Springer-Verlag, Berlin Heidelberg New York, p 31-38

Goodglick LA and Kane AG (1986) Role of reactive oxygen metabolites in crocidolite asbestos toxicity to mouse macrophages. Cancer Res 46:5558-5566

Hatch GE, Gardner DE and Menzel DB (1980) Stimulation of oxidant production in alveolar macrophages by pollutant and latex particles. Environ Res 23:121-136

Kang K, Bice D, D'Amato R, Ziskind M and Salvaggio JE (1979) Effects of asbestos and beryllium on release of alveolar macrophage enzymes. Arch Environ Health 34:133-140

Kaw JL and Zaidi SH (1975) In vitro studies on the cyto-toxic action of different varieties of asbestos dust on macrophages. Acta Pharmacol Toxicol 36:225-235

McGee MP and Myrvik, QN (1981) Collection of alveolar macrophages from rabbit lungs. In Herscowitz, HB (eds) Manual of macrophage methodology. Marcel Dekker, Inc., New York, p 17-22

McLemore T, Corson M, Mace M, Arnott M, Jenkins T, Snodgrass D, Martin R, Wray N and Brinkley BR (1979) Phagocytosis of asbestos fibers by human pulmonary alveolar macrophages. Cancer Letters 6:183-192

Miller K, and Harington JS (1972) Some biochemical effects of asbestos on macrophages. Brit J Exptl Pathol 53:397-405

Mossman BT and Marsh JP (1985) Mechanisms of toxic injury by asbestos fibers: Role of oxygen-free radicals. In Beck EG, Bignon J (eds) In vitro effects of mineral dusts. Springer-Verlag, Berlin Heidelberg New York, p 66-74

Myrvik QN, Knox EA, Gordon Mark, and Shirley PS (1985) Effects of asbestos on the random migration of rabbit alveolar macrophages. Environ Health Perspect 60:387-393

Shatos MA, Doherty JM, Marsh JP and Mossman BT (1987) Prevention of asbestos-induced death in rat lung fibroblasts and alveolar macrophages by scavengers of active oxygen species. Environ Research 44:103-116

Sirois P, Rola-Pleszcynski M and Begin R (1980) Phospholipase A activity and prostaglandin release from alveolar macrophages exposed to asbestos. Prostaglandins Med 5:31-37

PRODUCTION OF RADICALS ARISING FROM SURFACE ACTIVITY OF FIBROUS MINERALS

R. Zalma, J. Guignard and H. Pezerat
Laboratoire de Réactivité de Surface et Structure
Université P. et M. Curie, 4 Place Jussieu
75252 - Paris Cédex 05, France.

M.C. Jaurand
Institut National de la Santé et de la Recherche Médicale
Unité 139, Henri Mondor
94010 - Créteil Cédex, France

INTRODUCTION

While it is true that fiber morphology increases,amplifies, the carcinogenic activity of an insoluble inorganic particle, the mechanism of genotoxicity will be, very probably, the same as the one observed with organic materials, that is an attack of the genome by electrophilic entities.

Then, for poorly soluble inorganic fibrous materials, the concerned activity will be a redox surface activity and we can make the hypothesis that the electrophilic entities will be oxygen active species (by reference to the mechanisms of oxidative stress), formed on the surface of the fibers.

So, we have explored the formation of oxygen active species from the interaction of reducing surface sites and oxygen dissolved in an aqueous aerated medium, according to the following reaction

$$\text{Reducing surface sites} + O_2 \longrightarrow \left\{\begin{array}{l} OH^{\cdot} \text{ or equivalent} \\ \text{oxidizing species} \end{array}\right.$$

In the samples under study, the reducing surface sites (donor of one electron) can be linked to Fe^{2+}, or may be, on the MMMF surface, to some dangling bonds, such as $\geq Si^{\cdot}$ or other radical sites.

If the electrophilic entities are the $OH^{\cdot}$ radicals, then oxygen is reduced following the classical scheme with the transfer of one electron from the surface sites in each step :

$$O_2 \xrightarrow{+e^-} O_2^{\cdot-}$$

$$O_2^{\cdot-} \xrightarrow{+e^- + 2H^+} H_2O_2$$

NATO ASI Series, Vol. H30
Effects of Mineral Dusts on Cells
Edited by B. T. Mossman and R. O. Bégin

$$H_2O_2 \xrightarrow{+e^-} OH^{\cdot} + OH^-$$

With the Fe^{2+} rich materials the two last steps can be replaced by :

$$2\,Fe^{2+} + O_2^{\bar{\cdot}} \longrightarrow Fe^{3+}\text{-}OH + \begin{cases} Fe^{3+} - O^- \\ \quad\text{or} \\ Fe^{4+} = O \end{cases}$$

these latter entities having approximatively the same oxidizing power as the $OH^{\cdot}$ radicals.

The oxidizing entities are detected by radical reaction with the formate anion :

$$HCO_2^{\bar{\cdot}} \xrightarrow{\text{oxidizing species}} CO_2^{\bar{\cdot}}$$

Hence, our test allows us to classify, for the fibers under study, the oxidizing activity as the ability to generate electrophilic entities.

METHODS

To detect the formation of radical species ($OH^{\cdot}$ or $R^{\cdot}$ as $CO_2^{\bar{\cdot}}$) we use a spin trap agent, the 5-5'-dimethyl-1-pyrroline-N-oxide (DMPO), which, in aqueous medium, gives radical adducts such as ($DMPO, CO_2^{\bar{\cdot}}$) detected and quantitatively measured by EPR spectroscopy (Kirino, 1980). The intensity of the EPR signal allows us to know the quantity of $CO_2^{\bar{\cdot}}$ generated, and thus the importance of the electrophilic entities present in the medium or on the fiber surface.

1 ml of 1M sodium formate solution in aqueous phosphate buffer (1M in phosphate, pH = 7.4) and 1 ml of 100mM DMPO in distilled water are added to 45 mg of the solid sample. Aliquots of the suspension are withdrawn at 30 and 60 min., without modification of the relative concentrations, and then filtered through 0.65 µm porosity filters. Controls contained no solid particles.

MATERIALS

Asbestos samples

Rhodesian (A) and Canadian (B) chrysotile, crocidolite and amosite were obtained from the "Union Internationale contre le Cancer (UICC)".

Mineral contaminants of chrysotile

Magnetite (commercial sample from BDH) and nemalite (from Asbestos, Québec) are classical contaminants of chrysotile, but we have verified the absence of nemalite in Rhodesian chrysotile UICC. Nemalite is a fibrous brucite $Mg(OH)_2$ where about 8 % of Mg^{2+} is substituted by Fe^{2+}.

Other mineral fibers

Attapulgite from Sénégal is a fibrous clay. It does not contain Fe^{2+}.

Erionite from Oregon is a fibrous zeolite.

Wollastonite is a commercial sample (Expandite limited). It contains 0.57 % of Fe^{2+} expressed as FeO).

Man-made minerals fibers (MMMF)

Rock Wool, glass fibers and Slag Wool are commercial samples.

JM 104, a commercial sample of fine fibers, (Johns-Manville "Tempstran") is a borosilicate which contains only 0.1 % of iron (measured as Fe_2O_3, but probably present as Fe^{2+}).

DMPO is from Sigma. It was purified by filtering through activated charcoal (Buettner and Oberley, 1978).

RESULTS

In table 1, 2, 3, 4 we present the results obtained with the above samples. Seven categories of oxidizing activity are designated ranging from "non active" to "out of categories". The signal intensity (within brackets) were measured after 30 min. of incubation with DMPO and formate. The materials having been kept in contact with air, surface sites were oxidized (therefore passivated).

Table 1. Activities of asbestos

Materials	Activities
Chrysotile UICC (A) (Rhodesian) Crocidolite UICC not reground Amosite UICC not reground	non active
Chrysotile UICC (B) Canadian without grinding	extremely active (1200)
Chrysotile UICC (B) Canadian, grinding 3h30	non active
Crocidolite UICC grinding 1 min.	very active (640)
Amosite UICC grinding 1 min.	very active (850)
Control	(10)

Table 2. Activity of contaminants of chrysotile

Materials	Activities
Nemalite grinding 1 min.	out of categories (2400)
Nemalite grinding 1 min. Without phosphate buffer	moderatly active (160)
Magnetite commercial sample	non active

Table 3. Activities of natural fibers (without grinding) other than asbestos

Materials	Activities
Attapulgite	non active
Erionite	non active (45)
Erionite (exchange with $FeCL_2$)	very active (800)
Wollastonite (commercial sample)	little active (135)

Table 4. Activities of some MMMF after grinding for 5 min.

Materials	Fe^{2+}(%FeO)	Activities
Rock Wool	5.2 - 6.5	very active (600-900)
Fine Fibers JM 104	0.07	very active (400-700)
Glass Fibers	0.2 - 0.07	active (340-540)
Slag Wool	0.3	little active (200)

Table 5. Activities of Rock Wool : effect of grinding

Material		$(DMPO,CO_2^-)$ signal intensity
Rock Wool	without grinding	110
	grinding 1 min.	310
	grinding 2 min.	560
	mechanical grinding 3 h	990

Crocidolite and amosite - table 1 - become very active after a moderate grinding just before the test. This activation process implies only the emergence of fresh surfaces (not oxidized).

Canadian chrysotile UICC (B) without grinding was active. After a long grinding this sample became inactive.

Nemalite, after grinding in order to increase the surface area, was extremely active ("out of categories"), but this activity decreases if the test is realized without phosphate nuffer (table 2).

Materials with or without a poor content of Fe^{2+} were inactive or little active (table 3). The introduction of Fe^{2+} into the channels of erionite by ion exchange allows us to obtain a level of activity of 800.

The case of magnetite is discussed in the paper of D. Costa et al. (same volume).

In table 4, MMMF samples were ground for 5 min. There is no correlation between the percentage of FeO and activity, but the more active MMMF is the one which contains the greater quantities of Fe^{2+}.

Table 5 shows that the activity of Rock Wool increases with grinding. This results is different from the one obtained with chrysotile (table 1) where the activity decreases after a long period of grinding.

DISCUSSION

After air aging the reducing surface sites of the particles are oxidized and thus passivated. The activation process consists of discovering new reducing sites either by grinding or by dissolution of the oxidized coating :

- by grinding. A short duration of grinding allows, in general, to bring out new active sites, but a prolonged grinding can completely oxidize and irreversibly passivate some soft materials. In the case of hard materials - like the Rock Wool fibers the activity can increase with grinding for longer periods.

- by scouring of the coating of the particles, a process facilitated by the presence of phosphate in the medium.

After this activation process, two possibilities exist for the Fe^{3+} formed in the same process as for the oxidizing entities, either this Fe^{3+} is extracted from the surface of the particle, or it participates in the formation of a new oxidized coating. The example of the first case is the nemalite, which is easily soluble in our aqueous phosphate medium to form a new solid phase, an insoluble magnesian-potassium phosphate. In this case all the Fe^{2+} of the particles appear on the surface in a short time and are able to react with oxygen. Then the material has a very strong oxidizing activity towards the formate. An example of the second possibility is silicates such as amphiboles (crocidolite, amosite) where the silicic skeleton can accept in the octahedral sheets the Fe^{3+} instead of the Fe^{2+}. In this case, in our experimental conditions it is impossible to activate fibers which are very little soluble and which have remained several years in contact with air. But in biological medium, Fe^{3+} can be extracted from the fibers by complexation or by reducing molecules, and thus a surface activity can appear.

It was not possible in our experimental conditions to activate the Rhodesian chrysotile, free of nemalite. May be another mechanism is implied here or a peculiar process of activation may be required.

In the absence of formate certain active fibers exhibit only a very weak formation of the radical adduct $(DMPO,OH)^{\cdot}$. Therefore, we are led to envisage, in the place of $OH^{\cdot}$ as oxidizing species, rather the formation of surface entities linked to iron, such as Fe^{3+}-O^{-} or Fe^{4+}=O, with about the same oxidizing power (Mimoun, 1986, Che and Tench, 1982, 1983).

Nemalite is the better example of a fiber very easy to solubilize, and hence with a very weak durability in biological medium (probably no more than a few days).However, it is known to be very carcinogenic in intraperitoneal medium (Pott, 1987). On the contrary, crocidolite is a good example of a fiber with a long durability, but its carcinogenicity in the Pott studies is approximatively equivalent to that one of nemalite. These examples allow us to consider that the durability parameter is not very important to appreciate the carcinogenic risk.

CONCLUSIONS

1) For the majority of the asbestos fibers, it is certain that their surface activity enables the formation of oxidizing entities in the intracellular medium after phagocytosis.

2) It remains to be proved, from both in-vitro and in-vivo experiments by utilizing non fibrous materials possessing the same type of activity, that the production of oxidizing entities may be at the origin of the carcinogenic action.

3) For certain asbestos varieties (Rhodesian chrsysotile) and erionite it is necessary to look for other mechanisms.

4) The fibrous morphology intervenes principally at the level of intracellular interactions.

The intracellular interaction between the genome and the oxidizing species is a function of the mean distance between the particle and the nucleus. This mean distance, in turn, taking into account the relative movements of the particle within the cell, is a function of the length of this particle. Longer will be the particle, greater will be the probability of contact or approach between the genome and the particle. The intervention of diameter in the data from intrapleural injections can be explained by the fact that the same dose in weight is implanted. So, the number of particles, correlated with the probability of a carcinogenic event, varies as the inverse of the mean diameter.

REFERENCES

Buettner G.R. and Oberley L.W. (1978). Consideration in the spin trapping of superoxide and hydroxyl radicals in aqueous systeme using DMPO. Biochem. Biophys. Research Com., 83, p. 69-74.

Che M. and Tench A.J. (1982). Characterization and reactivity of molecular oxygen species on oxide surface. Adv. Calysis, 31, 77-133.

Che M. and Tench A.J. (1983). same ref., 32, 1-48.

Kirino Y., Ohkuma T. and Kwan T. (1981). Spin trapping with 5,5-dimethylpy-

rroline-N-oxide in aqueous solution. Chem. Pharm. Bull., 29, p. 29-34.
Mimoun H. (1986). Comprehensive coordination chemistry. The synthesis reaction, properties and applications of coordination compounds. Wilkinson G., Ed., Pergamon Press, pp. 317-340.
Pott F., Ziem U., Reiffer F-J., Huth F., Ernst H. and Mohr U. (1987). Carcinogenicity studies on fibers, metal compounds and some other dusts in rats Exp. Path. (JENA), 33(3), pp. 129-152.

POTENTIAL ROLE OF SILICON-OXYGEN RADICALS IN ACUTE LUNG INJURY[1,2]

N.S. Dalal, Xianglin Shi, V. Vallyathan[3]

Department of Chemistry
West Virginia University
Morgantown, West Virginia 26506, USA

INTRODUCTION

This paper summarizes our recent investigations of the potential role of silicon-oxygen based free radical species in the biochemical mechanism of the lung injury caused by the inhalation of freshly fractured silica. The impetus for this study came from the following observations. Inhalation of crystalline silica is associated with the development of acute and chronic silicosis. Chronic silicosis is characterized by the development of sharply marked concentric fibrotic nodules in the lung. This occurs over a period of several decades with the development of progressive respiratory impairment. In contrast, acute silicosis is manifested by the filling of the alveoli with an amorphous lipoproteinaceous exudate within a short period after exposure, and rapid development of respiratory disability often leading to fatality. Most studies on the pathogenesis of silicosis have focused on the elucidation of cellular mechanisms of cell injury and the development of chronic silicosis. Because pulmonary responses to crystalline silica are distinctly different for the acute and chronic silicosis, we hypothesized that the acute response is associated with some unique properties of freshly formed silica particle surfaces, due to the formation of some reactive chemical species caused by the rupturing of the silicon-oxygen bonds. In occupations involving drilling, tunneling and sandblasting operations acute silicosis can be correlated with this unique surface reactivity of silica. The formation of reactive species was demonstrated, using electron spin resonance (ESR) spectroscopy (Dalal et

[1]This research has been supported by the Department of the Interior's Mineral Institute Program administered by the Bureau of Mine through the Generic Technology Center for Respirable Dust under grand G1135142.
[2]We would like to thank Myhanh Nguyen, Julia Martin, Daniel Davies, and Hyacinth Paul for technical assistance.
[3]From the Division of Respiratory Disease studies, National Institute for Occupational Safety and Health, Morgantown, West Virginia, USA.

NATO ASI Series, Vol. H30
Effects of Mineral Dusts on Cells
Edited by B. T. Mossman and R. O. Bégin

al., 1986; Fuibini et al., 1987; Hochstrasser and Antononi, 1972). Grinding of quartz crystals under ultra high vacuum (10^{-10} torr) leads to the formation of Si· and SiO· radicals and these radicals are quenched by exposure to air and other gases (Hochstrasser and Antonini, 1972). ESR signals have also been reported from micronized quartz samples that were evacuated at 413 K (Bolis et al., 1983). In order to find a more direct relationship of such radicals to acute silicosis, we made ESR measurements on quartz dusts freshly ground in ambient air and found that indeed grinding under ambient environment leads to the formation of the Si· and SiO· type of radicals which decay on storing the dusts in air or in biological buffers (Dalal et al., 1986; Vallyathan et al., 1988). The Si· and SiO· radical formation on grinding has also been reported by Fubini et al. (1987). In addition, we recently found that ·OH radicals are generated in aqueous suspensions of freshly ground quartz (Dalal et al., 1988; Shi et al., 1988; Vallyathan et al., 1988). We have subsequently carried out detailed cytotoxicity and biological assays on freshly made as well as aged quartz dusts. As summarized below, the results provide new insight into the biochemical mechanism of acute silicosis.

MATERIALS AND METHODS

ESR spectra were obtained at X-band (~9.7 GHz) using a Bruker ER 200D ESR spectrometer, employing a self-tracking NMR gaussmeter and a microwave frequency counter. Crystalline quartz with particle sizes of 0.2 to 2.5 mm was obtained from the Generic Respirable Dust Technology Center, Pennsylvania State University, University Park, PA. Particles in the range of smaller than 25 microns were produced in an agate mortar ball machine grinder in air for 30 minutes and sieved through a 20-micron mesh filter before use. All chemicals and biochemicals used were obtained from Sigma, St. Louis, MO.

Alveolar macrophages were harvested from male pathogen-free Sprague-Dawley rats by bronchopulmonary lavage using a calcium- and magnesium- free Hank's balanced salts solution as described previously (Myrvia and Evans 1967). Microscopic estimates of purity indicated that 90 to 95% of the lavaged cells were alveolar macrophages (Phillips, 1973).

Silica-induced activation of respiratory burst in alveolar macrophages was monitored by measuring O_2^- and H_2O_2 release. O_2^- release was monitored by measuring the superoxide-dependent reduction of cytochrome c spectrophotometrically at 550 nm using a Gilford Spectrophotometer (Model 300-N) (Sweeney et al., 1981). H_2O_2 release was monitored by measuring the change in fluorescence of scopoletin in the presence of horseradish peroxidase (Van Scott et al., 1984). Fluorescence was monitored at an excitation wavelength of 350 nm and an emission wavelength of 460 nm using a Perkin-Elmer Fluorescence Spectrophotometer (Model MPG-36) equipped with a stirrer and temperature controlled at 37°C.

The cytotoxic potential of fresh or aged silica was monitored by determining the effects of these dusts on cellular membrane integrity, i.e., hemolysis of red blood cells and release of cytosolic lactate dehydrogenase (LDH) from alveolar macrophages. Hemolytic activity of freshly ground or aged quartz particles was measured in a 2% suspension of sheep erythrocytes as the amount of hemoglobin released after incubation in the presence of quartz particles (Nolan et al, 1981). The LDH release was measured by incubating the alveolar macrophages (2×10^6) for different time intervals in a shaking water bath at 37°C in the presence of quartz particles. After incubation, cell suspensions were centrifuged and LDH released from the macrophages was estimated in the supernate (Wroblewski et al., 1955).

RESULTS

A. ESR Measurements on Freshly Ground Quartz

Figure 1 shows some typical ESR spectra of fresh quartz particles ground by an agate ball machine grinder in air. The measured g-values are $g_{\parallel} = 2.0017$ and $g_{\perp} = 2.0007$. Such ESR spectra from quartz are characteristic of silicon-based radicals (Si·, SiO·) (Hochstrasser and Antonini, 1972; Dalal et al., 1986; Fubini et al., 1987; Vallyathan et al., 1988). It was found that these silicon-based radicals decay with a half-life of about 30 hours. We stress, however, that the decay kinetics followed "first order" only very approximately and that about 20% of the radicals were still detectable even after four weeks of storage in air. We note that the spectral resolution and lineshape depend somewhat on the grinding time and procedure as well as on temperature. Further studies are under progress to understand the details of these observations.

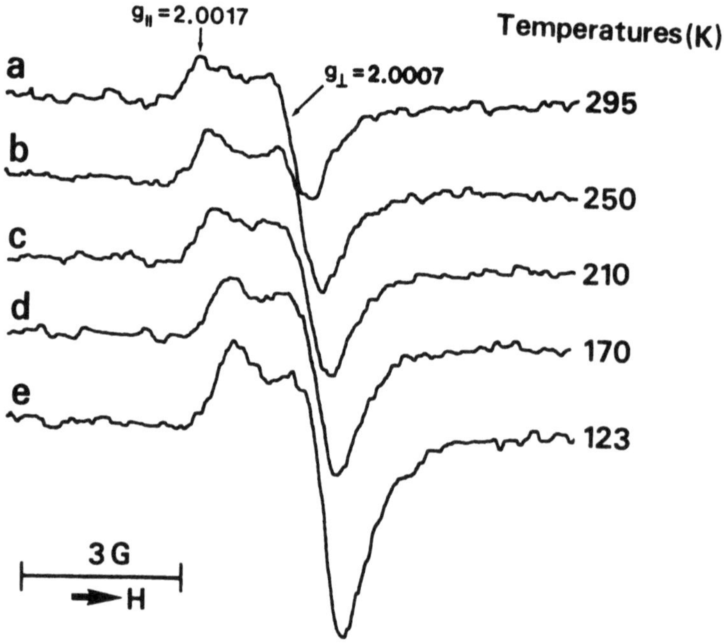

Figure 1. Typical ESR spectra from quartz dust freshly produced by an agate ball machine grinder in air. The spectra were recorded at different temperatures as indicated.

B. Activation of Alveolar Macrophages by Freshly Ground Quartz Particles

In order to investigate the possible role of silicon-based radicals and the oxygenated radicals generated upon reaction of freshly ground quartz particles with aqueous media (Dalal et al., 1988; Shi et al., 1988). we examined if freshly ground silica was a more potent stimulant of the respiratory burst in alveolar macrophages. The experiments consisted of measuring the O_2^- and H_2O_2 after *in vitro* exposure of alveolar macrophages to either fresh quartz particles or aged ones from the same stock. As shown in Figure 2, fresh quartz particle-induced O_2^- release decreased by 16% or 27% after storage in phosphate buffer for 24 or 96 hours, respectively. The fresh quartz particle-induced H_2O_2 release decreased by 65% after 24 hours of storage (Figure 3). From Figures 2 and 3, it may be noted that the half time for thedecrease in the ability of fresh quartz particles to activate alveolar macrophages was approximately 22 hours, which was comparable with the half-life for the silicon-based radicals and the fresh quartz dust's ability to generate oxygenated radicals (Dalal et al., 1988; Shi et al., 1988).

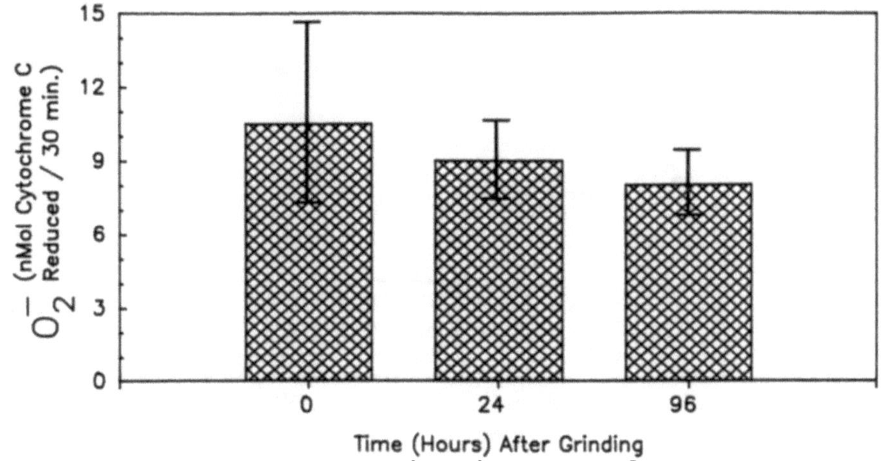

Figure 2. The effect of quartz particles on O_2^- release from alveolar macrophages.

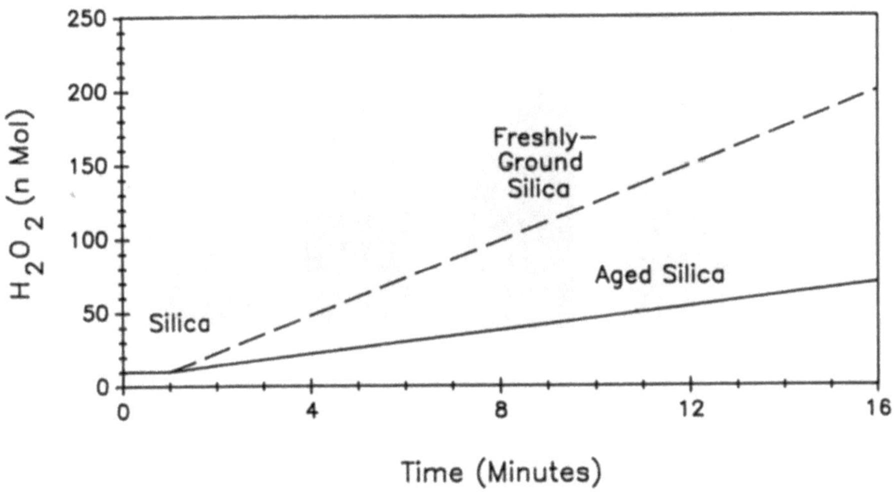

Figure 3. The effect of quartz particles on H_2O_2 release from alveolar macrophages.

C. Measurements of Cytotoxicity of Fresh Quartz Particles

Figure 4 shows that fresh quartz particles exhibit a greater disruptive effect on red blood cell membrane integrity than do particles stored in a phosphate buffer solution for several hours. Although there was a tendency for

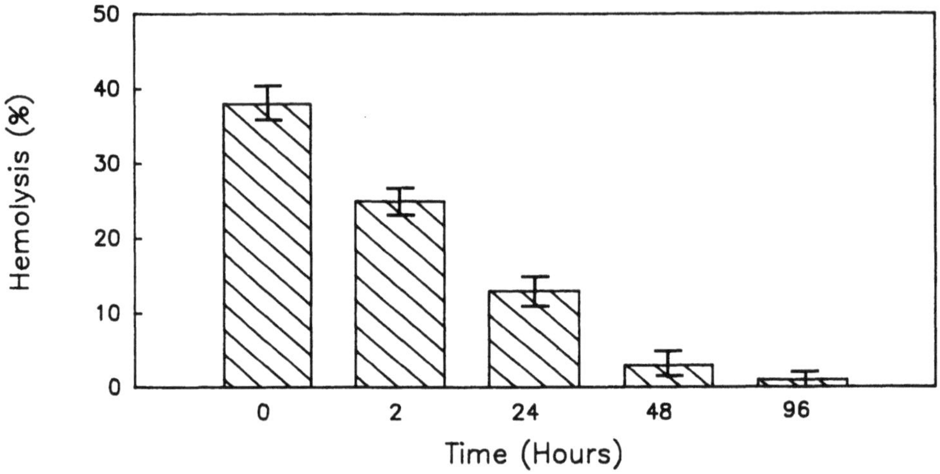

Figure 4. Hemolytic potential of quartz particles.

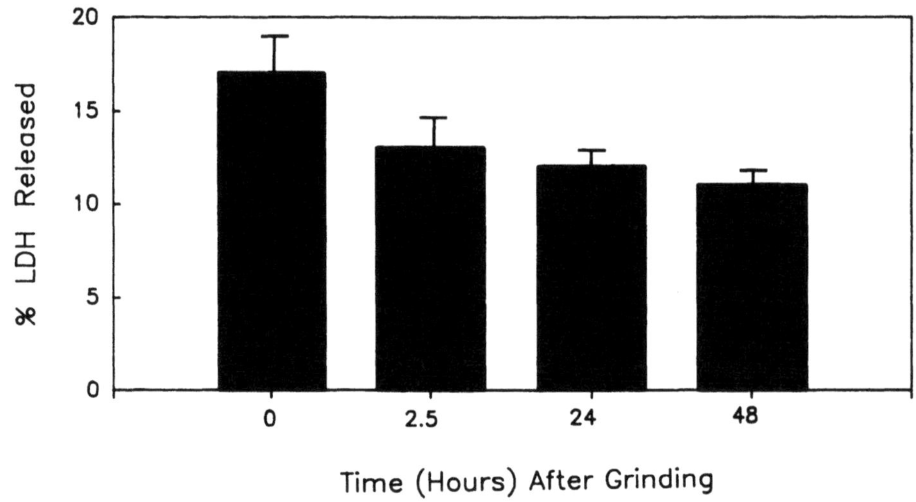

Figure 5. The effect of fresh ground quartz particles on release of cytosolic lactate dehydrogenase, LDH, from alveolar macrophages.

the fresh quartz particle-induced release of LDH to decline as silica aged, the differences were minimal (Figure 5) compared to hemolytic change.

DISCUSSION

Our results provide evidence that mechanical grinding of crystalline quartz under ambient laboratory conditions in air produces silicon-based radicals and that these radicals decay with time (Dalal et al., 1986). In our precedingpaper (Dalal et al., 1988) and elsewhere (Shi et al., 1988), we have reported that freshly ground quartz particles have the ability to generate oxygenated radicals and hypothesized that the silicon-based radicals and their associated oxygenated radicals might be involved in the initiation of the lipid peroxidation of the macrophage membrane, resulting into cell death (Dalal et al., 1988; Shi et al., 1988; Vallyathan et al., 1988). The results presented here show that fresh quartz particles are more biologically reactive than aged ones, i.e., fresh quartz particles induce a greater respiratory burst in alveolar macrophages. The generation of Si· and SiO· radicals and the formation of ·OH radicals in suspensions of fresh silica dusts might partially explain the enhanced reactivity of fresh quartz particles and the pathogenesis of acute silicosis which results from the inhalation of fresh quartz. We also find that the ESR spectra obtained from quartz particles ground by an agate ball machine grinder (Figure 1) are different from those ground by hand using agate mortar and pestle as reported elsewhere (Dalal et al., 1986; Vallyathan et al., 1988). In this connection we point out that the method of reducing the particle size for quartz has been found to have a marked effect on the toxicity. For example, the toxicity was found to increase when quartz was ground in a mill than when crushed in a press (Rassk and Schilling, 1980; Robock and Klosterkötter, 1973). Further investigation of the effect of different ways of grinding on the free radical formation and its relevance to fresh quartz dust's fibrogenic and cytotoxic potentials is currently underway.

REFERENCES

Bolis V, Fubini B, Venturello G (1983) Surface characterization of various silicas. A tentative correlation between the energy of adsorption sites and the different biological activities. *J Thermal Anal* **28**:249-258.

Dalal NS, Shi X, Vallyathan V (1988) Oxygenated radical formation by fresh quartz dust in a cell-free aqueous medium and its inihibition by scavengers. Paper in this volume.

Dalal NS, Suryan MM, Jafari B, Shi X, Vallyathan V, Green FHY (1986) Electron spin resonance detection of reactive free radicals in fresh coal dust and quartz dust and its implications to pneumoconiosis and silicosis. In: *Respirable Dust in the Mineral Industries: Health Effects, Characterization, and Control* (Frantz R.L., and Ramani R.V., eds.), pp. 25-29. American Conference of Governmental Industrial Hygienists (ACGIH) Publication, ISBN 0-936712-76-7.

Fubini B, Bolis V, Giamello E (1987) The surface chemistry of crushed quartz dust in relation to its pathogenicity. *Inorg Chim Acta* **138**:193-197.

Hochstrasser G, Antonini JF (1972) Surface states of pristine silica surfaces. *Surface Sci* **32**:644-664.

Surface Sci **32**:644-664.
Myrvia QN, Evans DG (1967) Metabolic and immunologic activities of alveolar macrophage. *Arch Biochem. Biophys.* **14**, 92-96.
Nolan RP, Langer AM, Harington JS, Oster G, Selikoff IJ (1981) Quartz hemolysis as related to its surface functionalities. *Environ Res* **26**:503-520.
Phillips HJ. (1973) Dye exclusion tests for cell viability. In:*Tissue Culture Methods and Applications*, eds., Kruse PR, Patterson MD, pp. 406-408.Academic Press, New York.
Raask E, Schilling CJ (1980) Research findings on the toxicity of quartz particles relevant to pulverized fuel ash. *Ann Occup Hyg* **23**:147-157.
Robock K, Klosterkötter W (1973) Specific toxicity of different silicon dioxide and silicate dust. *Staub* **33**:60-63.
Shi X, Dalal NS, Vallyathan V (1988) ESR evidence for the hydroxyl radical formation in aqueous suspension of quartz particles and its possible significance to lipid peroxidation in silicosis. *J Toxicol Environ Health* **25**:237-245.
Sweeney JD, Castranova V, Bowman L, Miles PR (1981) Factors which affect superoxide anion release from rat alveolar macrophages. *Exp Lung Res* **2**: 85-96.
Vallyathan V, Shi X, Dalal NS, Irr W, Castranova V (1988) Generation of free radicals from freshly fractured silica dust: potential role in acute silica-induced lung injury. *Am Rev Respir Dis* **138**:1213-1219.
Van Scott MR, Miles PR, Castranova V (1984) Direct measurement of hydrogen peroxide release from rat alveolar macrophages: artifactual effect of horseradish peroxidase. *Exp Lung Res* **6**:103-104.
Wroblewski F, LaDue JS (1955) Lactic dehydrogenase activity in blood. *Proc Soc Biol Med* **90**:210-214.

OXYGENATED RADICAL FORMATION BY FRESH QUARTZ DUST IN A CELL-FREE AQUEOUS MEDIUM AND ITS INHIBITION BY SCAVENGERS[1]

N.S. Dalal[2], Xianglin Shi, V. Vallyathan[3]

Department of Chemistry
West Virginia University
Morgantown, West Virginia 26506
USA

Introduction

This report summarizes our electron spin resonance (ESR) detection of oxygenated radical formation by freshly crushed quartz particles in a cell-free aqueous medium, and its inihibition by radical scavengers. This study was undertaken with to elucidate the primary biochemical mechanism involved in the development of silicosis (Farber, 1983; Reiser and Last, 1979; Silicosis and Silicate Disease Committee, 1988). In particular the mechanism by which the quartz particles exert their toxic action on cells and the process(es) by which these actions progress to fibrogenesis are still not well understood (Farber, 1983; Reiser and Last, 1979;). It is generally believed that the action of quartz particles on the cell membrane is the starting point of the silicotic process (Farber, 1983; Parazzi et al., 1968). Thus the surface characteristics of the quartz particles have been the subject of several recent studies (Bolis et al., 1983; Dalal et al., 1986; Fubini et al., 1987). Recently, We reported that mechanical crushing of quartz under normal atmosphere generates free radicals which decay with time, and that these free radicals are associated with a higher cytotoxicity of fresh quartz dust as compared to aged dust from the same stock (Dalal et al., 1986; Vallyathan et al., 1988). The formation of Si· and SiO·-type of radicals from quartz particles crushed under atmospheric conditions has also been reported by Fubini et al. (1987). In addition, Gulumian and van Wyk (1987) have reported that quartz particles react with hydrogen peroxide (H_2O_2) to generates hydroxyl radicals (·OH) and suggested that this process might contribute to

---

[1]This research has been supported by the Department of the Interior's Mineral Institute Program administered by the Bureau of Mine through the Generic Technology Center for Respirable Dust under grant G1135142.
[2]Author to whom correspondence should be addressed.
[3]From the Division of Respiratory Disease studies, National Institute for Occupational Safety and Health, Morgantown, West Virginia, USA.

NATO ASI Series, Vol. H30
Effects of Mineral Dusts on Cells
Edited by B. T. Mossman and R. O. Bégin

quartz's pathogenicity. However, these authors did not report the effects of metal chelators, so it was not clear as to whether the detected ·OH radicals simply resulted from the Fenton reaction involving exogenous metal ion contaminants. Earlier, Marasas and Harington (1960) reported that quartz particles could function as an oxidant in a number of *in vitro* oxidations, including the hydroxylations of proline and lysine. They postulated that silica dusts, on reacting with water, could release H_2O_2 which might have the potential to react with various biological constituents and thus cause tissue damage. In fact, Gabor and coworkers (Gabor and Anca, 1974; Gabor et al., 1975) suggested that the cytotoxicity of quartz particles might be associated with the generation of some factor or factors possessing the properties of a free radicals capable of promoting the peroxidative chain cleavage of polyunsaturated fatty acid moieties of the phospholipids in the cell membrane. More recently Weitzman and Graceffa (1984) reported that asbestos is able to catalyze the generation of ·OH radicals from H_2O_2. Their later work indicated that lipid peroxidation may be one of the mechanism by which asbestos produces tissue injury (Weitzman and Weitberg, 1985). This is significant because ·OH radicals are capable of causing peroxidation by abstracting hydrogen atoms from cell-membrane lipids and initiating lipid peroxidation in lysosomal membrane (Fong et al., 1973). In fact, the formation of oxygenated species in reactions involving quartz was suspected since Kolbanev et al. (1980) reported the detection of H_2O_2 in the reaction of fresh quartz particle with aqueous medium. The present work was thus undertaken to detect the anticipated formation of some oxygenated radical species by reactions of quartz in a cell-free media and its inhibition by radical scavengers. Because of the reactive nature of the oxygenated radicals and their short half-lives, ESR spin trapping methodology (Buettner, 1984) was employed in the present studies, as in the studies using asbestos (Weitmen and Graceffa, 1984) and quartz with H_2O_2 (Gulumian and van Wyk, 1987). The results show that ·OH and possibly O_2^- radicals are formed by fresh quartz dust in cell-free aqueous medium and their formation can be inhibited by the dust as it ages or by radical scavengers. Some preliminary results have been reported recently (Shi et al., 1988).

Materials and Methods

ESR spectra were obtained at X-band (~9.7 GHz) using a Bruker ER 200D ESR spectrometer, employing a self-tracking NMR gaussmeter and a microwave frequency counter. An ASPECT 2000 computer was used for data acquisition, analysis, and for ESR spectral simulations to obtain the splitting constants. Crystalline silica with particle sizes ranging from 0.2 to 2.5 mm was obtained from the Generic Respirable Dust Technology Center, Pennsylvania State University, University Park, PA. Particles in the range of smaller than 25 microns were produced by hand grinding in air, with an agate mortar and pestle because of the structural similarity of agate to that of quartz. Also, a rather mixed particle size rather than a specific range, was employed, to roughly approximate the random particle-size distribution in the mining atmosphere. 5,5-dimethyl-1-pyrroline-N-oxide (DMPO) was used as a spin trap for the ·OH radicals. The DMPO was purchased from Aldrich, Milwaukee, WI and used without further purification, because very weak or no ·OH radical signal was obtained from the purchased sample when used by itself. Superoxide dismutase (SOD) and catalase were obtained from Sigma, St. Louis, MO. Dimethyl sulfoxide (DMSO) was obtained from Fisher Scientific Company, Pittsburgh, PA. All experiments were performed at room temperature (22°C).

Results and Discussions

Figure 1 shows some typical results of the ESR spin trapping measurements. A 0.1 M aqueous solution of the spin trap DMPO with as received (unground) quartz particles did not give a detectable ESR spectrum (Figure 1a). When the quartz particles were crushed in a 0.1 M DMPO aqueous solution or when fresh quartz particles were mixed with 0.1 M DMPO aqueous solution, an ESR spectrum (g = 2.0059), consisting of a 1:2:2:1 quartet pattern with splittings of a_N = a_H = 14.9 G, was observed (Figure 1b). Based on the splitting constants and 1:2:2:1 lineshape this spectrum was assigned to the DMPO-OH adduct (Buettner, 1984). As supporting evidence for the ·OH radical formation, ESR spin tapping measurements were made in which 30% ethanol was added as a secondary trap (Buettner, 1984). It has been shown (Buettner, 1984; Shi et al., 1988) that in presence of ethanol, the intensity of the DMPO-OH spin-adduct signal decreases because ethanol scavenges some of the ·OH radicals to form theethanolyl radicals. The signals from the DMPO-trapped ethanolyl radicals are indicated by arrows in Figure 1b. The measured splitting constants, a_N = 15.8 G and a_H = 22.8 G, for these signals are indeed typical of those of the DMPO-$CHOHCH_3$ adducts (Buettner, 1984), thus supporting the ·OH radical formation in the quartz particle aqueous suspension, without any added H_2O_2.

An important observation was that the concentration of the ·OH radicals, as

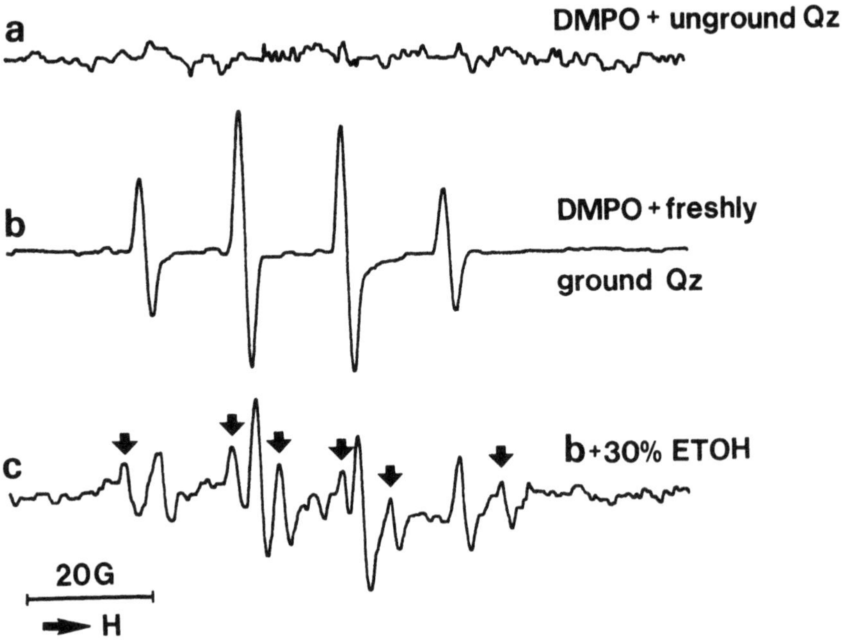

Figure 1. ESR spectra recorded 2 minutes after mixing a 100 mM DMPO aqueous solution with (a) unground quartz; (b) freshly ground quartz particles; (c) same as (b) but with 30% ethanol added. The arrows show the signals from the DMPO-ethanolyl radical adduct, attesting to the ·OH radical formation.

measured by the peak-to-peak height of ESR spin trapping adduct signal, which increased with time of grinding (Figure 2). This showed that the ·OH radical formation is related to some surface property of the fresh quartz particles (Kolbanev et al., 1980). The likely active sites may be the silicon-based radicals (Si· and SiO·) formed after grinding (Bolis et al., 1983; Dalal et al., 1986; Fubini et al., 1987; Vallyathan et al., 1988). Additional spin trapping experiments show that the ability of the fresh quartz dust to generate ·OH radicals decreases with the aging of the dust in air (Figure 3). It was found that, on an average, freshly ground quartz particles, when kept in air, lose their ·OH-generating capacity by about 50% in approximately a day.

To detect the possible formation of the O_2^- radicals in the fresh quartz particle suspension, SOD (50 ug/ml) and catalase (5000 units/ml) wereadded individually into the reaction medium. As noted in Figure 4b, SOD reduced the ·OH radical formation to about 65%, indicating that O_2^- radicals may be involved in the mechanism of ·OH radical formation (Weitzman and Graceffa,

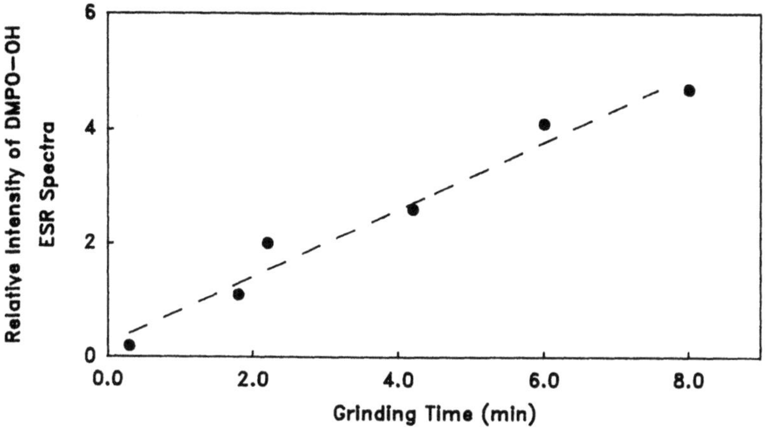

Figure 2. Dependence of the ESR intensity of the DMPO-OH adduct (i.e. ·OH radical production) on grinding time (i.e. surface) of quartz particles).

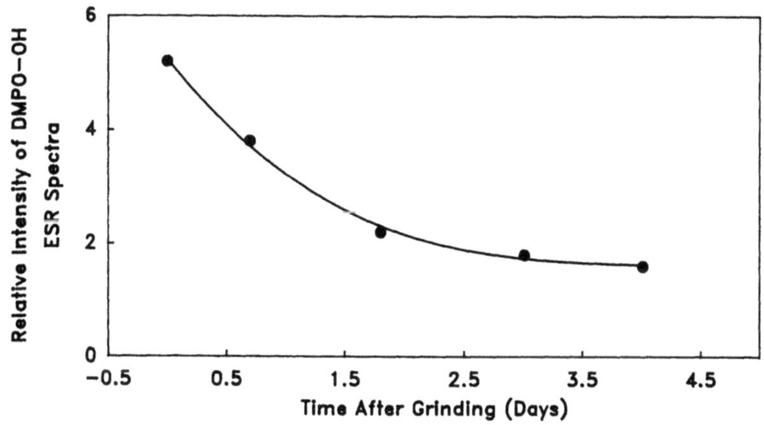

Figure 3. Dependence of the ESR intensity of the DMPO-OH on the aging of the quartz particles.

1984). Catalase, however, completely suppressed the ·OH formation (Figure 4c), indicating that H_2O_2 plays an important role in the ·OH radical formation as reported earlier (Vallyathan et al., 1988). The generation and detection of H_2O_2 was reported with wet analytical chemistry, the $KMnO_4$ reduction (Kolbanev et al., 1980). Using the same methodology, we did confirm the reducing activity of fresh quartz particle suspension with respect to $KMnO_4$, although the H_2O_2 yield was measured to be an order of a magnitude smaller for our sample. Further experiments showed that DMSO, commonly used radical

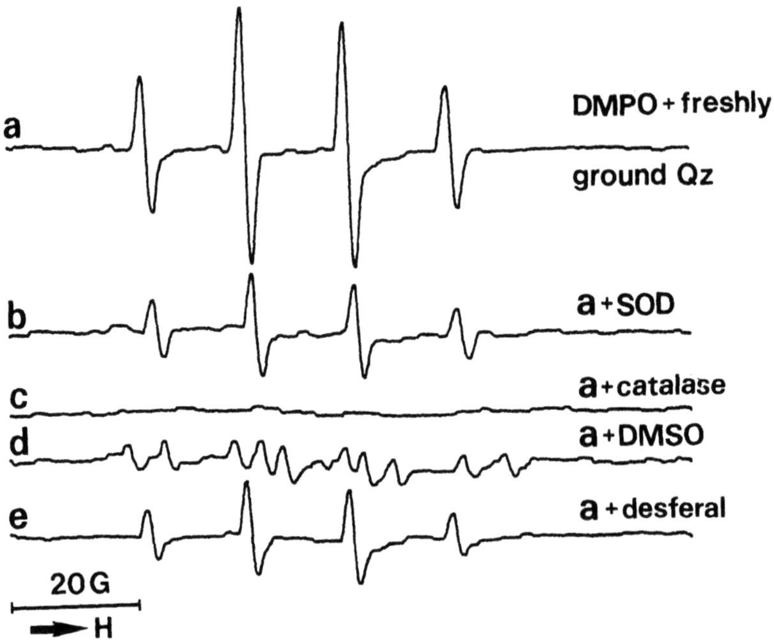

Figure 4. ESR spectra recorded 2 minutes after mixing 100 mM DMPO with (a) fresh quartz particles; (b) same as (a) but with SOD (50 ug/ml); (c) same as (a) but with catalase (5000 units/ml); (d) same as a but with DMSO (25%); (e) same as (a) but with desferal (1 mM).

scavenger (Britigan et al., 1986), suppresses the ·OH radical formation by more than 70% (Figure 4d). We noted that addition of desferal (1 mM) reduced the ·OH radical formation by more than 80%. Thus the Fenton reaction, Fe^{2+} + $H_2O_2 \longrightarrow Fe^{3+} + \cdot OH + OH^-$, may play some role in the ·OH radical formation (Morehouse and Mason, 1988). The Fe^{2+} may be present as a trace impurity or impregnated in the chemical structure of quartz dust.

The above results lead us to suggest the following model of the initial events in the reaction of the quartz dust with a cell membrane. The free radicals (Si· and SiO·) on the surface of silica dusts and their associated oxygenated radicals (·OH, O_2^-, and H_2O_2) are involved in the interaction of cell membrane with quartz dust. The results of this interaction would be the release of reactive oxygenated species (H_2O_2, O_2^-, ·OH, R·, and RO·). These reactive oxygenated species would further react with the cell membrane leading to additional release of these species and to lipid peroxidation (Fantone and Ward, 1982). As to the sites of the reactions between the cell membranes and the quartz dust, we noted that the reaction of quartz particles with H_2O produces silanol (SiOH) groups on the particle surface, as detected by infrared spectroscopy (Tsuchiya, 1982). These silanol groups could form hydrogen bonds with the nitrogen or oxygen atomic sites on the cell membrane. This is supported by the report of a weak (hydrogen) bond formation between a secondary amide (peptide) of proteins and silanol moieties (Summerton et al., 1977). Such hydrogen bonding could bring the silica surface and cell membrane close enough to provide a favorable environment for the initiation of lipid peroxidation by the silicon-based radicals and their associated oxygenated radicals.

Conclusions

The present ESR spin trapping experiments show that freshly generated, micron-size quartz particles have an ability to generate the ·OH and possibly the O_2^- radicals in a cell free media. Moreover, the ability of such particles to generate these oxygenated radicals decreases with aging of the particles. This result implies that fresh quartz particles should be more fibrogenic than the stale dust (Dalal et al., 1986 Shi et al., 1988; Vallyathan et al., 1988). Indeed, we have reported recently that fresh quartz dust activates a greater respiratory burst in alveolar macrophages than aged dust and that fresh dust exhibits a 50% greater ability for inducing lipid peroxidation (Vallyathan et al., 1987, Vallyathan et al., 1988). We hypothesize that silicon-based radicals and their associated oxygenated radicals may be involved in the initiation of the lipid peroxidation of the cell membrane which results in cell death (Vallyathan et al., 1988).

References

Bolis V, Fubini B, Venturello G (1983) Surface characterization of various silica: a tentative correlation between the energies of absorption sites and the different biological activities. *J Thermal Anal* 28:249–257.

Britigan BE, Rosen GM, Thompson BY, Chai Y, Cohen MS (1986) Do human neutrophils make hydroxyl radicals? *J Biol Chem* 261:17026–17032.

Buettner GR (1984) The spin trapping of superoxide and hydroxyl radicals. In: *Superoxide Dismutase*, vol. 2, ed., Oberly LW, pp. 63–81. CRC press, Boca Raton, Florida.

Dalal NS, Suryan MM, Jafari B, Shi X, Vallyathan V, Green FHY (1986) Electron spin resonance detection of reactive free radicals in fresh coal dust and quartz dust and its implications to pneumoconiosis and silicosis. In: *Respirable Dust in the Mineral Industries: Health Effects,*

Characterization, and Control (Frantz R.L., and Ramani R.V., eds.), P. 25-29. American Conference of Governmental Industrial Hygienists (ACGIH) Publication, ISBN 0-936712-76-7.
Fantone JC, Ward PA (1982) Role of oxygen-derived free radicals and metabolites in leukocyte-dependent inflammatory reactions. *Am Assoc Pathol* 107:397-418.
Farber JL (1983) How do mineral dusts cause lung injury? *Lad Invest* 49:379-390.
Fong KL, McCay PB, Poyer JL, Keel BB, Misra H (1973) Evidence that peroxidation of lysosomal membranes is initiated by hydroxyl free radicals produced during flavine enzyme activity. *J Biol Chem* 248:7792-7797.
Fubini B, Bolis V, Giamello E (1987) The surface chemistry of crushed quartz dust in relation to its pathogenicity. *Inorg Chim Acta* 138:193-197.
Gabor S, Anca Z (1974) Effect of silica on lipid peroxidation in the red cells. *Int Arch Arbeitsmed* 32:553-558.
Gabor S, Anca Z, Zugravu E (1975) *In vitro* action of quartz on alveolar macrophage lipid peroxides. *Arch Environ Health* 30:499-501.
Gulumian M, van Wyk A (1987) Free radical scavenging properties of polyvinyl -pyridine-N-oxide: a possible mechanism for its action in pneumoconiosis. *Med Lav* 78:124-128.
Kalbanev IV, Berestetskaya IV, Butyagin PU (1980) Mechanochemistry of quartz surface. *Kinet Katal* 21:1154-1158.
Marasas LW, Harington JS (1960) Some oxidative and hydroxylative actions of quartz: their possible relationship to the development of silicosis. *Nature* 188:1173-1174.
Morehouse KM, Mason RP (1988) The transition metal-mediated formation of the hydroxyl free radical during the reduction of molecular oxygen by ferredoxin-ferredoxin: $NADP^+$ oxidoreductase. *J Biol Chem* 263:1204-1211.
Parazzi E, Secchi GC, Pernis B, Vigliani E (1968) Cytotoxic action of silica dusts on macrophages *in vitro*. *Arch Environ Health* 17:850-859.
Reiser KM, Last JA (1979) Silicosis and fibrogenesis: fact and artifact. *Toxicology* 13:51-72.
Shi X, Dalal NS, Vallyathan V (1988) ESR evidence for the hydroxyl radical formation in aqueous suspension of quartz particles and its possible significance to lipid peroxidation in silicosis. *J Toxicol Environ Health* 25:237-245.
Silicosis and Silicate Disease Committee (1988) Diseases associated with exposure to silica and nonfibrous silicate minerals. *Arch Pathol Lab Med* 112:673-720.
Summerton J, Hoenig S, Butler C, Chvapil M (1977) The mechanism of hemolysis by silica and its bearing on silicosis. *Exp Mol Pathal* 26:113-128.
Tsuchiya I (1982) Infrared spectroscopic study of hydroxyl groups on silica surface. *J Phys Chem* 86:4107-4112.
Vallyathan V, Shi X, Dalal NS, Irr W (1987) Role of reactive oxygen radicals in silica toxicity. *4th International Congress on Oxygen Radicals (extended abstract)*, pp. 98-99, La Jolla, California.
Vallyathan V, Shi X, Dalal NS, Irr W, Castranova V (1988) Generation of free radicals from freshly fractured silica dust: potential role in acute silica-induced lung injury. *Am Rev Respir Dis* 138:1213-1219.
Weitzman SA, Graceffa P (1984) Asbestos catalyzes hydroxyl and superoxide radical generation. *Arch Biochem Biophys* 228:373-376.
Weitzmab SA, Weitberg AB (1985) Asbestos-catalyzed lipid peroxidation and its inhibition by desferrioxiamine. *Biochem J* 225:259-262.

INTERACTION OF QUARTZ WITH BOVINE ALVEOLAR MACROPHAGES: STUDIES ON CYTOTOXICITY, ELECTROPHORETIC MOBILITY AND SUBCELLULAR COMPONENTS

A. Seidel, E. Drosselmeyer, V. Hofheinz, W.G. Kreyling[a], F. Krombach[b], S. Patzold, G. Polzer, J. Schimmelpfeng, G. Wurtemberger[c]
Kernforschungszentrum Karlsruhe
Institut fur Genetik und Toxikologie
Postfach 3640
D-7500 Karlsruhe 1, F.R.G.

Introduction

Bovine alveolar macrophages have been used for testing the influence of air pollutants on cells (Fisher et al. 1983; Hooftman et al. 1988). They are also an ideal source of large quantities of subcellular components for studies on the mechanisms of toxic actions at the subcellular level (Seidel et al. 1988). These studies are of interest for example, for investigations of the influence of quartz particles on macrophage lysosomes, the damage to which appears responsible for the cytotoxicity of quartz (Allison et al. 1967; Allison et al. 1979).

Lysosomal enzyme release after incubation of lysosomal preparations from rat or rabbit liver with quartz particles has been observed (Dehnen et al. 1967, Weissman et al. 1972). Further information might be gained by working with isolated macrophage lysosomes which had internalized quartz particles, prior to isolation.

The aim of our study was to describe the effects of quartz on bovine alveolar macrophages and to compare the reactions of bovine alveolar macrophages with those of cells from other species, including man, in order to provide a sound basis for the extrapolation of results. A further main aim of our research is to quantify by biochemical methods the reaction of the lysosomal apparatus after quartz phagocytosis and to relate it to the cytotoxic effects.

Materials and Methods

Alveolar macrophages were obtained from bovine and rat lung by post mortem lavage, from dog (beagle), monkey (Macaca fascicularis) and man (one

[a]Gesellschaft fur Strahlenschutz, Institut fur Strahlenschutz, D-8042 Neuherberg, F.R.G.
[b]Klinikum Grobhadern, Institut fur chirurgische Forschung der Universitat Munchen, D-8000 Munchen 70, F.R.G.
[c]Universitatsklinikum Freiburg, Robert-Koch-Klinik, D-7800 Freiburg, F.R.G.

NATO ASI Series, Vol. H30
Effects of Mineral Dusts on Cells
Edited by B.T. Mossman and R.O. Bégin

chronic smoker, two persons with sarcoidosis, one of them under steroid treatment) by segmental bronchoalveolar lavage using routine procedures. In the case of bovine cells they were purified by Ficoll-Paque centrifugation. For all experiments the culture medium was RPMI 1640 with 100 U/ml penicillin and 100 μg/ml streptomycin without serum. The interspecies comparison was performed with monolayer cultures after removal of the nonadherent cells 1-2 hours after plating. The macrophages were incubated with 250 μg/ml DQ12 quartz (AMAD<5 μm) for 20 hours, thereafter the viability of the adherent cells and the release of enzymes were measured. In additional experiments the incubations were done in the presence of 333 μg/ml L-α-dipalmitoyl-lecithin (DPL). Subpopulations of bovine alveolar macrophages were separated by carrier-free electrophoresis (VaP 11, Bender-Hobein, Munich, FRG) in a 1 mM acetic acid/triethanolamine buffer containing 306 mM sucrose at $5^{o}C$ with a field strength of 1430 mV/cm and an electrical current of 185 mA (10^7 cells/ml, 15 ml/hour). For electrophoresis the cells were analyzed after 20 hours in suspension culture (Polypropylene) without or with 250 or 500 μg/ml native or DPL pretreated quartz (Sikron F600, Palas, Karlruhe, FRG; Geometric diameter 2-5 μm). In this case, DPL was removed from the quartz suspension before incubation of the particles with the cells according to Wallace et al. (1985). After electrophoresis cell number and viability were determined in each of the fractions.

In order to test the effects of quartz that can or cannot be phagocytosed, Sikron F600 quartz was immobilized by fixing it on cover-slips which had been coated with Eukitt (Acryl resin in xylene, Kindler, Freiburg, F.R.G.). The quartz powder was applied to the Eukitt surface shortly before it had completely been dried and non-fixed particles were removed by blowing air over the layer. This resulted in a surface occupation of 50-100 ug quartz per cm^2. In our monolayer cultures exposed to 500 μg/ml quartz in suspension the surface density after settling of the particles was 90 $μg/cm^2$. Whereas in this case the macrophages were plated before adding the quartz suspension, the cells were plated directly to the cover-slips with the immobilized quartz. The plating efficiency of bovine alveolar macrophages on the immobilized quartz surface is the same as on normal materials (>90% within 15 min). Viability and enzyme release were tested after 20 hours incubation, $O_2^{\cdot-}$-production after 60 or 90 minutes.

To analyse the influence of quartz on the intracellular distribution of lysosomal enzymes, bovine alveolar macrophages were incubated in monolayer cultures for 2 or 16 hours with native or DPL pretreated DQ12 quarts (250 µg/ml). The cells were scraped off, homogenized with 0.306 M sucrose containing 1 mM Na_2-EDTA (ethylenediaminetetraacetate) and the homogenate was centrifuged for 10 min at 3500 g_{av} (resulting in the nuclei pellet N), for 20 min at 16.000 g_{av} (yielding the mitochrondria-lysosome fraction ML) and for 45 min at 105.000 g_{av} (cytosol, S). The activity of N-Acetyl-β-glucosaminidase (NAc) was determined in these fractions as well as in the culture medium, the sum of all activities was taken as 100% and the content in each fraction expressed as a percentage thereof. In addition the release of this enzyme from a suspended N- or ML-fraction (in 0.306 M sucrose) was tested by recentrifugation of the fractions after 2 hours of incubation at 37°C and determination of the free enzyme content in the supernatant.

Viability was tested by trypan blue exclusion, N-Acetylglucosaminidase (NAc) according to Andersen et al. (1982), $O_2^{\cdot-}$-production according to Johnston (1981) by measuring the cytochrome c reduction. Opsonized Zymosan (Los Alamos Diagnostics, Los Alamos, USA) was used as a control stimulus.

Results and Discussion

In Fig. 1 the interspecies comparison of the effects of DQ12 quartz on alveolar macrophages is shown. Though differences in the response of the cells certainly exist, the general similarity of the reactions can be assumed, except for NAc in bovine macrophages and in one of the three humans, which was a chronic smoker. There was no difference between the results from the steroid treated and non-treated patient. The exceptionally high release of NAc by bovine cells cannot yet be explained, the release of lactate dehydrogenase was again more uniform between the species (Schimmelpfeng et al., unpublished). Preliminary results with chrysotile asbestos (250 µg/ml) have shown that the viability was reduced to values between 50 and 30% independent of the species. Although an exact statement must be based on the detailed comparison of the dose-effect-relationships, which is under way in our laboratory, we can conclude already that bovine alveolar macrophages can be used as a representative model for testing the effects of mineral dusts on macrophages. Concerning the protective effect of DPL, the data for bovine alveolar macrophages and rat cells are significantly different. A possible explanation might be differences in the activities of phospholipases. However, Wallace et al. (1985) obtained an almost 100% protection of rat alveolar macrophages by

DPL treatment. We cannot yet explain the discrepancies between their and our results.

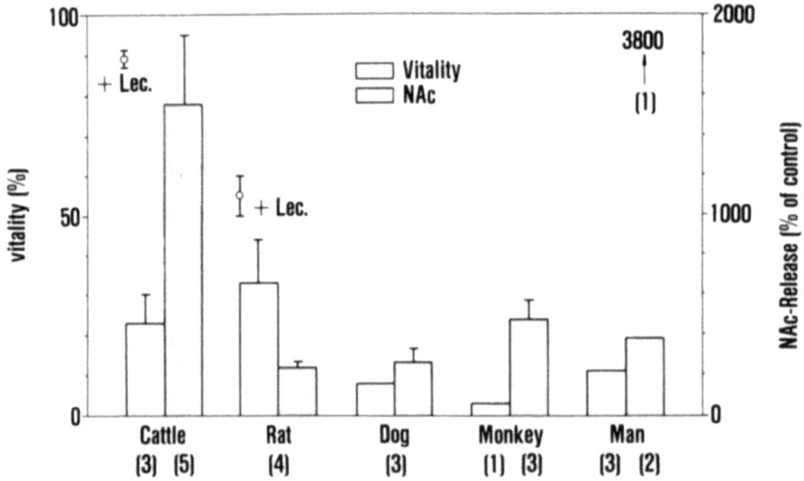

Fig.1: Vitality of adherent cells and release of N-Acetylglucosaminidase into the medium after 20 hours incubation of alveolar macrophages from different species with 250 µg/ml DQ12 quartz. Lec: Incubation in presence of Lecithin (DPL). Arithmetic means with S.E. (vertical bars), number of experiments in brackets. Note the exceptionally high enzyme release in one human experiment (smoker). Enzyme release is expressed as percentage of amount released by the respective untreated control cells.

Bovine alveolar macrophages can be separated by carrier-free electrophoresis into three or four subpopulations according to electronegativity of the cell surface (see control curves in Fig. 2). It can be seen from the three control curves in Fig. 1 that the relative contribution of the subpopulations varies considerably between different individuals, however, three or four peaks can always be distinguished. Incubation with Dikron F 600 quartz prior to electrophoretic separation leads to the disappearance of the two slowly migrating fractions in case of the 500 µg/ml concentration. Under these conditions DPL pretreatment also completely abrogates the effects of quartz (Fig. 2, lowest graphs). It should be mentioned that we could not detect differences between the subpopulations by light or electron microscopy, α-Naphthylacetate staining or Nitroblue tetrazolium reduction. However, the specific activity of lysosomal enzymes was higher in the most rapidly migrating fraction (Seidel et al., submitted to J. Leucocyte Biol.).

When quartz particles cannot be phagocytosed by bovine alveolar macrophages an inverse effect on vitality, enzyme release and O_2^{-}-production is observed as compared to the standard exposure with suspended quartz (Fig.

3): The influence on viability and enzyme release is reduced, whereas

Fig.2: Results after carrier-free electrophoresis of bovine alveolar macrophages after 20 hours of incubation with native or lecithin (DPL)-pretreated Sikron F 600 quartz. Values correspond to number of living cells per fraction as percentage of total number of all cells (sum of living and dead in all fractions). Anode left side. Data for 500 µg/ml quartz representative for three experiments, other data one experiment each (Graph taken from Schimmelpfeng et al. in: J. Aerosol. Med., in press).

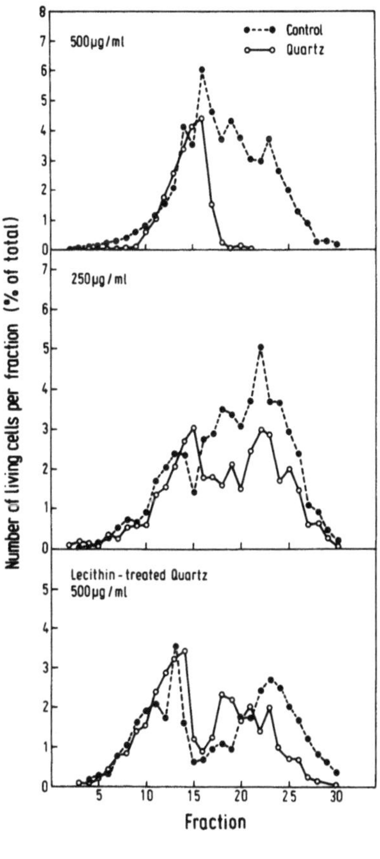

the $O_2^{\cdot}$-production is distinctly increased when quartz cannot be phagocytosed. Obviously, complete phagocytosis is important for the full development of the deleterious quartz effects whereas the higher $O_2^{\cdot}$-production can probably be interpreted as a result of "frustrated phagocytosis" where a comparatively large part of the cell surface is continuously in contact with the toxic particles.

Phagocytosis of DQ12 quartz is followed by a release of NAc into the culture medium which is paralleled by a decrease of the relative content in the cells, especially in the N- and ML-fraction (Fig. 4). This can clearly be observed after 16 hours incubation with quartz, but the change already becomes visible after two hours. It should be noted that DPL-pretreated particles had no such effect which also means that the enzyme redistribution is not an artifact due to damage of the quartz-filled lysosomes during homogenization. This would also have resulted in a high enzyme content in S. After a two hours of incubation of the N- or ML-fractions, the amount of lysosomal enzyme released is the same for fractions from untreated control cells as for such from quartz exposed cells (Fig. 5), indicating that most of the quartz-containing lysosomes were not grossly damaged during this period. The possibility of readsorption of released enzyme onto the quartz has been excluded by control experiments with free enzyme. From these results it cannot yet be concluded whether the redistribution of the lysosomal enzyme from the ML-fraction into the medium is the primary or a secondary event.

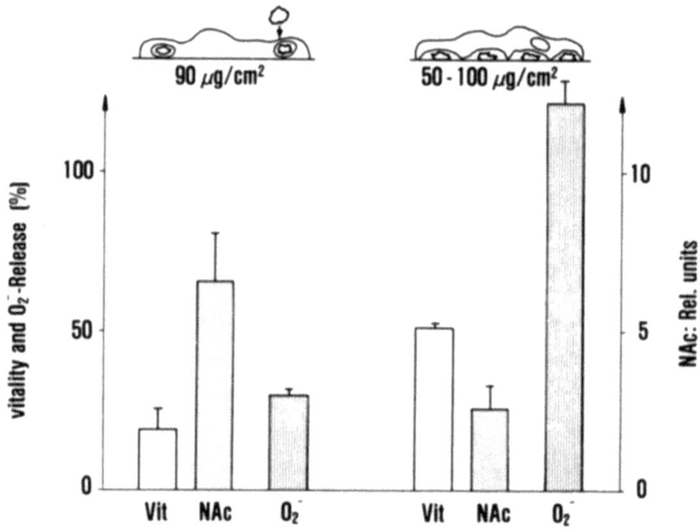

Fig.3: Influence of Sikron F 600 quartz on bovine alveolar macrophages. Left side: Standard monolayer culture with quartz in suspension. Right side: Quartz fixed on cover-slips (see Methods section). Data for vitality and N-Acetylglucosaminidase (NAc) release obtained after 20 hours of incubation. O_2^- production per 90 min, left side per 60 min incubation (data are expressed as pecentage of O_2^- produced after Zymosan stimulation of the same cells). Arithmetic means with S.E. (vertical bars), vitality data from three, others from four determinations.

It is surprising that we were not able to detect any lysosomal enzyme release, when isolated, quartz-filled lysosomes were incubated in vitro. It must be expected that lysosomes within cells which had been incubated with 250 µg/ml quartz for two hours, become damaged within the subsequent two hours and release their enzyme at least to some degree. One explanation for this failure might be that lysosomal damage by quartz particles requires additional intracellular conditions which are lacking after breaking up the cells and isolating the lysosomes.

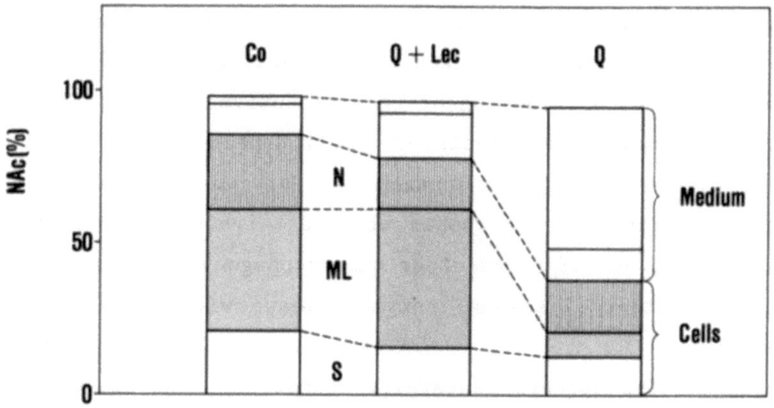

Fig.4: Relative distribution of N-Acetylglucosaminidase within subcellular fractions of bovine alveolar macrophages and in the culture medium after 16 hours incubation with 250 µg/ml native or lecithin (DPL) pretreated DQ12 quartz. Data expressed as percentage of sum of enzyme activity in cells plus medium, representative for four experiments. For explanation of abbreviations see Methods section.

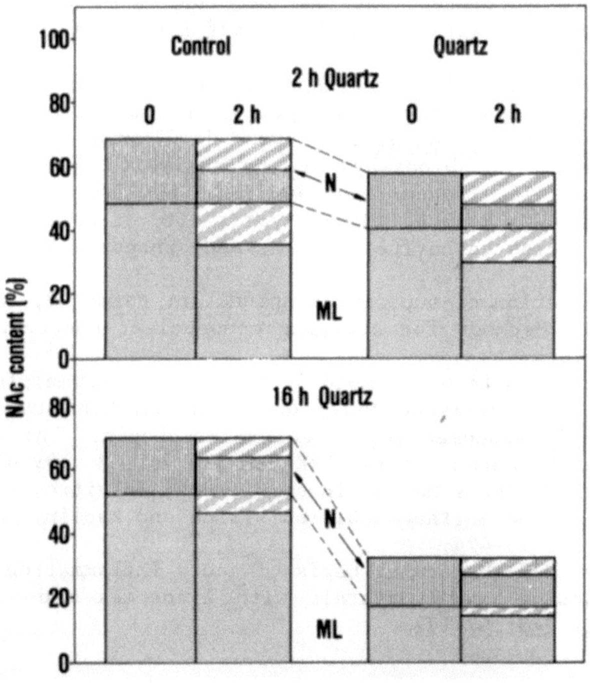

Fig.5: Release of N-Acetylglucosaminidase after two hours incubation of the N- or ML-fraction from bovine alveolar macrophages (control cells and those incubated two or 16 hours with 250 μg/ml DQ12 quartz). Left columns: centrifugation immediately after mixing of quartz with the fractions, right columns: after two hours incubation. Shaded areas in the two hour columns represent the fraction released. Further explanations see Fig. 4, representative for 2 experiments.

Summing up we can conclude from our results that bovine alveolar macrophages react to mineral dusts qualitatively similarly to those of several other species, including man. They consist of three to four subpopulations according to electrical surface charge, quartz has a distinct influence on the relative contribution of the respective subpopulations to the total cell population. The effects of quartz depend on whether it can be phagocytosed by bovine alveolar macrophages or not. According to cell fractionation experiments, the number of intact lysosomes decreases drastically between two and 16 hours after quartz phagocytosis, but it was not possible to demonstrate lysosomal enzyme release with isolated lysosomes from these cells.

Acknowledgement: The authors gratefully acknowledge the technical assistance of Mrs. E.Kruger, Mrs. U. Bergner and Mrs. R. Mauser.

References:

Allison AC, Morgan DML (1979) Effects of silica, asbestos and other particles on macrophage and neutrophil lysosomes. In: Dingle JT, Jacques PJ, Shaw IH (ed) Lysosomes in applied biology and therapeuics, North-Holland Publ. Comp. Amsterdam, p. 149.

Andersen OK, Stenvold SE, Volden G (1982) Optimalized assay conditions of 3T3 cell lysosomal hydrolases. Eur Rev Med Pharmacol Sci IV 265-272.

Dehen W, Fetzer J (1967) Zur Wirkung von Quarz auf Lysosomen in vitro. Naturwissenschaft 54:23.

Fisher GL, McNeill KL, Democko CJ (1983) Application of bovine macrophage bioassays in the analysis of toxic agents in complex environmental mixtures. In: Waters MD, Sandhu SS, Lewtas J, Claxton L, Chernoff N, Nesnow S (ed) Short-term bioassays in the analysis of complex environmental mixtures III, Plenum Press New York, p. 257.

Hooftman RN, Arkesteyn CWM, Roza P (1988) Cytotoxicity of some types of welding fume particles to bovine alveolar macrophages. Ann Occup Hyg 32:95-102.

Johnston RB (1981) Secretion of superoxide anion. In: Adams DO, Edelson PJ, Koren HS (ed) Methods for studying mononuclear phagocytes, Academic Press New York, p. 489.

Seidel A, Drosselmeyer E, Hotz G, Patzold S, Polzer G, Schimmelpfeng J, Wilczek W (1988) Biochemische Untersuchungen zum Einfluß inhalierbarer Partikel auf Lysosomen von Alveolarmakrophagen. In: Report of Karlsruhe Nuclear Research Centre, KfK-PEF 35, Vol. 2, 399-410.

Wallace WE, Vallyathan V, Keane MJ, Robinson V (1985) In vitro biological toxicity of native and surface-modified silica and kaolin. J Toxicol Environm Health 16:415-424.

Weissmann G, Rita GA (1972) Molecular basis of gouty inflammation: Interaction of monosodium urate crystals with lysosomes and liposomes. Natur-New Biology 240:167-172.

Degradation of an Extracellular Matrix Component by Bronchoalveolar Leukocytes *In Vitro*: Modulation by Mineral Dust

Geraldine M. Brown, Kenneth Donaldson and Joan Slight
Institute of Occupational Medicine
8 Roxburgh Place, Edinburgh EH8 9SU, Scotland.

Introduction

Occupational exposure to harmful mineral dusts is frequently associated with development of chronic fibrotic lung disease e.g. silicosis, asbestosis or coalworkers' pneumoconiosis (Morgan and Seaton, 1984). Alveolitis is a characteristic feature of such diseases (Begin *et al.* 1986; Voisin *et al.* 1985) and although the aetiology of the fibrosis is not yet fully established, such an accumulation of inflammatory leukocytes is firmly implicated in the pathogenesis of the occupational lung diseases.

Proteolytic enzymes are released from inflammatory leukocytes at sites of inflammation and some "bystander injury" to the surrounding tissue is an inevitable consequence (Ryan and Majno, 1977). Where the inflammation is short-lived, acute injury is quickly resolved and normal tissue architecture is restored. Persistent lung inflammation, however, leads to severe tissue derangement resulting in fibrosis; the mechanisms involved in the pathogenesis of fibrotic lung disease have recently been reviewed by Hunninghake *et al.* (1984). In occupational lung disease there are two likely causes of persistent inflammation, not mutually exclusive, and these are:- (a) the inherent harmfulness of the dust and (b) persistent exposure. This study has evaluated the role of inherent harmfulness in causing persistent inflammation by assessing the inflammation developing in response to a single intratracheal dose of the fibrogenic mineral dust quartz, compared with TiO_2, a mineral which is virtually inert in experimental studies *in vivo* (Donaldson *et al.* 1987) and *in vitro* (Donaldson and Brown, 1988) and is not associated with pathology in occupationally-exposed groups. We compared these responses with those elicited by two other model inflammogens which reflect the normal scope of the lung to mount an inflammatory response without causing irreversible tissue damage. *C. parvum* was chosen to represent a normal bacterial challenge in the lung and since mineral dusts can activate complement, zymosan was also used as it is a very efficient activator of complement. We have investigated one of the aspects of persistent exposure by assessing the effect of a second dust challenge (*in vitro*) on the functional status of inflammatory bronchoalveolar leukocytes obtained following *in vivo* dosing with dust. We measured two key functional

NATO ASI Series, Vol. H30
Effects of Mineral Dusts on Cells
Edited by B.T. Mossman and R.O. Bégin

parameters of the bronchoalveolar leukocytes relevant to proteolytic injury and its consequences by (a) assessing the ability of bronchoalveolar leukocytes to degrade fibronectin, a connective tissue component of major importance in maintaining tissue integrity and (b) measuring plasminogen activator production. Plasminogen activator is a leukocyte protease that is frequently cited as being central to the inflammatory response (Robinson, 1988).

Materials and Methods

Animal injection and lavage

Rats used were syngeneic, SPF, males of the PVG strain.

Zymosan and both mineral dusts (DQ_{12} standard quartz; titanium dioxide (rutile) Tioxide Ltd., Stockton were injected directly into the trachea as 1mg in 0.5ml saline, *C. parvum* as 1.4mg in 0.2ml saline; animals were killed at selected time points thereafter. The lavage procedure was performed as previously described (Donaldson *et al.* 1988).

Fibronectin proteolysis

The ability of leukocytes or leukocyte supernatants to degrade fibronectin was assessed by measuring breakdown of a solid phase [^{125}I] fibronectin matrix as previously described (Brown and Donaldson, 1988). Proteolytic activity was measured as counts per minute of degraded [^{125}I] fibronectin released into the supernatant medium over a 4 hour period.

Plasminogen activator assay

Plasminogen activator activity was measured as previously described (Donaldson *et al.* 1987) by measuring plasminogen-dependent breakdown of [^{125}I] fibrin. Results are shown as counts per minute of [^{125}I] fibrin released into the supernatant medium in a 4 hour period.

Statistical analyses

Differences between means of treatment groups were tested using Student's 't' test.

Results and Discussion

Alveolitis is a characteristic feature of chronic lung disease associated with exposure to harmful mineral dusts but there is little information regarding the role of the inflammatory leukocytes in the pathogenesis of the disease process. We have addressed this question by investigating inflammation developing in the lung in response to a single dose of the fibrogenic dust, quartz, compared with other non-fibrogenic stimuli. We have shown here that the acute inflammatory response to quartz was less than that caused by a bacterial preparation (*C. parvum*); it was, however, greater than that produced by TiO_2 or the particulate activator of complement, zymosan, in terms of total numbers of leukocytes recruited (Figure 1) and the percentage of those cells that were neutrophils (Figure 2).

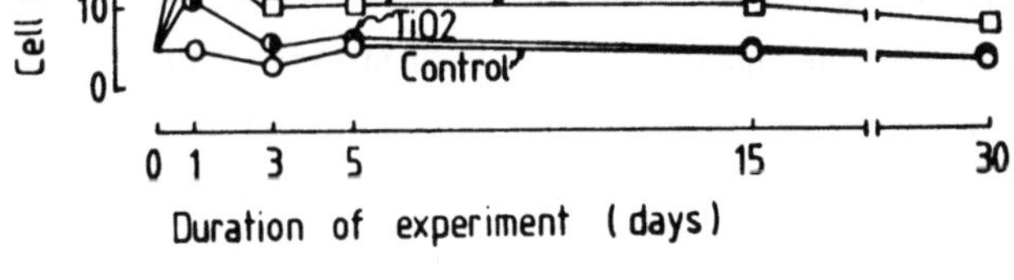

Fig. 1. Total number of leukocytes retrieved in bronchoalveolar lavage. Results represent the mean of 3 rats per treatment group.

The neutrophil is the classical leukocyte of acute inflammation and the neutrophil response to *C. parvum*, TiO_2 and zymosan was a typical acute one with the percentage of neutrophils falling off rapidly after a peak at 1 day and returning to around control levels by 5 days.

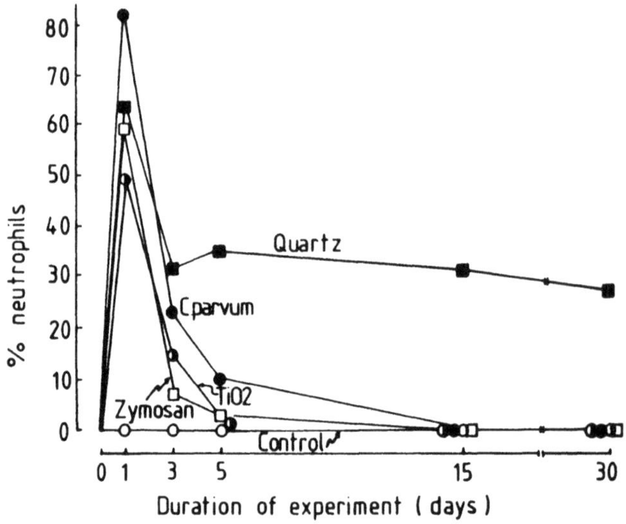

Fig. 2. Percentage of neutrophils in the bronchoalveolar lavage cell population. Results represent the mean of 3 rats per treatment group.

By 3 days, the difference between quartz and the other inflammogens was apparent – the percentage of neutrophils in the quartz lavage declined less than in the others and thereafter persisted at around 30%. This result is in agreement with the finding of approximately 6–8 fold more neutrophils in bronchoalveolar lavage in silicosis, asbestosis and coalworkers' pneumoconiosis compared with the normal population (Begin *et al.* 1986; Voisin *et al.* 1985 and of the neutrophil alveolitis found in rats inhaling quartz (Donaldson *et al.* 1988). The macrophage alveolitis which persisted up to 30 days in the *C. parvum* and zymosan animals was not present in the TiO_2 rats. This suggests that the leukocyte influx in TiO_2-exposed lungs (which was less than any other group) may occur largely in response to chemotactic stimuli released during phagocytosis of the TiO_2 particles by macrophages. Subsequent to phagocytosis, there is no further reaction between the macrophage and the TiO_2 particle and thus no means of prolonging the inflammation, so no pathological change results.

The proteolytic activity of the inflammatory leukocytes reflected the severity of the inflammation, peaking at 1 day with all of the inflammogens (except quartz which was maximal at 3 days) and declining rapidly thereafter in all but the quartz-elicited group (Figure 3).

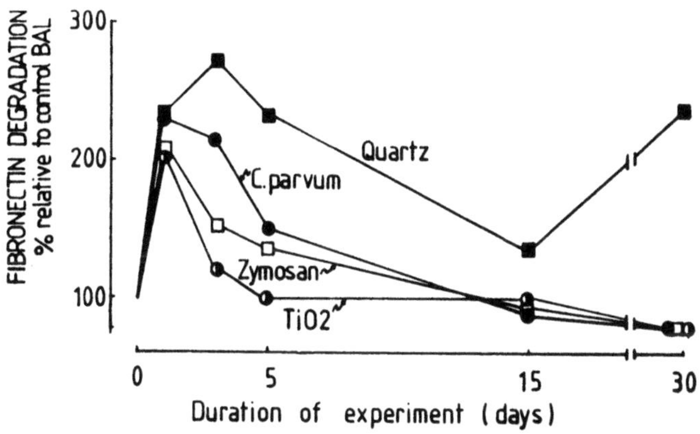

Fig. 3. Protease activity as the percentage of fibronectin degraded in relation to control leukocyte proteolytic activity.

The proteolytic activity of the quartz-elicited leukocytes remained at least double that of the other inflammatory leukocyte populations from 1 day onwards and although there was a slight drop in the percentage of neutrophils at day 30, total leukocytes in the bronchoalveolar lavage increased and the proteolytic activity of these leukocytes also increased. The increase in proteolytic activity at day 30, concomitant with a fall in the percentage of neutrophils in that population, suggests that the macrophages were involved in the increased proteolysis either directly by secreting more proteolytic enzymes or indirectly by triggering increased neutrophil-mediated proteolysis. Neutrophil elastase has previously been considered to be of major importance in leukocyte-mediated injury in the lung (Cohen and Rossi, 1983) but this enzyme is susceptible to α1-protease inhibitor, the major source of antiprotease activity in the alveoli. If, as we suggest, macrophage proteases were secreted at high levels during chronic inflammation in the quartz-treated rats, this might be more likely to result in tissue damage since macrophage proteases are not inhibited by α1-protease inhibitor (Fels and Cohn, 1986).

Though we confirmed that we could measure increased plasminogen activator activity by using bronchoalveolar leukocytes elicited by thioglycollate broth (results not published) we found no evidence of enhanced plasminogen activator activity, on a cell for cell basis, (Figure 4) in leukocytes elicited by dusts or bacteria in rat lungs. This, however, does not preclude a role for plasminogen activator in the inflammatory response since enhanced levels of this enzyme may occur in the alveoli due to the increased accumulation of leukocytes.

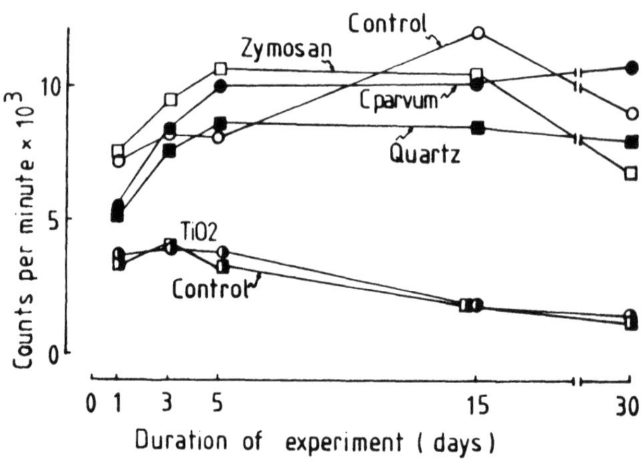

Fig. 4. Plasminogen activator secretion as counts per minute of degraded fibronectin released. TiO_2 was tested independently of the other samples and thus has a separate control.

In the key parameter we measured, proteolysis of fibronectin, the differential between controls and the different inflammatory leukocyte populations remained constant, despite addition *in vitro* of the triggers quartz, TiO_2, zymosan or PMA. In the case of control leukocytes, this suggests that in addition to the dust, one or more *in vivo* signals are required for leukocyte activation; mediators such as interleukin 1, interferon or tumour necrosis factor suggest themselves as likely priming agents. The dust-elicited leukocytes may have been maximally triggered *in vivo* and thus would be incapable of further stimulation *in vitro*. The assay of *in vitro* triggering was carried out within a four hour period which may have been too short for measurable stimulation to occur; experiments are currently in progress to measure *in vitro* stimulation over several days.

In summary, we have demonstrated marked differences in the inflammation engendered in rat lungs by a dust of known fibrogenic activity, quartz, compared with the inert dust, TiO_2. TiO_2 elicited a transient acute inflammatory response but alveolar inflammation in response to quartz persisted throughout the 30 days of the study and secretion of proteolytic enzymes by these leukocytes increased. Thus, tissue injury mediated by leukocyte proteases may be a major arbiter of the permanently deranged tissue structure and functional impairment found in the chronic mineral dust-related lung diseases. Work is in progress to characterize the cellular origin of the connective tissue degrading activity described in inflammatory leukocyte populations.

References

Begin R, Bisson G, Boileau R, Massé S (1986) Assessment of disease activity by Gallium ^{67} scan and lung lavage in the pneumoconioses. Sem Resp Med 7:271-280

Brown GM, Donaldson K (1988) Degradation of connective tissue components by lung-derived leukocytes *in vitro*: role of proteases and oxidants. Thorax 43:132-139

Cohen AB, Rossi M (1983) Neutrophils in normal lungs. Am Rev Resp Dis 127:53-59

Donaldson K, Bolton RE, Jones A, Brown GM, Robertson MD, Slight J, Cowie H, Davis JMG (1988) Kinetics of the bronchoalveolar leukocyte response in rats during exposure to equal airborne mass concentrations of quartz, chrysotile asbestos or titanium dioxide. Thorax 43:525-533

Donaldson K, Slight J, Bolton RE (1987) *In vitro* fibrinolytic activity of rat alveolar macrophages treated with inflammation-generating dusts. Agents and Actions 20:87-92

Fels AOS, Cohn ZA (1986) The alveolar macrophage. J Appl Phys 60:353-369

Hunninghake GW, Garrett KC, Richerson HB, Fantone JC, Ward PA, Rennard SI Bitterman PB, Crystal RG (1984) Pathogenesis of the granulomatous lung diseases Am Rev Resp Dis 130:476-496

Morgan WKC, Seaton A (1984) Occupational Lung Diseases. Saunders, Philadelphia

Robinson BWS (1988) Production of plasminogen activator by alveolar macrophages in normal subjects and patients with interstitial lung disease. Thorax 43:508-515

Ryan GB, Majno G (1977) Acute inflammation. A review. Am J Path 86:1:183-276.

Voisin C, Wallaert B, Aerts C, Grosbois JM (1985) Bronchoalveolar lavage in coalworkers' pneumoconiosis: oxidant and antioxidant activities of alveolar macrophages. In: Beck EG, Bignon J (eds) *In Vitro* effects of mineral dusts. Springer, Berlin, pp 93-100

Silica Induced Alveolar Leucocytosis: Modulation by Mepacrine*

J. Bernard L. Gee, M.D. and Patricia S. Mikes, M.D.
Section of Pulmonary Medicine, Yale University School of Medicine, New Haven, CT, USA

The initial response of the lung to inhalation of many dusts frequently includes the mobilization of polymorponuclear leucocytes (PMN) into the alveolar spaces. Lugano et al. (1982) demonstrated such a response following intra-tracheal instillation of silica particles in guinea pigs. The response occurred within the first 24 hours and was followed by fibrosis at the seventh post-instillation day. We used a somewhat similar model employing adult Fisher 344 rats which received 0.5 mg of autoclaved endotoxin free Minusil in 0.35 ml of saline by the intratracheal route-a full report is now published (Mikes, P.S., et al. 1988a). Briefly this model showed that broncho-alveolar lavage (BAL) fluid showed a twofold increase in cell yield, of

---

*Supported by Grant No. 1667A from The Council for Tobacco Research.

NATO ASI Series, Vol. H30
Effects of Mineral Dusts on Cells
Edited by B.T. Mossman and R.O. Bégin

which 35% were PMN. This contrasts with the 1% PMN yield in control animal BAL fluid. This represents a 200 fold increase in the number of lavage recovered PMN. The PMN response was not accompanied by much increase in either total protein or albumin concentration in the BAL fluid from silica treated animals, thus no great inflamatory response occurred at least as judged by this measure of capillary permeability.

We used this model to examine the potential pharmacologic modulation of PMN mobilization. We considered three different groups of drugs.

First, colchicine was examined since this agent, acting on microtubules, impairs chemotactic responses by PMN (Caner, J.E.Z., 1965; Fruhman, G.J., 1960; Malawista, S.E.; Andriole, V.T., 1968). Intraperitoneal colchicine (0.5mg/kg body weight), 48 hours and 4 hours prior to silica instillation, produced variable results. Our initial series showed a sharp suppression of PMN mobilization. Unfortunately these results were not reproducible and we cannot, at present, explain this. We suspect this agent still warrants further study.

The second series of drugs were those acting on mast cells, notably reserpine and cyproheptadine. The former acts by depleting mast cell histamine stores and the latter is a competitive inhibitor of serotonin. We considered these agents for two reasons. First, in a murine model of skin delayed hypersensitivity, Askenase et al. (1982) have shown that PMN migration through endothelial gaps can be prevented by these agents. Second, Wagner et al (1984) have reported that rats exposed to asbestos dust both show the expected fibrosis and also an increase in the number of pulmonary interstitial mast cells, a feature also present in the BAL fluid from patients with idiopathic pulmonary fibrosis. We employed doses in excess of those required to block the PMN responses in the murine model but failed to modulate PMN intra-aveolar mobilization following silica instillation, so suggesting vasoactive amines produced by mast cells are not involved in the silica induced PMN diapedesis.

The third approach was much more fruitful. Canham et al. (1983) showed that mepacrine sharply diminished PMN mobilization into alveoli induced by intravenous phorbol myristate acetate (PMA). These workers and ourselves (Mikes, P.S., et al.1988b) also showed that mepacrine dimished both the lysosomal enzyme release and $O_2^{\cdot}$ generation evoked in human PMN in vitro by PMA. We used quinacrine hydrochloride (Mepacrine, Sigma; 10 mg/kg body weight). This was given I.V. 15 minutes prior to the intratracheal silica instillation. The data (Mikes, P.S., et al. 1988a) indicated that: (1) the total cell yield induced by silica was halved by mepacrine; (2) the PMN% of the total cell yield fell by 75% from 32 to 8% with mepacrine and (3) the total PMN recovery by BAL was diminished sevenfold by mepacrine pretreatment of silica exposed animals. Thus to the extent that such PMN mobilization is a component of the lung's inflammatory response, mepacrine is an anti-inflammatory agent in this experimental model.

We devoted considerable effort to examining the loci of this action of mepacrine (Mikes, P.S., et al. 1988a). The mechanisms were shown to be multiple, specifically including an action on PMN mobility and more importantly on the alveolar macrophage (AM) release of chemotaxin(s). Random movement of PMN and also PMN chemotactic responses to formyl-methionine-leucine and to AM derived PMN chemotaxin(s) was sharply diminished by mepacrine. Thus there is an action in the effector limb of the inflammatory response, namely PMN mobility and the release from PMN of tissue injuring agents, e.g. elastase and $O_2^{\bar{\cdot}}$ (Mikes, P.S., et al. 1988b). Of perhaps greater interest is the effect of mepacrine on the AM, one component of the afferent limb inflammatory mechanism causing PMN mobilization. We were able to show that mepacrine diminished the release of PMN chemotaxin(s), from rat AM exposed _in vitro_ to silica. Even more importantly, we could demonstrate this effect of mepacrine in the _in vivo_ system employing intra-tracheal silica. This was apparent from study of the PMN chemotactic activity of BAL fluid following silica instillation. Mepacrine diminished such activity fourfold when either human or rat PMN were employed to assess the chemotactic activity of BAL fluids. That this effect, manifest in BAL fluid, was at least partially AM derived was demonstrated by comparison of short term (4 hour) cultures of lavaged AM from _in vivo_ silica and mepacrine treated rats. Thus _in vivo_ mepacrine diminished the AM culture media PMN chemotactic activities. The cultured AM from in vivo mepacrine and silica exposed animals showed one third of the PMN chemotactic activity observed with in vivo silica exposure alone. While silica may also generate PMN chemotaxin(s) by non-AM pathways, e.g. complement activation, there is clearly also a mepacrine sensitive AM activity induced by intratracheal silica.

SIGNIFICANCE

We have reviewed several reasons to inculpate PMN in acute lung injury and also perhaps in the disordered repair mechanisms causing lung fibrosis (Mikes, P.S., et al. 1988a; Gee, J.B.L.; Lwebuga-Mukasa, J., 1984). Neutrophil involvement in both human adult respiratory distress syndrome (ARDS) and experimental inflammatory pulmonary edema (Gee, J.B.L., Lwebuga-Mukasa, J. 1984) is generally recognized. Neutrophil mobilization is a feature both of ARDS and also of experimental and human pulmonary fibrosis. Excess PMN are found in the BAL fluids in human idiopathic pulmonary fibrosis, asbestosis and also in the fibrosis stages of collagen vascular disorders affecting the lung. It has been suggested that PMN collagenase plays a role in this lung fibrosis (Gee, J.B.L., Lwebug-Mukasa, J. 1984). Thus agents diminishing both PMN flux and secretion, such as mepacrine, could conceivably play an important role in the management of ARDS and the tissue injury features of pulmonary fibrosis.

Additionally, the demonstration, in this model, that mepacrine affects AM within the broncho-alveolar spaces suggests mepacrine can act to modify other secretory functions than PMN chemostaxin(s) generation. Among such functions are the secretion of growth factors by AM. The latter promote fibroblast mobility, replication and thus in turn engender a fibrotic process. An effect on AM growth factor production by mepacrine is at present only suggested by analogy with the observations described above. However, the yellow drug mepacrine, or rather the related colorless 4 amino-quinoline, chloroquin can be given to man in doses that do not cause ocular toxicity. Aside from the initial use of mepacrine as a World War II winning anti-malarial, chloroquin has been used in the management of rheumatoid arthritis and of cutaneous sarcoidosis. Further studies of chloroquin in lung disease may be therefore of clinical importance.

REFERENCES

Askenase, P.W.; Metzler, C.M.; Gershon, R.K. (1982), Localization of leukocytes in sites of delayed type hypersensitivity and in lymph nodes: dependence on vasoactive amines, Immunology 47: 239-246.

Caner, J.E.Z. (1965), Colchicine inhibition of chemotaxis, Arthritis Rheum. 8: 757.

Canham, E.M.; Shoemaker, S.A.; Tate, R.M.; Harada, R.N.; McMurty, I.F.; Repine, J.E. (1983), Mepacrine but not methylprednisolone decreases acute edematous lung injury after injection of phorbol myristate acetate in rabbits, Am. Rev. resp. Dis., 127: 594-598.

Fruhman, G.J. (1960), Inhibition of neutrophil mobilization by colchicine, Proc. Soc. exp. biol. Med. 104: 284-286.

Gee, J.B.L.; Lwebuga-Mukasa, J. (1984), Cellular and matrix mechanisms in occupational lung disease, ed. W.K.C. Morgan, A. Seaton, Occupational Lung Disease, 2nd edition, W.B. Saunders Co., Philadelphia.

Lugano, E.M.; Dauber, J.H.; Daniele, R.P. (1982), Acute experimental silicosis, Am. J. Path., 109: 27-36.

Malawista, S.E., Andriole, V.T. (1968), Colchicine: antiinflammatory effect of low doses in a sensitive bacterial system, J. Lab. clin. Med., 72: 933-942.

Mikes, P.S.; Polomski, L.L.; Gee, J.B.L. (1988a), Mepracrine Impairs Neutrophil Response after Acute Lung Injury in Rats: Effects on Neutrophil Migration, Am. Rev. resp. Dis., 138: 1464-1470.

Mikes, P.S.; Polomski, L.L.; De Musis, M.R.; Gee, J.B.L. (1988b), The Neutrophil, Friend or Foe: Pharmacologic Manipulation, Respiration, 54: 9-15.

Wagner, M.M.F.; Edwards, R.E.; Moncrieff, C.B.; Wagner, J.C. (1984), Mast cells and inhalation of asbestos in rats, Thorax, 39: 539-544.

ALTERED FUNCTIONAL EXPRESSION OF ALVEOLAR MACROPHAGE SUBPOPULATIONS AFTER SERPENTINE AND AMPHIBOLE ASBESTOS EXPOSURE

Elliott Kagan, Tamio Inamoto and M. Mala Georgian

Department of Pathology
Georgetown University School of Medicine
3900 Reservoir Road, N.W.
Washington, D.C. 20007, U.S.A.

INTRODUCTION

There is considerable evidence that alveolar macrophages play a pivotal role in the pathogenesis of asbestosis (Kagan 1985, Kagan 1988), since these phagocytic cells are a notable feature of histologic lesions at sites of asbestos deposition in the lungs (Davis 1963, Brody et al. 1981, Oghiso et al. 1984). Multinucleated giant cells also are a prominent component of the macrophage response to inhaled asbestos (Kagan et al. 1983a, Kagan 1988). Conceivably, the accumulation of alveolar macrophages in evolving asbestotic lesions may reflect a localized tissue response to the release of asbestos-induced pulmonary chemoattractants. Support for this concept is provided by experimental studies which have demonstrated that inhalation of, asbestos fibers can evoke the release of chemotactic stimuli for alveolar macrophages (Kagan et al. 1983b, Warheit et al. 1985). Asbestos inhalation also is associated with enhanced expression of immune-associated (Ia) surface antigens on alveolar macrophages and with augmented production of the immunostimulatory monokine, interleukin-1 (Hartmann et al. 1984a, Hartmann et al. 1984b). Since the surface expression of Ia antigens on macrophages is a useful morphologic correlate of macrophage activation (Garrett et al. 1984, Kunkel et al. 1984), these findings collectively suggest that contact with asbestos fibers may induce alveolar macrophages to become functionally activated.

The use of continuous or discontinuous density gradients in several animal species has demonstrated that pulmonary alveolar macrophage populations are heterogeneous in nature. Thus they comprise subpopulations of both young and functionally more mature macrophages (Murphy and Herscowitz 1984, Shellito and Kaltreider 1984, Chandler et al. 1986, Ferro et al. 1987). Alveolar macrophages which aggregate at sites of intrapulmonary asbestos deposition conceivably may reflect increased percentages of newly-recruited, immature cells as well as increased proportions of activated phagocytic cells. The present study was undertaken to test this hypothesis, by examining the effects of chrysotile (i.e. serpentine) and crocidolite (i.e. amphibole) asbestos inhalation on the distribution of rat alveolar macrophage subpopulations defined by continuous, iso-osmotic Percoll gradients.

MATERIALS AND METHODS

Experimental Animals and Exposure Protocol: The study entailed a comparison of three matched groups of adult male, inbred Fischer-344 rats. One group of rats were exposed to a time-weighted average concentration of 10.23 mg/m^3 of chrysotile asbestos. A second group of rats

NATO ASI Series, Vol. H30
Effects of Mineral Dusts on Cells
Edited by B.T. Mossman and R.O. Bégin

received a time-weighted average concentration of 10.52 mg/m^3 of crocidolite asbestos. A control group of rats were sham-exposed to filtered, clean air only. The exposures were performed under contract at The University of Vermont. The rats were exposed by intermittent inhalation for 6 hours/day on 5 days/week over 4 weeks. Some animals were sacrificed on days 7 or 17, respectively, of the exposure regimen, whereas others were sacrificed at 2-3 weeks or 2-5 months, respectively, after the cessation of exposure.

Bronchoalveolar Lavage Studies: Bronchoalveolar cells were obtained by bronchoalveolar lavage, as described previously (Kagan et al. 1983a).

Fractionation of Alveolar Macrophages on Percoll Gradients: A continuous, iso-osmotic Percoll (Pharmacia, Piscataway, New Jersey) density gradient was generated, using a two-chamber gradient mixer (Chandler et al. 1986). The density of the gradient varied from 1.006-1.123 g/ml (Table 1). Unfractionated alveolar macrophage populations (containing at least 1 x 10^7 macrophages) then were layered on the linear density gradient and subsequently centrifuged at 800 x g for 30 minutes at 4° C. Ten Percoll density-defined alveolar macrophage fractions were collected in 6 ml volumes in plastic tubes, using a sterile, dense sucrose solution (1.45 g/ml) to displace the gradient. Each Percoll fraction was then washed twice in Ca^{++}- and Mg^{++}-free phosphate-buffered saline.

Macrophage Morphologic Assessment: The composition of unfractionated and density-defined cell fractions (including the proportion of multinucleated macrophages) was enumerated by proportional analysis on Giemsa-stained cytospin preparations. In virtually all instances, the various fractions comprised approximately 90-99% macrophages, as defined by morphologic criteria and by non-specific esterase positivity. Determinations also were made of the macrophage yield (i.e. the proportion of unfractionated macrophages recoverable) in each density-defined fraction.

Assessment of Ia Surface Antigen Expression on Macrophages: The surface expression of Ia antigen on rat alveolar macrophages was determined by an immunoperoxidase technique on cytospin preparations, as described previously (Hartmann et al. 1984a). A monoclonal mouse IgG anti-rat Ia antiserum (Accurate Chemicals, Westbury, Connecticut) was used as the primary antibody for this purpose. Thereafter, a biotin-labelled horse anti-mouse IgG antiserum (Vector Laboratories, Burlingame, California) as the secondary antibody.

Functional Interleukin-1 Assays: For the assessment of interleukin-1 production, various density-defined alveolar macrophage fractions (2.5 x 10^6 macrophages/ml) were allowed to attach to 24-well microtiter plates (Costar Plastics, Vineland, New Jersey) for 2 hours at 37° C. After washing twice, the macrophages were then cultured in serumless Neuman and Tytell medium (GIBCO, Grand Island, New York) for 24 hours at 37° C. Thereafter, the macrophage culture supernatants were assayed for interleukin-1 activity by a standard functional assay, using C_3H/HeJ mouse thymocytes (Mizel 1981). The presence of interleukin-1 was assayed in parallel cultures containing either unstimulated macrophages, or macrophages stimulated with 100 μg/ml of bacterial lipopolysaccharide (LPS, Difco Laboratories, Detroit, Michigan.).

Interpretation of Data: Results were expressed as the mean $\pm$ SEM of 3-5 experiments. Statistical comparisons were made using Student's *t*-test.

RESULTS

Density-Defined Distribution of Alveolar Macrophages: When unfractionated alveolar macrophages from the 3 groups of exposed rats were layered on continuous, iso-osmotic Percoll gradients, the majority (i.e. > 70%) of the cells were recoverable from density fractions 6 and 7. There were, nevertheless, distinct differences between sham-exposed and asbestos-exposed groups of rats with respect to the density-defined distribution of alveolar macrophage subpopulations. Thus, when rats were sacrificed on day 7 and day 17 during exposure, significantly greater proportions of alveolar macrophages were recoverable in the highest density fractions (i.e. fractions 8-10) after crocidolite exposure than after sham exposure ($p < 0.01$). Similar, albeit less striking, changes were detectable after chrysotile inhalation ($p < 0.05$). The crocidolite-related effect was persistent, being detectable in rats sacrificed at 2-3 weeks and 2-5 months, respectively, after the cessation of exposure ($p < 0.05$).

Proportions of Multinucleated Alveolar Macrophages: When compared with sham-exposed controls, both chrysotile and crocidolite exposure were associated with significantly increased numbers of multinucleated alveolar macrophages recoverable by bronchoalveolar lavage. It should be noted that the proportion of multinucleated macrophages was significantly increased in virtually all density-defined fractions after both types of asbestos exposure (Table 2). This effect occurred within 7 days of the commencement of asbestos exposure and persisted throughout the entire period of the study.

Ia Surface Antigen Expression on Alveolar Macrophages: The proportion of Ia antigen-bearing alveolar macrophages also was noted to be significantly increased after asbestos inhalation in multiple density-defined fractions (Table 3). This finding was noted in the context of both types of asbestos exposure. However, it was more prominent after crocidolite than after chrysotile exposure. Notably, the effect could be detected within 7 days of the initiation of asbestos inhalation and, in the case of crocidolite, persisted throughout the entire course of the study. Moreover, when consideration was given to the presence of Ia-positive multinucleated alveolar macrophages in the various density defined fractions, both types of asbestos inhalation were associated with relatively large percentages of such cells. Thus, approximately 2-32% of multinucleated macrophages expressed the Ia surface marker in different density-defined fractions. By contrast, Ia-positive multinucleated cells were virtually never detected in any Percoll fraction after sham exposure.

Interleukin-1 Production by Alveolar Macrophages: Both types of asbestos exposure were associated with considerably augmented production of interleukin-1 by alveolar macrophages when compared with sham exposure. These differences were already evident within 7 days of the commencement of exposure and were most striking in LPS-stimulated cultures. However, differences between asbestos-exposed and sham-exposed rats also were evident in unstimulated cultures. Thus, LPS-directed functional interleukin-1 activity was usually less than 100 units/ml, and never exceeded 400 units/ml, in culture supernatants from any density-defined fraction. On the other hand, similar culture supernatants from chrysotile- and crocidolite-exposed animals frequently contained functional interleukin-1 levels ranging from 500-1,600 units/ml. In general, these asbestos-related effects were most pronounced in the higher density-defined macrophage fractions (i.e. fractions 6-10).

DISCUSSION

This study has demonstrated that the deposition of asbestos fibers in the lungs produces striking alterations in the distribution and composition of alveolar macrophage subpopulations defined by continuous, iso-osmotic Percoll gradients. These asbestos-related changes are characterized by increased proportions of multinucleated macrophages and higher density fraction-defined subsets of cells. Alveolar macrophages in the higher density fractions have been shown previously to be developmentally immature (Shellito and Kaltreider 1984, Elias et al. 1985). Thus, the finding of enriched numbers of higher density phagocytes in bronchoalveolar lavage cell subpopulations from asbestos-exposed rats may reflect the accumulation of immature, newly-recruited monocytes and/or macrophages at sites of intrapulmonary asbestos deposition. Studies demonstrating the presence of asbestos-related chemoattractants for alveolar macrophages have provided support for this notion (Kagan et al. 1983b; Warheit et al. 1985).

Because alveolar macrophage populations are heterogeneous in nature, their overall function is dependent on mixtures of cells with varied functional capabilities. A disproportionate increase in the number of functionally activated phagocytic cells can result in macrophage-mediated lung injury, as has been shown for sarcoidosis and other granulomatous pulmonary diseases (Garrett et al. 1984; Kunkel et al. 1984). A similar mechanism may pertain to asbestosis, since enhanced interleukin-1 production was detected in several density-defined fractions after asbestos exposure in the present study. This finding is particularly noteworthy, since this effect was noted both spontaneously and in LPS-stimulated cultures. The enhancement of Ia antigen expression in most density-defined alveolar macrophage fractions provides an additional morphologic correlate of functional macrophage activation induced by asbestos. This phenomenon was especially prominent among multinucleated alveolar macrophage subpopulations. Although the significance of these multinucleated cells in asbestosis has not been established, they may represent a distinct subset of functionally activated phagocytes. Their appearance may possibly result from the fusion of macrophage membranes induced by γ-interferon (Nagasawa et al. 1987). The fact that bronchoalveolar lavage cells from patients with asbestosis recently have been shown to manifest enhanced *in vitro* production of γ-interferon lends credence to this hypothesis (Robinson et al. 1988).

Both serpentine and amphibole exposure are capable of inducing the changes seen in this study and these effects develop shortly (i.e. by day 7) after the commencement of asbestos exposure. Nevertheless, the effects of crocidolite exposure tend to be more prominent and to persist for a longer period after the cessation of inhalational asbestos exposure. These variations in the biologic effects of the two asbestos fiber types may relate to the relatively more efficient clearance and greater solubility of chrysotile compared with crocidolite (Becklake 1982; Kagan 1988).

In conclusion, this study has demonstrated that asbestos inhalation can induce striking alterations in the composition and distribution of alveolar macrophage subpopulations. Given the pivotal role played by the alveolar macrophage in asbestos-related lung injury, these findings may have biologic relevance to the pathogenesis of asbestosis.

ACKNOWLEDGEMENT

This work was supported by U.S. Environmental Protection Agency Grant # R812036-01-0.

REFERENCES

Becklake MR (1982) Asbestos-related diseases of the lungs and pleura. Am Rev Respir Dis 127:187-194

Brody AR, Hill LH, Adkins B, Jr, O'Connor RW (1981) Chrysotile asbestos inhalation in rats: deposition pattern and reaction of alveolar epithelium and pulmonary macrophages. Am Rev Respir Dis 123:670-679

Chandler DB, Fuller WC, Jackson RM, Fulmer JD (1986) Studies of membrane receptors and phagocytosis in subpopulations of rat alveolar macrophages. Am Rev Respir Dis 133:461-467

Davis JMG (1963) An electron microscopy study of the effect of asbestos dust on the lung. Br J Exp Pathol 44:454-464

Elias JA, Schreiber AD, Gustilo K, Chien P, Rossman MD, Lammie PJ, Daniele RP (1985) Differential interleukin 1 elaboration by unfractionated and density fractionated human alveolar macrophages and blood monocytes: relationship to cell maturity. J Immunol 135:3198-3204

Ferro TJ, Kern JA, Elias JA, Kamoun M, Daniele RP, Rossman MD (1987) Alveolar macrophages, blood monocytes, and density fractionated alveolar macrophages differ in their ability to promote lymphocyte proliferation to mitogen and antigen. Am Rev Respir Dis 135:682-687

Garrett KC, Richerson HB, Hunninghake GW (1984) Pathogenesis of the granulomatous lung diseases. II. Mechanism of granuloma formation. Am Rev Respir Dis 130:477-483

Hartmann DP, Georgian MM, Kagan E (1984a) Enhanced alveolar macrophage Ia antigen expression after asbestos inhalation. J Immunol 132:2693-2695

Hartmann DP, Georgian MM, Oghiso Y, Kagan E (1984b) Enhanced interleukin activity after asbestos inhalation. Clin Exp Immunol 55:643-650

Kagan E, Oghiso Y, Hartmann DP (1983a) The effects of chrysotile and crocidolite asbestos on the lower respiratory tract: analysis of bronchoalveolar lavage constituents. Environ Res 32:382-397

Kagan E, Oghiso Y, Hartmann DP (1983b) Enhanced release of a chemoattractant for alveolar macrophages after asbestos inhalation. Am Rev Respir Dis 128:680-687

Kagan E (1985) Current perspectives in asbestosis. Ann Allergy 54:464-473

Kagan E (1988) Current issues regarding the pathobiology of asbestosis: a chronologic perspective. J Thorac Imag 4(3):1-9

Kunkel SL, Chensue SW, Plewa M, Higashi GI (1984) Macrophage function in the *Schistosoma mansoni* egg-induced pulmonary granuloma: role of arachidonic acid metabolites in Ia antigen expression. Am J Pathol 114:240-249

Mizel SB (1981) Production and quantitation of lymphocyte-activating factor (interleukin 1). In: Herscowitz HB, Holden HT, Bellanti JA, Ghaffar A (eds) Manual of Macrophage Methodology: Collection, Characterization and Function. Marcel Dekker, New York, p 407

Murphy MA, Herscowitz HB (1984) Heterogeneity among alveolar macrophages in humoral and cell mediated immune responses: separation of functional subpopulations by density gradient centrifugation on Percoll. J Leukocyte Biol 35:39-54

Nagasawa H, Miyaura C, Abe E, Suda T, Horiguchi M, Suda T (1987) Fusion and activation of human alveolar macrophages induced by recombinant interferon-γ and their suppression by dexamethasone. Am Rev Respir Dis 136:916-921

Oghiso Y, Kagan E, Brody AR (1984) Intrapulmonary distribution of inhaled chrysotile and crocidolite asbestos: ultrastructural features. Br J Exp Pathol 65:467-484

Robinson BWS, Rose AH, Hayes A, Musk AW (1988) Increased pulmonary gamma interferon production in asbestosis. Am Rev Respir Dis 138:278-283

Shellito J, Kaltreider HB (1984) Heterogeneity of immunologic function among subfractions of normal rat alveolar macrophages. Am Rev Respir Dis 129:747-753

Warheit DB, George G, Hill LH, Snyderman R, Brody AR (1985) Inhaled asbestos activates a complement-dependent chemoattractant for macrophages. Lab Invest 52:505-514

TABLE 1:

CONTINUOUS ISO-OSMOTIC PERCOLL GRADIENTS

Fraction No.	Density Range (gm/ml)
1	< 1.018
2	1.018 - 1.028
3	1.028 - 1.037
4	1.037 - 1.046
5	1.046 - 1.055
6	1.055 - 1.064
7	1.064 - 1.072
8	1.072 - 1.081
9	1.081 - 1.090
10	> 1.090

TABLE 2:

PROPORTIONS OF MULTINUCLEATED ALVEOLAR MACROPHAGES (%)

	Fraction Number	Rat Exposure Group: Sham-Exposed	Chrysotile-Exposed	Crocidolite-Exposed
	Unfractionated	1.31 ± 0.47	6.04 ± 1.10**	8.03 ± 0.85**
	1	-	1.67 ± 1.67	8.89 ± 6.48
	2	0.74 ± 0.53	3.55 ± 1.58	3.08 ± 0.98*
	3	1.04 ± 0.50	6.87 ± 1.52**	5.10 ± 1.42**
	4	1.09 ± 0.37	5.99 ± 1.06**	6.29 ± 1.33**
During Exposure	5	1.02 ± 0.41	5.34 ± 0.95**	5.77 ± 1.40**
(Days 7-17)	6	0.86 ± 0.33	4.90 ± 0.93**	5.84 ± 0.61**
	7	0.91 ± 0.40	5.90 ± 0.98**	5.97 ± 1.37**
	8	0.77 ± 0.26	7.88 ± 1.30**	8.51 ± 1.11**
	9	0.95 ± 0.57	6.95 ± 1.11**	10.64 ± 1.35**
	10	0.23 ± 0.23	7.45 ± 1.61**	12.84 ± 1.41**
	Unfractionated	1.71 ± 0.41	5.31 ± 1.60*	5.27 ± 1.28*
	1	1.00 ± 1.00	3.94 ± 3.70	1.89 ± 1.89
	2	1.83 ± 1.32	3.23 ± 1.06	1.19 ± 0.70
	3	0.74 ± 0.40	7.21 ± 0.97**	3.93 ± 1.02**
	4	0.95 ± 0.31	5.20 ± 1.08**	4.65 ± 0.80**
2 Weeks-5 Months	5	1.56 ± 0.43	6.14 ± 0.80**	4.70 ± 1.09**
Post-Exposure	6	1.02 ± 0.31	4.92 ± 0.70**	4.01 ± 0.53**
	7	0.73 ± 0.39	5.26 ± 1.03**	5.02 ± 0.93**
	8	1.73 ± 0.71	7.65 ± 2.19**	6.58 ± 1.32**
	9	1.13 ± 0.44	8.39 ± 2.45*	11.77 ± 2.06**
	10	0.51 ± 0.45	6.78 ± 0.90**	13.28 ± 1.20**

* $P < 0.05$ (Sham-Exposed vs. Asbestos-Exposed)

** $P < 0.01$ (Sham-Exposed vs. Asbestos-Exposed)

Results are expressed as Mean ± SEM of 10 experiments

TABLE 3:

PROPORTIONS OF Ia-POSITIVE ALVEOLAR MACROPHAGES (%)

	Fraction Number	Rat Exposure Group: Sham-Exposed	Chrysotile-Exposed	Crocidolite-Exposed
	Unfractionated	3.12 ± 1.58	7.28 ± 1.25**	9.90 ± 2.01**
	1	-	-	-
	2	0.58 ± 0.58	0.80 ± 0.51	0.21 ± 0.21
	3	0.98 ± 0.65	3.44 ± 1.19*	2.15 ± 0.64*
	4	1.39 ± 0.75	4.19 ± 1.25*	3.04 ± 1.32
During Exposure	5	2.19 ± 1.19	6.47 ± 2.31*	6.06 ± 2.01*
(Days 7-17)	6	1.51 ± 0.78	4.98 ± 2.12	7.04 ± 1.74**
	7	1.55 ± 0.89	4.82 ± 1.34*	6.02 ± 2.03*
	8	1.66 ± 0.51	2.47 ± 0.71*	4.26 ± 1.51*
	9	1.31 ± 1.31	2.14 ± 0.88	6.51 ± 3.16*
	10	1.42 ± 1.27	2.96 ± 1.69	4.01 ± 1.17**
	Unfractionated	2.83 ± 1.21	9.88 ± 2.81*	16.88 ± 3.62**
	1	0.68 ± 0.68	1.16 ± 1.16	2.50 ± 2.50
	2	0.96 ± 0.96	1.51 ± 1.14	4.21 ± 3.44
	3	2.43 ± 0.40	2.87 ± 2.02	4.67 ± 2.00
	4	2.67 ± 1.10	6.60 ± 2.89	8.38 ± 2.25*
2 Weeks-5 Months	5	2.64 ± 1.12	6.24 ± 3.27	10.80 ± 2.07**
Post-Exposure	6	1.31 ± 0.50	6.39 ± 4.02	9.70 ± 3.81*
	7	1.60 ± 0.43	7.34 ± 4.22	8.97 ± 2.30**
	8	3.67 ± 0.68	5.14 ± 2.43	7.46 ± 1.62**
	9	3.72 ± 1.83	6.21 ± 3.41	9.81 ± 3.53
	10	5.38 ± 1.93	10.44 ± 4.68	14.09 ± 4.74*

* $P < 0.05$ (Sham-Exposed vs. Asbestos-Exposed)

** $P < 0.01$ (Sham-Exposed vs. Asbestos-Exposed)

Results are expressed as Mean ± SEM of 10 experiments

PARTICLE INDUCED *IN VITRO* CHEMOTAXIS OF ALVEOLAR MACROPHAGES (AM) AND NEUTROPHILS AND *IN VIVO* EFFECTS.

G. Oberdorster, J. Ferin, N. Corson, S. Gavett and D. Hemenway,[1]
Environmental Health Sciences Center,
University of Rochester,
Rochester, NY 14642
USA

The initial contact of AM with deposited particles could occur by chance or by a directed movement of AM towards the site of deposition. Activation of chemotactic factors for AM in serum by particles has been suggested as a mechanism for particle-macrophage encounter in the lungs thereby initiating the mechanical clearance of those particles (Warheit *et al.*, 1985; 1988). Indeed, Warheit *et al.* (1988) could show that there was a very good correlation between the sites of deposition in the transitional zone of the lungs of rats and the appearance of AM shortly thereafter. No neutrophils appeared at these sites. Particles of different shapes and composition have been identified by these investigators as being able to activate serum components. They also found, that particles from Mt. St. Helens ash did not activate serum C5a and suggested that these particles may be cleared less effectively than other particles from the lung. Activation of complement followed by generation of C5a from the complement cascade is the likely source of the chemotactic factor (Warheit *et al.*, 1985).

Since C5a is a very potent chemoattractant for neutrophils as well (Toews *et al.*, 1984), the question arises as to why neutrophils are not always attracted to the site of particle deposition, like AM. A second relevant question is: Is there a correlation between the potency of a particle to activate serum attractants and its lung clearance *in vivo*? We hypothesize that 1. particles in general have a potential to activate chemotactic factors in serum, and 2. that the alveolar epithelial barrier is of importance for the effectivity of induced chemotactic factors like C5a: If this barrier is intact, then a relocation of AM within the alveolar space occurs primarily; however, if this barrier is damaged, neutrophils and macrophages from interstitium and the blood compartment are recruited. Solutes of the molecular weight of C5a (12.5 Kd) will not easily penetrate the normally tight alveolar epithelium (Jones *et al.*, 1984) but require more time than smaller solutes for penetration into the interstitium. However, a damaged epithelium is more easily permeable to solutes of larger molecular sizes, and we expect in such case to see a larger influx of neutrophils if a factor like C5a is involved.

This paper presents a preliminary report on studies investigating these hypotheses.

---

[1]University of Vermont, Dept. of Pathology, Burlington, VT 05405

NATO ASI Series, Vol. H30
Effects of Mineral Dusts on Cells
Edited by B. T. Mossman and R. O. Bégin

Methods:

The following particles were used: Carbon-black (Cabot, Regal 660); PVC-dust (Chem. Works Huls, F.R. Germany). TiO_2 (Fisher Scientific, Anatase); amorphous silica (J.M. Huber Co., Havre de Grace, Md.); α-quartz (Thermal American Fused Quartz Co., Montville, NJ); cristobalite (C.N. Mineral Co., King of Prussia, PA).

For the *in vitro* studies, rat serum from freshly drawn blood (Fisher 344 rats) was incubated for 45 min at 37°C (particle concentrations: 2.5-200 mg/ml serum), followed by heat inactivation (567°C) for 30 min. After centrifugation, chemotactic and chemokinetic activity of the particle free serum was determined in a 48 micro-well chemotaxis chamber (Neuroprobe) by placing the activated serum (10% in HBSS) in the lower wells, four wells per particle type. 50 µl AM suspensions, 1.5×10^6 cells/ml, or 50 µl neutrophil suspensions, 0.5×10^6 cells/ml, were placed into the upper part of the wells separated from the chemoattractant by a PVP-free nucleopore filter (5 µm pores for AM; 3 µm pores for neutrophils). AM were obtained from control Fisher 344 rats by exhaustive lung lavage, and neutrophils were separated from the blood of control rats by dextran sedimentation followed by Ficoll gradient centrifugation. After 2 h (AM) or after 30 min (neutrophils) at 37°C, the nucleopore filters were prepared for counting by light microscopy of the cells that had migrated through the pores towards the chemoattractant. Positive and negative controls, respectively, included zymosan activated serum (10 mg/ml serum, 5% activated serum in HBSS) and heated (decomplemented) serum activated with particles after heating.

For the *in vivo* studies, three male Fisher 344 rats per dust, b.w. 200-225 g, were intratracheally instilled with a suspension of the various particles in 0.2 ml normal saline. Doses ranged from 2.5-400 µg/rat, depending on the *in vitro* chemotactic activity (see below). 24 hrs after instillation, the lungs were exhaustively lavaged (10 x with 5 ml) for determination of cell parameters and protein content of lavage fluid (as indicator of alveolar epithelial permeability).

Since it is conceivable that the specific surface area of the particles may contribute to its properties to chemotactically activate serum, the surface areas were determined by the nitrogen absorption method (Flowsorb 2300, Micromeritics) after drying the particles at 200°C (except PVC dust, 357°C). The results were correlated with the chemotactic activity.

The sequence of the experiments was as follows: First, the ability to chemotactically activate serum for AM was measured over a wide range of concentrations for each particle type. Second, chemotactically equipotent doses were calculated and instilled into the animals for the *in vivo* studies. Third, chemotactically equipotent serum concentrations (for AM) were tested for their capacity to activate *in vitro* chemotactic factors for neutrophils.

Results:

a) chemotactic activation of serum for AM:

The results of the chemotactic in vitro assay for the different particles are shown in Fig. 1, left panel. For all particles, a typical increase in chemotactic activation of serum factors with increasing concentration up to a maximum was found, followed by a decrease. Incubation of decomplemented serum with the dust particles did not lead to activation of chemotactic factors. Carbon was found to be the most active chemotactic particle (lowest concentration needed), followed by PVC dust, then amorphous silica and α-quartz, and then cristobalite; the least active particle which needed the highest particle concentration (100 mg/ml) for equivalent activation was TiO_2.

Determination of specific surface area gave the following results (in m^2/g): Carbon = 10.1±9.65; amorphous silica = 79.0±0.03; cristobalite = 8.8±0.19; TiO_2 = 6.45±0.04; PVC dust = 3.69±0.08; α-quartz = 2.9±0.05. Fig. 1, right panel, depicts the results of the chemotactic assay correlated to the surface area of the different doses of the dusts. The order of chemotactic activation is now quite different, with α-quartz and PVC-dust showing the highest activity and carbon and amorphous silica showing the lowest activity (in terms of their surface area required to activate serum).

b) intratracheal instillation:

The analysis of the lung lavages 24 h after instillation of chemotactically equipotent doses is shown in Table 1. Two of the dusts, amorphous silica and cristobalite, led to an extremely high influx of neutrophils. Instillation of cristobalite also elicited a significant influx of peroxidase positive macrophages, a possible indicator of newly arrived macrophages. A statistically significant increase in protein content of the lavage fluid was seen for both particle types as

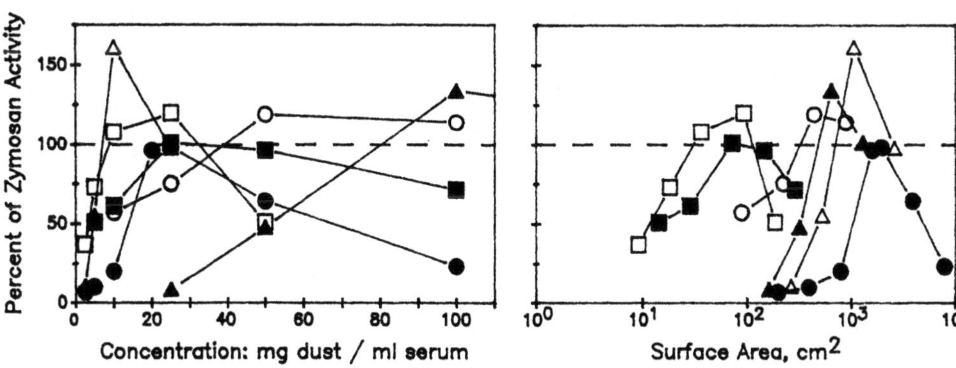

Fig.1: Results of chemotaxis assay with AM after incubation with 10% serum activated with different dust concentrations. 5% Zymosan A activated serum (10 mg/ml) served as positive control (=100%). Left panel: chemotactic activity vs dust concentration; right panel: chemotactic activity vs dust surface area. Δ = carbon; □ = PVC-dust; ● amorphous silica; ■ α-quartz; ○ = cristobalite; ▲ TiO_2.

well. Although a significant decrease in relative numbers of AM after cristobalite occurred, the absolute numbers of AM showed an increase, which, however, was statistically insignificant. α-quartz led also to a significant increase in numbers of neutrophils. Although after both α-quartz and amorphous silica a significant decrease in relative numbers of AM occurred (Table 1), this was not significant in terms of absolute numbers. Both TiO_2 and Carbon black induced a slight, yet significant increase of neutrophils to about 3% and 2%, respectively, whereas the response after PVC-dust and zymosan instillation was not significantly different from control values. A linear correlation between epithelial permeability, P, and numbers of neutrophils, N, was found in this study ($P = 100.5 + 2.62 N$; $r = 0.91$).

c) chemotactic activation of serum for neutrophils:

Serum concentrations of the different particles with equal peak chemotactic activation potency for AM (Fig. 1) were used to determine the chemotactic activation of serum for neutrophils. The results show that serum activation with TiO_2 and carbon resulted in the strongest response, closely followed by amorphous silica and cristobalite whereas PVC-dust and α-quartz showed a significantly lower response compared to the Zymosan control ($p < 0.05$). When comparing these results with those in Table 1, it appears that the two dusts, which *in vivo*

Table 1: Analysis of alveolar lavage of after intratracheal instillation of different particles into rats (n=3 per compound, mean ±S.D.)

	i.tr. dose µg	Total cell numbers x 10^{-7}	AM %	Neutrophils %	Peroxidase positive AM %	Rel. Protein Content, % (Control= 100%)
Control	0.2ml saline	1.16 ±0.10	99.4 ±0.8	0.5 ±0.4	0.1±0.2	100±15
Zymosan	25	1.12 ±0.12	94.3* ±2.6	5.3 ±2.6	0.2±0.2	96±5
PVC-dust	50	1.08 ±0.05	99.6 ±0.2	0.4 ±0.4	0	102±3
Carbon	40	1.07 ±0.21	98.2 ±0.6	1.8* ±0.6	2.3±1.2	86±5
TiO_2	400	1.12 ±0.11	96.6 ±1.3	3.3* ±1.3	0.3±0.2	113±10
Amorphous Silica	100	1.91* ±0.27	55.3* ±8.6	44.7* ±8.6	0.2±0.2	142±16*
α-Quartz	125	1.12 ±0.09	90.3* ±1.6	9.3* ±1.9	0.4±0.2	100±9
Cristobalite	200	4.19* ±0.42	38.9* ±6.9	60.7* ±6.8	4.6±0.2*	161±1*

*Significantly different from control ($p < 0.05$, two tailed t-test.

showed no change in alveolar epithelial permeability yet gave rise to a slight increase in neutrophil numbers (TiO_2 and carbon black), induced now the highest in vitro chemotactic activity for neutrophils.

Discussion:

The results of these experiments show that the different particles studied can activate serum chemotactic factors for AM, albeit with various potency. Some particulate materials -like carbon, PVC-dust - led to a maximum response at low concentrations in serum whereas others - like cristobalite and TiO_2 - required up to 10 fold higher concentrations in serum to achieve the same effect. Warheit *et al.* (1988) reported that Mt. St. Helens ash did not induce chemotactic activation of serum. This may be related to the concentration of volcanic ash tested in those studies, which was 25 mg/ml serum, since we did not observe a significant chemoattractant activation at this concentration with TiO_2 either. Only when we tested higher concentrations of 50 and 100 mg/ml did we see an effect with TiO_2.

Our studies indicate that the specific surface area of the particles is not the primary factor for the activation of serum components. However, a pilot study with two TiO_2 compounds (both anatase) of different particle sizes (not described in this paper) showed that within a given particulate compound the surface area of the particles - or other parameters related to surface area - may well be an important determinant of the activation of chemotactic serum factors.

The well-known cytotoxicity of some of the particles studied - e.g. cristobalite and α-quartz - is not reflected in their potential to activate serum factors: Carbon-black and TiO_2, which are both nuisance dust particles, showed the highest and lowest chemotactic activation potential, respectively. Among the well-known cytotoxic dusts, α-quartz and cristobalite showed also both relatively high and low chemotactic activation potentials. Thus, there is no correlation between cytotoxicity of dust and its activation of chemotactic factors in serum.

Our preliminary results are surprising, showing that the different dusts may activate serum factors for neutrophils differently than for AM. It appears that α-quartz and possibly also PVC-dust have a lower potential to activate neutrophil chemotactic factors in serum than the other dusts at chemotactically equipotent doses for AM. Obviously, this finding has to be confirmed before we can be certain that different chemotactic factors can be activated in serum with different potencies for AM and neutrophils.

When evaluating the results of the *in vitro* and *in vivo* studies together it is intriguing to speculate that changes in the alveolar epithelial barrier function are necessary to recruit neutrophils and macrophages into the alveolar space. Both amorphous silica and cristobalite induced a large influx of neutrophils 24 hrs after intratracheal instillation which was accompanied by a change in alveolar epithelial permeability (increased protein in lavage fluid) and by the appearance of peroxidase positive macrophages (possibly representing newly arrived macrophages). TiO_2 and carbon black induced only a mild influx of neutrophils without permeability changes, and these two particle types were most effective in evoking a chemotactic response for neutrophils in serum.

Based on these results, we propose the following hypothetical sequence for cell mobilization in the lung by deposited particles (fig. 2): Upon deposition of the inhaled particles in the lung a chemotactic factor (e.g. C5a) is generated locally which attracts AM to the site of deposition (relocation of AM within alveolar space). The alveolar epithelium is normally tight enough that C5a will not penetrate quickly, and only if it is generated to a larger degree will sufficient amounts reach the interstitium and the blood compartment and attract neutrophils (TiO_2 and Carbon-black). Phagocytosis of the deposited particles by AM will lead to activation of these cells (Roos, 1981) which in turn release mediators, including reactive oxygen species, proteases and chemoattractants. The degree of mediator release depends on the particle type; however, a known damaging effect of released oxygen radicals or proteases on the epithelium (Freeman and Crapo, 1982 Donaldson *et al.*, 1987) will make its tight junctions more leaky, thus allowing chemotactic factors to penetrate more easily and attract interstitial macrophages, blood monocytes and neutrophils (cristobalite, amorphous silica). At a first glance the results after α-quartz instillation - no permeability changes but increase in neutrophil numbers - do not fit into this scheme. The explanation could be that instillation of α-quartz represents an acute cytotoxic event to AM. Possibly, chemotactic factors released from injured or dying AM are capable of penetrating the alveolar epithelium to attract neutrophils. A number of chemotactic factors have been identified which are released from

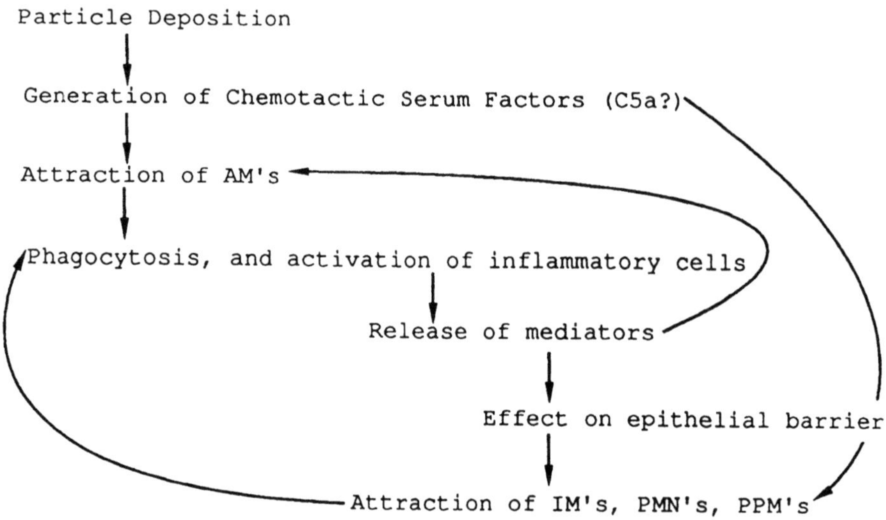

Fig. 2: Hypothetical sequence of particle induced cell mobilization in the lung.

AM and monocytes (Martin *et al.*, 1987; Yoshimura *et al.*, 1987; Tanaka *et al.*, 1988). A complex interaction between different inflammatory monokines may be involved in the influx of neutrophils into the alveolar space, and the different degree of cytotoxicity of the particles probably plays a crucial role in the quantity and quality of released factors. The present results (α-quartz, TiO_2; carbon) support conclusions from our previous studies that the influx of neutrophils into the alveolar space can be independent of changes in alveolar epithelial permeability (Oberdorster *et al.* 1986). Apparently, such influx of inflammatory cells depends very much on the degree to which the various chemotactic factors generated in the alveolar space can penetrate the intact alveolar epithelium or whether a damaged, leaky epithelium has to be present for penetration of certain factors.

With regard to the importance of chemotactic factors generated at the site of particle deposition, there is no evidence at present that AM mediated particle clearance is accelerated for a particle with high potential for chemotactic activation, or depressed by a particle with a lower potential. The two particles of our study with the most extreme potencies to activate chemotactic serum factors, carbon black on the one hand, and TiO_2 on the other, are not cleared differently *in vivo*: Pulmonary retention halftimes for both are about 60 days in rats (Ferin, 1981; Lee *et al.*, 1987). Even a particle which was found to elicit no chemotactic activation of serum factors in another study - Mt. St. Helens ash (Warheit *et al.*, 1988) - was found to be cleared from the lungs of rats with a retention halftime of about 39 days (Wehner *et al.*, 1984). In one study reporting virtually no lung clearance of Mt. St. Helens ash (Sanders *et al.*, 1982) very high lung doses of 40-77 mg/rat lung were used which clearly led to the development of lung overload induced cessation of lung clearance as discussed by Morrow (1988) for many inert particles.

The clearance studies cited above refer to long term lung clearance of the respective particles, and although there may not be a correlation between activation of chemotactic serum factors and long term lung clearance it does not mean that chemotactic factor generation at the site of particle deposition is unimportant for AM mediated particle clearance. As demonstrated by Warheit *et al.* (1985), the very early AM - particle encounter is highly influenced by the potential of particles to activate chemotactic factors. This event possibly prevents particles from entering interstitial spaces protecting these more sensitive sites. Thus, the generation of serum chemotactic factors by deposited particles may prevent lung injury and may support - but is not essential - for AM-mediated long term particle clearance from the lung.

References

Donaldson K, Slight J, Brown GM, Bolton RE (1988) The ability of inflammatory bronchoalveolar leucocyte populations elicited with microbes or mineral dust to injure alveolar epithelial cells and degrade extracellular matrix *in vitro*. Br. J. Exp. Path. 69:327-338.

Ferin J (1981) Alveolar macrophage mediated pulmonary clearance suppressed by drug-induced phospholipidosis. Exp. Lung. Res. 4:1-10.
Freeman BA, Crapo JK (1982) Biology of disease. Free radicals and tissue injury. Lab. Invest. 47:412-426.
Jones JG, Minty BD, Royston D (1982) The physiology of leaky lungs. Br. J. Anaesth. 54:706-721.
Lee P, Gorski R, Hering W, Chan T (1987) Lung clearance of inhaled particles after exposure to carbon black generated from a resuspension system. Environ. Res. 43:367-373.
Martin I, Raugi G, Merrit T, Henderson W (1987) Relative contribution of leukotriene B4 to the neutrophil chemotactic activity produced by the resident human alveolar macrophage. Clin. Invest. 80:1114-1124.
Morrow PE (1988) Possible mechanisms to explain dust overloading of the lungs. Fund. Appl. Tox. 10:369-384.
Oberdorster G, Anderson GL, Mavis RD, Marcello NL, Drago SR (1986) Time sequence of inflammatory reaction in the lung induced by cadmium compounds. I: Cellular events and epithelial permeability. The Toxicologist 6:197.
Roos D (1981) The metabolic response to phagocytosis. In: Handbook of Inflammation, Vol. 2. Ed. G. Weissman Elsevier/North Holland, pp. 337-385.
Sanders CL, Conklin AW, Gelman RA, Adee RR, Rhoads K (1982) Pulmonary toxicity of Mt. St. Helens volcanic ash. Environ. Res. 27:118-135.
Tanaka S, Robinson EA, Yoshimura T, Matsushima K, Leonard EJ, Appella E (1988) Synthesis and biological characterization of monocyte-derived neutrophil chemotactic factor. FEBS Letters 236:467-470.
Toews GB, Vial WC (1984) The role of C5 in polymorphonuclear recruitment in response to streptococcus pneumoniae. Am. Rev. Resp. Dis. 129:82-86.
Warheit DB, George G, Hill LH, Snyderman R, Brody AR (1985) Inhaled asbestos activates a complement dependent chemoattractant for macrophages. Lab. Invest. 52:505-514.
Warheit DB, Overby LH, George G, Brody AR (1988) Pulmonary macrophages are attracted to inhaled particles through complement activation. Exp. Lung Res. 14:51-66.
Wehner AP, Wilerson CL, Stevens DL (1984) Lung clearance of neutron activated Mt. St. Helens volcanic ash in the rat. Environ. Res. 35:211-217.
Yoshimura T, Matsushima K, Oppenheim JJ, Leonard EJ (1987) Neutrophil chemotactic factor produced by lipopolysaccharide (LPS)-stimulated human blood mononuclear leukocytes: partial characterization and separation from interleukin-I (IL1). J. Immunol. 139:788-793.

Acknowledgement: This paper was partially based on work supported by NIH grants ES01247 and ES04872.

CONSEQUENCES OF MMMF INHALATION ON LUNG AND PLEURAL CAVITY CELL POPULATIONS

K. Miller, F. Lawrence, R.A. Riley
Department of Immunotoxicology and Inhalation
British Industrial Biological Research Association
Woodmansterne Road, Carshalton, Surrey SM5 4DS, UK

INTRODUCTION.

The basic mechanisms by which inhalation of mineral fibres initiate a series of events which lead to diffuse pleural fibrosis, pleural plaques or pleural mesothelioma are unknown. Furthermore the importance of the physical characteristics of fibres in inducing pleural pathology is unclear. Several experimental investigations have shown, however, that shorter fibres translocate from the alveoli to the interstitium within a few days of inhalation (Brody et al. 1981) whereas long fibres appear to be preferentially retained in the lung, suggesting that a role for migrating short fibres in inducing pleural events should be considered. Supportive evidence for such a postulate comes from work by Sebastian (1979) who observed a drift by asbestos and other dust particles toward the pleura in human lungs and experimental studies by Morgan and co-workers (1977) who exposed rats to radiolabelled asbestos and found the radioactivity concentrated in subpleural foci within a few months. In addition, Holt (1983) has provided evidence that inhalation of chrysotile asbestos leads to the accumulation of dust-laden macrophages near the pleura as well as observing dust-containing cells penetrating into the pleura. Inflammatory mediators and growth factors elaborated by such macrophages could therefore diffuse into the pleural cavity, and cause both tissue injury and cellular proliferation.

In order to address these issues, studies were designed to investigate the changes inhalation of man-made mineral fibres may induce in pleural cell populations. Since pulmonary macrophages would play a central role we combined, the characterization of lavagable pleural cells with studies on lavagable lung cells. In a second component of the study we also monitored the time required for fibres to reach the pleural cavity.

NATO ASI Series, Vol. H30
Effects of Mineral Dusts on Cells
Edited by B.T. Mossman and R.O. Bégin

MATERIALS AND METHODS

Experimental and exposure protocol. Inbred male PVG rats were used and exposed to either glass fibres (John Manville Reference C100) or ceramic material (Kaowool Zirconia, Blanket, Morganite Ceramic Fibres) six hours a day for 5 days and sacrificed 3, 30, 90 and 180 days later. Control rats were sham-exposed to clean air. Particulate atmospheres were generated by a dust dispenser designed to produce consistent concentrations throughout the exposure period. Briefly, the material is packed in pellet form in a precision bore tube which is moved at a constant rate against a rotating knife edge. The loosened dust is then introduced into a calibrated air steam to give the desired atmospheric concentration. The mean mass concentrations during exposures was 13.9 µg/l for glass fibres (GF) and 11.8 µg/l for ceramic fibres (CZF). It must be noted, however, that because of the blowing process used to manufacture refractive ceramic fibres exposure to this dust is mixed, with the number of particles greater than the fibrous component (Fig.1).

Collection of cells. Lavage of the pleural space was performed essentially as described by Oberdörster et al (1983). The abdomen was opened, a cannula was inserted through the diaphragm and 8 ml of cold 0.9% saline (NS) installed into the thoracic space, after which several washes were performed. For lavaging the lung, NS was installed through an intra-tracheal cannula and several washes were performed in situ. Both the lavaged pleural and lung cells were kept on ice until centrifuged at 1200 rpm for 20 minutes and resuspended. Two cytocentrifuge preparation were made and stained with Diff-Quick for cell differentiation, and for non-specific esterase activity using β-naphthol butyrate as a substrate.

Phenotypic analysis. A panel of 3 rat monoclonal antibodies each defining specific differentiation antigens was used in immunofluorscent studies on cell suspensions to characterise broncho-alveolar and pleural populations. The antibodies MRC OX41, OX42 and Class II (Ia) were obtained commercially (SeraLabs, Crawley Downs, Sussex, UK). The individual specificities and basic properties of MRC OX41 and OX42 have been determined by Robertson, White and Mason (1986).

RNA Dot Blot and Hybridisation. Total cellular RNA obtained from freshly lavaged or cultured alveolar macrophages was denatured with formaldehyde (White and Bancroft, 1982), serially diluted and spotted onto nitrocellulose. After baking the blots at 80°C for 2 h, hybridisation was performed with either alpha actin or inter-leukin one (IL-1) ^{32}P DNA probes prepared by random priming and autoradiographs analysed by scanning video densitometry. The alpha actin (Ac-3) was a gift Dr J. Warwick, ICRF, London and the IL-1 kindly provided by Dr S. Volsen, Eli Lilly, Bagshot.
Scanning electron microscopy. Pleural and lung cells suspended in RPMI 1640 containing 10% FCS were added to glass cover slips (1 x 10^6 cells/ml), allowed to adhere for 2 h at 37°C, washed vigorously and cultured overnight. They were prepared for SEM as described previously (Miller & Kagen, 1976).

RESULTS

Inhalation of GF or CZF did not result in significant differences in the number or population characteristics of cells recovered by broncho-alveolar lavage compared to air-exposed animals. More than 95% of cells were alveolar macrophages and 98% of cells were viable at all time points tested. When alveolar macrophages from fibre-exposed animals were examined by SEM they showed increased spreading on glass and increased numbers of cytoplasmic processes compared to air-exposed animals and many cells contained fibres at all time points examined (Fig.1).

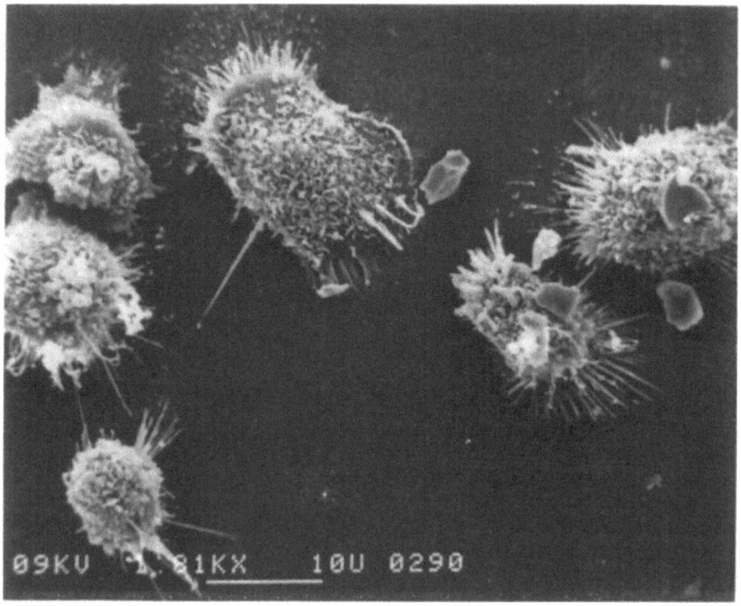

Fig.1. Scanning electron micrograph of alveolar macrophages obtained 30 days after exposure to ceramic fibres.

However, we were not able to demonstrate transcription of the IL-I gene (although high levels of alpha actin expression could be easily demonstrated) with cytoplasmic lysates prepared from freshly lavaged cells. Only when cells were cultured for 3 hours prior to lysate preparations could IL-I transcription be demonstrated suggesting that a phagocytic stimuli such as adherence is necessary for expression of IL-I.

Results of the pleural lavage cell counts are given in Table 1. In all groups the number of total lavaged cells increased with age and the population recovered was heterogenous. Typical distributions from air-exposed rats are 15-25% mast cells, 10-15% eosinophils, 5-10% lymphocytes, 2-8% mature macrophages and 45-60% macrophage-like and "other" cells. The majority of cells classified as macrophage-like have eccentric, uniform nuclei and are smaller than alveolar macrophages. The other mononuclear phagocyte (mature macrophages) are morphologically similar to alveolar macrophages. Cells not easily distinguishable ("other") appear to have "doughnut" shaped nuclei and could have been juvenile neutrophils or immature monocytes. Although the total numbers of cells harvested from the pleural spaces of air- or fibre-treated rats are not significantly different, the number of mature macrophages and macrophage-like cells appear to have increased in CZF fibre-exposed compared to control animals. However, whereas there was no discrepancy between morphology and esterase staining (considered a reliable cytochemical marker for macrophage identification) with alveolar macrophages the majority of pleural macrophage-like cells did not demonstrate any level of esterase activity.

Table 1. Differential counts of pleural cell populations from air of Man Made Mineral Fibre exposed rats

		% Cell type					
		Mast	Eosin	Lymph	Mature Macs	Macrophage-like	Other
Glass Fibres							
3 day	C	12	13.5	8	7	29.5	30
	T	17.5	13.5	9.5	11.5	27.5	20.5
90 day	C	27.5	14.5	1	3	20	34
		26	12.5	5.5	2	21	32
Ceramic Fibres							
3 day	C	15.5	13	10	7	26	28.5
	T	15.5	9	7.5	12	44	13
90 day	C	27	14	7	3	21	28
	T	30.5	9.5	4	7.5	32	16.5

In order to better characterise the pleural cell population, pleural and alveolar cells from untreated rats were identified by moncolonal antibodies against cell surface antigens. Phenotypic analysis revealed wide differences between cell populations with respect to two anti-rat peritoneal macrophage antibodies, MRC OX-41 and MRC OX-42

(Table 2). Alveolar macrophages are 95% OX-41 postive, are labelled with OX-42, whereas only 2% of pleural cells are OX-41 positive and 60% label with OX-42, demonstrating that with respect to surface phenotype they resemble one subset of alveolar macrophages. Unlike peritoneal macrophages, however, they bind only that antibody (OX-42) which recognizes a surface antigen associated with complement receptor functions.

Table 2. Phenotypic analysis of alveolar and pleural cell populations

Market	Alveolar cells % +ve	Pleural cells % +ve
OX-41	95	2
OX-42	37	60
Class II (Ia)	12	18
Non-specific esterase	75	15

Finally, examination of pleural cell cultures obtained after fibre exposure, demonstrated the presence of cells containing partly phagocytosed fibres as early as 3 days post-exposure of both GF and CZF fibres (Fig.2). Studies are

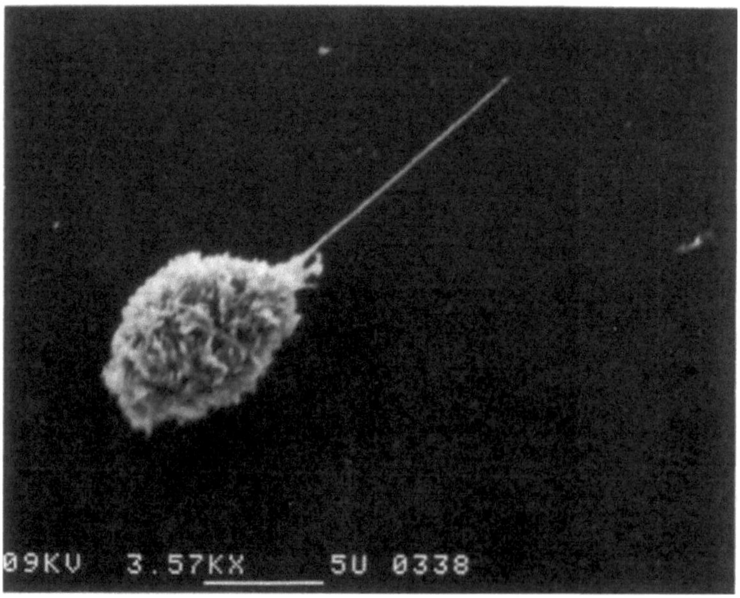

Fig.2. Screening electron micrograph of pleural cell obtained 3 days after exposure to ceramic fibres.

presently in progress to quantify the number of fibres present in the pleural lavage supernatants. The majority of fibres observed by SEM appeared shorter than 8 µm.

CONCLUSIONS

These early results, which form part of a larger investigation into the interaction of alveolar macrophages with pleural cavity cell populations during and after mineral fibre inhalation, raise several basic issues of conceptual importance. The migration of fibres into the pleura observed in this study appears to be faster than reported by Holt (1983) or Morgan <u>et al</u> (1977). Studies by Oberdoester <u>et al</u> (1983) also suggest an early arrival of fibres at pleural sites. Whatever the path of migration by which fibres reach the pleura, the possible role of short fibres in inducing pleural events ought therefore to be considered.

High numbers of eosinophils and mast cells were found in the pleural cavity of the rats used in this study and have also been found by other investigatiors who make use of the rat model (Lehnert, Valdez & Bolmalaski, 1985). Mast cells in particular have been shown to play a role in the development of fibrotic responses and it is not known to what extent pathological changes induced by pleural installation of fibres may be influenced by these inflammatory cells. Further research is needed to clarify this point.

Perhaps more importantly, it is evident from the above studies that pleural macrophages are unlike both peritoneal and alveolar macrophages. Furthermore there is an obvious lack of knowledge with respect to both the functional properties and phenotypic expression of these cells. Further studies into potential cellular responses and production of cytokines in the pleural cell population that may follow the inhalation of mineral dusts are needed to elucidate the underlying mechanisms involved in the pathogenesis of pleural responses.

References

Brody AR, Hill LH, Adkins B, O'Connor RW (1981) Chrysostile asbestos inhalation in rats: Deposition pattern and reaction of alveolar epithelium and pulmonary macrophages. Am Rev Respir Dis 123:670-679.

Holt PF (1983) Translocation of inhaled dust to the pleura. Environment Res. 31:212-220.

Lehnert BE, Vldez YE, Bomalaski SH (1985) Lung and pleural free- cell responses to the intrapulmonary deposition of particles in the rat. J Toxicol Environ Health 16:823-839.

Miller K, Kagan E (1976) The in vivo effects of asbestos on macrophage membrane structure and population characteristics of macrophages. J Reticuloendothel Soc. 20:159-171.

Morgan A, Evans JC, Holmes A (1977) Deposition and clearance of inhaled fibrous minerals in the rat. IN: Walton WH (ed) Inhaled Particles IV, Pergamon Press Oxford pp.259-274.

Oberdoerster G, Ferin J, Marcello NL, Meinhold SH (1983) Effect of intrabronchially instilled amosite on lavagable lung and pleural cells. Environm Health Perspect 51:41-48.

Robinson AP, White TM, Mason DW (1986) Macrophage heterogeneity in the rat as delineated by two monoclonal antibodies MRC OX-41 and MRC OX-42. Immunol 57:239-247.

Sebastien P, Fondimare A, Bignon J, Monchaux G, Desbordes J, Bonnand G (1977) Topographic distribution of asbestos fibres in human lung. In: Walton WH (ed) Inhaled Particles IV, Pergamon Press Oxford pp.435-446.

This work is supported by the UK Health and Safety Executive to whom our thanks are due.

THE ROLE OF ALVEOLAR MACROPHAGE ACTIVATION IN THE RESPONSE TO MINERAL FIBERS

G. J. Rosenthal, R. P. Stranahan,III, M. M. Fort, and M. I. Luster
Immunotoxicology Section
Systemic Toxicology Branch
National Institute of Environmental Health Sciences
Research Triangle Park, NC 27709
USA

Introduction

Alveolar macrophages (AM) perform a variety of functions important to the host which go beyond their role as phagocytic cells. These include, but are not limited to, the processing of complex antigens, secretion of reactive oxygen intermediates, and elaboration of bioactive soluble products (i.e. cytokines) (Fels and Cohn, 1986). Certain mineral fibers added to AM cultures can modulate a number of these cellular functions (Cantin et al., 1988; Gardner, 1984). By virtue of their immunoregulatory activity, their role as an early defense mechanism against inhaled particulates, and their intimate association with lung parenchyma, these cells are considered major elements in acute and chronic inflammation as well as in the proliferation of many cell types, including interstitial lung fibroblasts (Fels and Cohn, 1986).

AM recovered from healthy, nonsmoking individuals or normal animals have been characterized as 'down-regulated', a term describing the desirable homeostatic mechanism involved in protecting the host from the potential damaging effects of these cells under normal conditions (Holt, 1986). Such cells are in a 'resident' state with limited immunologic reactivity. A variety of stimuli, including lymphokines, monokines, serum proteins, lipopolysaccharide, muramyl dipeptide as well as xenobiotics can raise the activation level of the AM. Pathologic states in humans and experimental animals, including pulmonary infections, fibrosis, cancer, sarcoidosis, asbestosis, and inflammation also result in AM with varying levels of activation and often enhanced levels of immune reactivity (Brain 1980; Adams et al., 1988). The role that the stage of AM activation plays as a determinant in the ultimate biological response to fibers is an area that has not been adequately addressed. This is particularly important considering that aberrations in AM functions can provoke host damage via impaired host resistance or overactive destructive activity. If in vitro studies on the effects of mineral dusts are to be used to predict likely human disease, then a more thorough understanding of the relationship between the activation stage and the biological response to mineral fibers is needed. Employing in vitro culture conditions that elicit AM in various stages of activation defined by operational terms (resident, responsive, primed and activated) previously used to characterize peritoneal macrophages (Adams and Hamilton, 1984) and a spectrum of immune

NATO ASI Series, Vol. H30
Effects of Mineral Dusts on Cells
Edited by B.T. Mossman and R.O. Bégin

parameters, we have defined potentially susceptible or resistant stages of AM activation. This review will describe the activation profile of AM from F344 rats and focus on fiber induced modulation of three biologically active AM products that may play an important role in tissue damage and fibroblast proliferation; hydrogen peroxide (H_2O_2), interleukin -1 (IL-1) and tumor necrosis factor (TNF).

METHODS

Mineral Fibers Chrysotile and crocidolite asbestos were obtained from the National Institute of Environmental Health Sciences (NIEHS) repository. The physical and chemical properties of the asbestos fibers have been previously described (Campbell et al., 1980). Wollastonite was a gift from Dr. Bernard Adkins of Northrop Services, Inc. Before use, the mineral fibers were resuspended in RPMI 1640 to a concentration of 1 mg/ml, vortexed and immediately added to cultures at the desired concentration.

Alveolar macrophage collection and Culture Conditions AM were isolated from female F344 rats as previously described (Warheit et al., 1986). Recovery ranged from 6-10 x 10^6 per animal of which >95% were macrophages. Once washed and resuspended, AM were allowed to adhere to plastic microplates in serum free RPMI 1640 containing 25mM Hepes, 2 mM L-glutamine, and 50 ug/ml gentamicin sulfate (Sigma) hereafter refered to as RPMI. Following adherence for 1 to 2 hours at 37° C in 5% CO_2, the plates were washed 2X with warm RPMI to remove nonadherent cells and cultured for 24 hrs in RPMI with or without various stimulators to obtain operationally active cells. *Resident* cells were obtained by culturing in RPMI alone while *responsive* cells were obtained by culturing in the presence 5% FBS. *Primed* cells were cultured in either RPMI with 100 units/ml rat gamma interferon (γIFN) [TNF and IL-1 assays] or RPMI, 5% FBS and 100 units/ml γIFN [H_2O_2 assay]. To obtain cells in the *activated* stage, *primed* AM were cultured with lipopolysacharride (LPS) at 1 ng/ml to induce TNF production or 10 μg/ml for IL-1 or H_2O_2 production.

IL-1 assay Alveolar macrophages were added in 1 ml aliquots (10^6/ml) to 12-well culture plates and allowed to adhere. The plates were then washed 2X with RPMI to remove nonadherent cells. The cells were cultured for 24 hrs in 1 ml of the respective activation media as described above. Following culture, supernates were passed through 0.44 μ filters and stored at -20° C until assayed. Supernates that did not already contain serum were adjusted to 10% FCS prior to filtering to prevent loss of IL-1 on the filter.

IL-1 was assayed using a thymocyte co-stimulation assay (Mizel, 1981). Serial dilutions of supernates were made in RPMI media with appropriate supplements and added in 100 μl aliquots to flat bottom microtiter plates (Linbro). Thymocytes from 5 - 9 week old female B6C3F1 mice were suspended at 2.0 x 10^7/ml in media containing RPMI 1640, 10% FCS, and 2.5 x 10^{-5} M 2-mercaptoethanol (Sigma). These cells were added in 50 μl aliquots to each well along with 50 μl of PHA (Burroughs Wellcome) at 2.5 μg/ml. Cultures were incubated for 64 hours at 37° C in 5% CO_2 . Cells were harvested onto glass fiber filters following a 6 hr pulse with ^{3}H-thymidine (New England Nuclear, Boston, MA) and DNA incorporated activity determined.

TNF Assay TNF in conditioned media was measured by quantitating the cytolytic activity against the L929 target cell line. Briefly, 2.5 x 10^4 L929 cells in 0.1 ml of RPMI 1640 containing 10% FBS were dispensed into wells of a microtiter plate and grown to confluency. The medium was decanted and samples of conditioned media were added to the wells in quadruplicate, in the presence of 2 µg/ml of actinomycin D. After a further 18 hr incubation, the medium was removed and the cells were simultaneously fixed and stained with 0.2% crystal violet in 2% ethanol for 10 min, washed with water, and air dried. Cytolysis was calculated from the reduction in mean absorbance at 595 nm measured with a Titertek Multiskan Reader (Flow Labs), relative to control wells incubated with medium only. Reference wells containing known amounts of recombinant murine TNF (2 x 10^7units/mg: Genyme, Boston, MA) where used to generate a standard curve.

Hydrogen Peroxide Quantitiation H_2O_2 was assessed using a previously described microassay (Pick 1986). Briefly, $2x10^5$ AM were allowed to adhere to microtiter plates for 2 h, washed, and incubated in respective media for 24 h to generate cells in various stages of activation as described above. Cells were exposed to chrysotile, crocidolite or wollastonite at 10 µg/10^6 cells and cultured for another 24 hr period after which basal or opsonized zymosan (0.25 mg/ml) stimulated H_2O_2 production was measured following an additional 2 hr incubation.

Results

Generation of AM in Different States of Activation Table 1 qualitatively demonstrates that the culture conditions employed elicited AM with distinct immunologic activity. Those cells termed *resident* had no detectable Ia antigen on their cell surface and were poor producers of H_2O_2, TNF, and IL-1 . *Responsive* cells acquired substantial capacity to produce H_2O_2, as well as low levels of TNF without LPS stimulation, while IL-1 production and Ia expression were still not detectable. *Primed* AM continue to produce large quantities of H_2O_2, while their ability to produce TNF and IL-1 remain within the same range as seen within the *responsive* stage. A central characteristic observed with *primed* AM was the expression of Ia antigen on a majority of the cells (>50%). *Activated* AM were characterized by a decreased ability to produce H_2O_2, while TNF and IL-1 production were produced in much greater quantities than in other stages. These results are summarized below:

Stages	Conditions	Functions			
		H_2O_2	TNF	IL-1	Ia
Resident	24 h in serum free media	+	-	-	-
Responsive	24-48 in serum containing media	++++	++	-	-
Primed	24-48 in serum containing media with γIFN	+++	-	-	+++
Activated	24-48 in serum containing media with γIFN and LPS	++	++++	++++	ND

Table 1 - In vitro generation of AM in different stages of activation. Data are expressed qualitatively and based upon a minimum of 3 separate experiments for each endpoint. ND = not done

Effects of Mineral Fiber Exposure on Hydrogen Peroxide Production Figure 1 depicts basal H_2O_2 production following mineral fiber addition to cultures of AM. ***Resident*** and ***responsive*** cells were refractory to fiber induced alterations while *primed* cells appeared to be sensitive only to chrysotile asbestos addition. In contrast, while inherently producing less H_2O_2 than *primed* or *responsive* cells, *activated* AM cultures were the most sensitive to modulation by fiber exposure, showing significantly increased production not only with the fibrogenic asbestos dusts but also with the non-fibrogenic fiber, wollastonite.

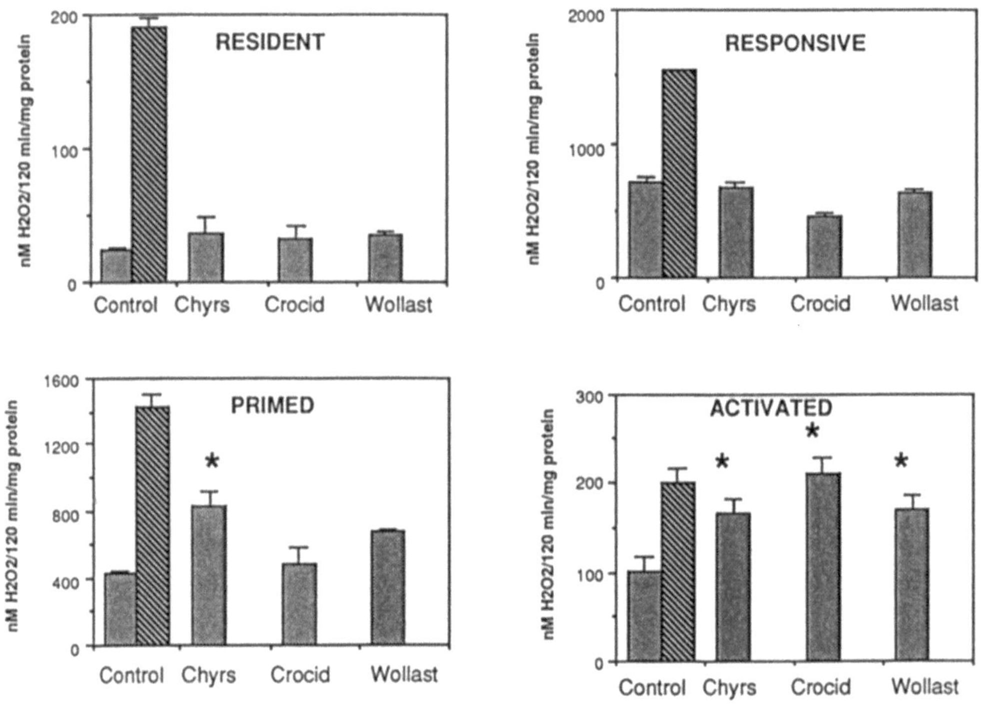

Figure 1 - Basal H_2O_2 production by AM in different stages of activation. chrys=chrysotile, crocid = crocidolite, wollast = wollastonite. Shaded bars represent basal H_2O_2 production while striped bar represents opsonized zymosan stimulated release from control cultures, provided for reference. Asterisks denote significant difference compared to vehicle control.

Effects of Mineral Fiber Exposure on Interleukin 1 and TNF Production IL-1 production in the absence of LPS stimulation was not observed in control or fiber exposed cultures (data not shown). Consequently, to determine whether mineral fibers modulated IL-1 production, we employed suboptimal concentrations of LPS (1 µg/ml) following the 24 hr incubation in respective activation media. LPS was added 30 minutes after mineral fiber addition. Under these conditions, chrysotile and crocidolite asbestos markedly stimulated IL-1 production when cells were pre-incubated in media resulting in *resident* and *responsive* cells (Figure 2). However, when cells were pre-incubated in media resulting in *primed* cells (cultured in the presence of IFN with or without serum), AM were refractory to mineral fiber induced modulation of IL-1 production. Wollastonite did not modulate IL-1 production from cells in any stage of activation even at concentrations 5X those used for the chrysotile or crocidolite asbestos.

In the absence of LPS stimulation TNF production was minimal in control or fiber exposed cultures after a 24 hr preincubation with respective media. AM produced small amounts of TNF during the initial preincubation period (generally < 8 units/ml; data not shown). Consequently, to determine whether mineral fiber exposure modulated TNF production, AM cultures were treated with suboptimal concentrations of LPS (1 ng/ml) following the 24 hr incubation in activation media. As in IL-1 assays, LPS was added 30 minutes after mineral fiber addition. Cultures treated with FCS, but not IFN, were highly effective in inducing TNF secretion (Figure 3). Cells cultured in RPMI alone or RPMI with IFN were unresponsive to 1 ng/ml LPS while cells cultured in RPMI with FCS produced small but detectable amounts of TNF. A marked synergistic effect was observed when cells were cultured in RPMI with both FCS and IFN present, eliciting >25X the concentration of TNF compared to cells cultured in FCS or IFN alone.

The addition of 30 µg/ml crocidolite asbestos to AM pre-incubated with RPMI and serum resulted in a significant increase in TNF production when compared to vehicle control (69 vs 6 units/ml, respectively). No other culture condition resulted in a fiber-sensitive stage of activation with respect to TNF production. Wollastonite similarly induced a significant increase in TNF production in AM pre-incubated with RPMI and serum although the magnitude of this increase was much smaller than that seen with crocidolite asbestos.

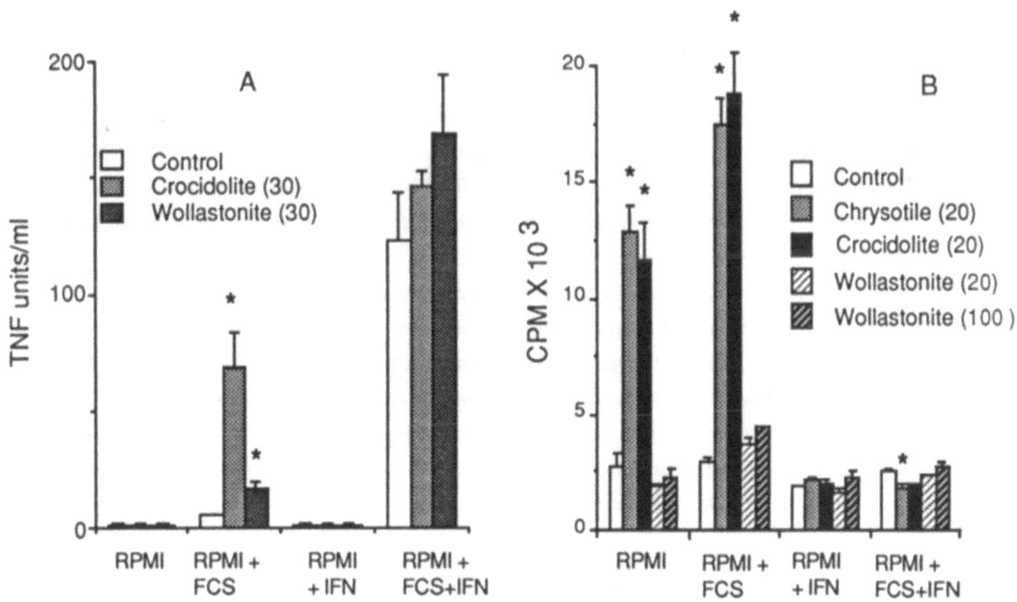

Figure 2: A - Modulation of TNF production by fibers following 24 h pre-culture in respective activation media. Fibers were added 30 minutes prior to suboptimal concentrations of LPS (1ng). Cells were then cultured for an additional 24 hr and conditioned media assayed for TNF. B -Modulation of IL-1 production by fibers following 24 h pre-culture in respective activation media. Fibers were added 30 minutes prior to suboptimal concentrations of LPS (1 µg). Cells were then cultured for an additional 24 hr and conditioned media assayed for IL-1. Values in parentheses ( ) represent the concentration of fibers employed in µg/ml

Discussion

With the increasing use of in vitro studies to understand and predict biological responses to mineral fibers, it is essential to understand the role of culture conditions in such responses and how these conditions may relate to the in situ pulmonary environment. The AM is a dynamic cell that responds to its external environment via activation, acquiring some capacities and functions while losing others. The culture conditions incorporated here can be considered, in a general sense, to represent conditions that may be found in the pulmonary environment. The *resident* cell demonstrated limited immunologic reactivity, as is purported for AM in normal healthy humans (Holt, 1986). The *responsive* cell is an AM that has been exposed to serum proteins, a situation not uncommon in smokers or conditions associated with diseases including fibrosis, inflammation or other pneumoconiosis. Exposure of cells to IFN (*primed*) can be classically seen in sarcoidosis, but may also occur in other pathogenic states considering the intimate involvement of interstitial lymphocytes with the pulmonary parenchyma. Triggers such as LPS used in the *activated* state may be manifested in vivo by bacterial antigens, or a variety of serum proteins such as fibronectin or complement components which have also been shown to stimulate release of TNF or IL-1 (Okusawa et al., 1988; Beezhold and Lause, 1987)

The present data demonstrate that the stage of activation is a major determinant in the ultimate response of the AM to mineral fiber exposure. Of the fibers examined, chrysotile, crocidolite, and wollastonite, the fibrogenic and carcinogenic fibers (chrysotile and crocidolite) generally modulated AM immune function while the non-fibrogenic wollastonite was relatively inert, even at much higher concentrations. Thus, it seems likely that fiber induced modulation of the AM is most likely a function of both the chemical and physical properties of the fiber and does not appear to be a nonspecific response to any particulate.

Is any one stage more susceptible to fiber induced immunologic modulation than another? The answer to this appears to be yes, however, this depends upon the endpoint examined. In other words, a stage of activation that is most susceptible with respect to one parameter, (e.g. IL - 1 production) may be relatively resistant with respect to another parameter, such as H_2O_2 production. This is not unusual if one considers that various immunologic functions or capacities are optimal at various stages (i.e. Ia is optimal in the primed stage while H_2O_2 production may be maximal in the responsive stage).

Using this multiple endpoint and multiple activation stage approach to understanding immunologic changes associated with mineral dust exposure should allow us to define activation not by a single parameter but by a complex and characteristic pattern of gene expression involving changes in secretory products, structural molecules, plasma membrane components and functional characteristics. Once the pattern of changes associated with known lung toxicants becomes understood, it may become possible to better understand, and possibly predict, the hazards of unknown particulates by comparing their spectrum of immune modulation with those compounds producing known human diseases.

References

Adams DO, Hamiltton T (1984) The cell biology of macrophage activation. Ann Rev Immunol 2:282-310

Adams DO, Lewis JG, Dean JH (1988) Activation of mononuclear phagocytes by xenobiotics of environmental concern: Analysis and host effects. In: Crapo JD, Gardner DE, Massaro EJ (eds) Target Organ Toxicology: Lung. Raven Press, New York p 351

Beezhold DH, Lause DB (1987) Stimulation of rat macrophage interleukin 1 secretion by plasma fibronectin. Immunol Invest 16:437-439

Brain JD (1980) Macrophage damage in relationship to the pathogenesis of lung disease. Env Hlth Persp 35:21-28

Campbell WJ, Higgens CW, Wylie AG (1980) Chemical and physical characterization of amosite, chrysotile, crocidolite and nonfibrous tremolite for oral ingestion studies by the NIEHS. Bureau of Mines Report of Investigations 8452

Cantin A, Dubois F, Begin R (1988) Lung exposure to mineral dusts enhance the capacity of lung inflammatory cells to release superoxide. J Leuk Biol 453:299-303

Fels AO, Cohn ZA (1986) The alveolar macrophage. J Appl Physiol 60:353-369

Gardner DE (1984) Alterations in macrophage function by environmental chemicals. Env Hlth Persp 55:343-349

Holt PG (1986) Down regulation of immune responses in the lower respiratory tract: The role of alveolar macrophages. Clin Exp Immunol 63: 261-270

Mizel SB (1981) Production and quantitation of lymhocyte activating factor (IL1). In: Herscowitz HB, Holden HT, Bellanti JA, and Ghattar A, (eds) Manual of Macrophage Methodology Marcel Dekker New York, p407

Okusawa S, Yancy KB, van der Meer JM, Endres S, Lonnemann G, Hefter K, Frank MM, Burke JF, Dinarello CA, Gelfand JA (1988) C5a stimulates secretion of TNF from human mononuclear cells in vitro J Exp Med 168:443-448

Pick E (1986) Microassays for superoxide and hydrogen peroxide production and nitroblue tetrazolium reduction using an enzyme immunoassay plate reader. In: Di Sabato G, Everse, J, (ed) Methods in Enzymology, vol 132 Academic Press, Orlando FL p 407

Warheit DB, Hill LH, George G, Brody AR (1986) Time course of chemotactic factor generation and the corresponding macrophage response to asbestos inhalation. Am Rev Resp Dis 134:128-133

In Vitro Studies of Leukocytes Lavaged from the Lungs of Rats Following the Inhalation of Mineral Dusts.

J.M.G. Davis, R.E. Bolton, D.M. Brown, G.M. Brown, K. Donaldson,
A.D. Jones, M.D. Robertson and J. Slight.
Institute of Occupational Medicine
8 Roxburgh Place
Edinburgh EH8 9SU

Introduction

Much has been learned concerning the harmful effects of mineral dusts and other toxic materials by both *in vivo* and *in vitro* techniques. Both approaches have disadvantages, however. With *in vivo* studies exposure can be conducted under physiologically normal conditions but it is more difficult to unravel the complex sub-cellular and molecular events. *In vitro*, target cells can be examined in much greater detail but dose levels may have to be excessive to produce effects in an acceptable time scale and the absence of the whole of the body's defensive systems and factors such as recruitment of new populations limit this approach. An approach now being adopted more frequently where target cells are readily available by simple techniques such as pulmonary lavage is to undertake exposure to harmful substances *in vivo* followed by examination of target cells and their products *in vitro*. This paper reports studies in which rats were exposed to three coalmine dusts, two varieties of asbestos, quartz and titanium dioxide administered by inhalation. Subsequently, their lungs were lavaged and a series of studies undertaken with the cell populations obtained.

Materials and Methods

Syngeneic rats of the PVG strain were exposed to dust clouds of chrysotile or amosite asbestos, quartz, titanium dioxide or three coalmine dusts from collieries mining anthracite, high rank coal and low rank coal. Dose levels were either $10 mg/m^3$ or $50 mg/m^3$ of respirable dust although chrysotile asbestos was used only at the $10 mg/m^3$ dose. Dust exposure continued for up to 75 days followed by a recovery period of 64 days. Animals were killed at intervals between 1 day of dusting and the end of the study and their lungs subjected to lavage with 0.85% NaCl at 37°C. The procedure was repeated

NATO ASI Series, Vol. H30
Effects of Mineral Dusts on Cells
Edited by B.T. Mossman and R.O. Bégin

4 times with massage of the lungs between each wash to increase the yield of cells. The first 10ml of saline recovered from the lungs was used to assess levels of enzymes present while the cell yield from all washes was centrifuged and pooled. Total cell numbers were obtained using a Neubauer chamber and the percentage viability was assessed using Trypan blue exclusion. Differential counts were undertaken following May Grunwald-Giemsa staining and these preparations were also used to examine the proportion of cells containing dust particles. Lactate dehydrogenase was estimated in the first saline wash by the method of Wroblewski and Ladue, (1955). The levels of macrophage activation was studied using the spreading assay described by Donaldson *et al*, (1984) and the ability of lavaged cells to phagocytose additional material was examined by treating coverslip cultures with fluorescinated latex beads. The production of the reactive oxygen intermediates (ROI) by lavaged cells was estimated by the method of Johnstone, (1981) for superoxide anion and the method of Pick and Keisari, (1980) for hydrogen peroxide. The effect of dust exposure on the ability of leukocytes to chemotact towards serum, either activated with zymosan or treated with dust was examined using Blindwell chambers and Nuclepore filters and the ability of lavage fluids to cause chemotaxis was studied using the same techniques with indicator cell populations of control bronchoalveolar cells or a neutrophil-rich bronchoalveolar cell population obtained following the intratracheal injection of *Corynebacterium parvum*.

Results

Pulmonary lavage of dusted rats showed a variable response in respect of cell numbers depending on the dust type. At a dose level of $10mg/m^3$ macrophage numbers did not increase significantly overall compared to controls, with any dust except quartz. With quartz macrophage numbers remained at normal levels for approximately 7 weeks but thereafter increased dramatically. For chrysotile, macrophage numbers were below control levels at 2 days, raised between 8 and 32 days of dusting before falling again by 75 days. At dose levels of $50mg/m^3$ the increase in macrophage numbers following treatment with quartz began sooner, at approximately 4 weeks and rose to much higher levels. With quartz at this dose, over 100×10^6 macrophages could be lavaged from each rat after 10 weeks of exposure. At the $50mg/m^3$ dose levels the three coalmine dusts all caused a significant increase in macrophage numbers by 75 days of exposure although the figures were approximately 5 times less than with quartz. Only amosite asbestos was used at the $50mg/m^3$ dose level. This caused some increase in macrophage numbers by 32 days but less than found with the coalmine dust samples. Macrophage numbers were not increased following treatment with titanium dioxide at 50mg/m^3.

An additional difference between the dusts in relation to macrophage numbers was found during the 64 day recovery period following dusting for either 32 or 75 days at both dose levels. With coalmine dusts, macrophage numbers, where raised, returned to control levels. With chrysotile at 10mg/m^3 and amosite at 50mg/m^3 macrophage numbers fell to below control levels during recovery after 32 days of dusting but remained slightly raised after 75 days of dusting. With quartz, macrophage numbers continued to increase, more than doubling during the recovery period at both dose levels. Examples of macrophage numbers lavaged from rat lungs following various dust treatments are illustrated in Figures 1 and 2. (see p. 343)

While several million pulmonary macrophages can be lavaged from the lungs of control rats, neutrophils are not found in these cell populations. The presence of neutrophils, therefore, represents a sensitive indication of pulmonary inflammation. At 10mg/m^3, all dust except titanium dioxide resulted in a significant increase in neutrophils by 32 days although once again the response to quartz was much greater than with coalmine dusts or asbestos. A similar pattern was found at 50mg/m^3 although with much higher neutrophil numbers reached and a significant increase detectable by 16 days or earlier. At 50mg/m^3 dose level the coalmine dust produced large numbers of neutrophils, although less than quartz and titanium dioxide produced a significant increase in neutrophil numbers between 52 and 75 days. Amosite asbestos although producing raised levels of neutrophils actually produced fewer than titanium dioxide at these later time points.

During the recovery period following treatment at 10mg/m^3, neutrophil numbers returned largely to control levels with all dusts except quartz. The number of lavageable neutrophils increased during this period with quartz although much less than the increase found with macrophages. A similar pattern was seen with quartz at the 50mg/m^3 dose level. At this dose, however, the number of lavageable neutrophils found during recovery had not returned to control levels by 75 days in animals treated with coalmine dusts, amosite or even titanium dioxide. Numbers of neutrophils lavaged from rat lungs in representative experiments are illustrated in Figures 3 and 4. (see p. 343)

The liberation of lactate dehydrogenase from cells is a very good indicator of cell damage and LDH levels in lavage fluid during the present study were found to parallel the neutrophil response very closely with most dusts. With a few dusts, however, a different pattern was seen. The low rank coalmine dust produced an early increase in LDH levels but at both 10mg/m^3 and 50mg/m^3 this fell back between 52 and 75 days to levels close to the controls. Chrysotile at 10mg/m^3 produced the most rapid rise in LDH of all dusts with very significant increases by 8 days of dusting. Thereafter the chrysotile figures were the highest recorded until 52 days when they were overtaken by quartz. Amosite at 10mg/m^3 also produced higher LDH figures in the later stages of dusting than the neutrophil figures would have indicated. With amosite, however, there was little variation in LDH release between 10mg/m^3 and 50mg/m^3 of dust. During the recovery period at both dose levels, LDH release fell with the coalmine dusts towards control figures. With amosite at 10mg/m^3 a reduction in LDH release was found during the recovery period but at 50mg/m^3 and with chrysotile at 10mg/m^3 the LDH figures remained at their high levels. With quartz at both dose levels the release of LDH continued to increase during the recovery period. Levels of LDH present in pulmonary lavage fluids from representative experiments are illustrated in Figures 5 and 6. (see p. 344)

Cells in the lavaged populations always showed over 90% viability when newly isolated. Where the dust was of a type clearly visible by light microscopy (quartz and chrysotile are not) the proportion of macrophages containing dust was also over 90% by as little as 8 days of dusting. The proportion of neutrophils containing dust was much lower and seldom exceeded 50%. Over 90% of macrophages were also capable of further phagocytosis of latex spheres.

The macrophage spreading assay indicated increased levels of macrophage activation with all dust types. However, no consistent pattern of increase was found and levels did not differ between the 10 and 50mg doses. There was no overall difference in the production of ROI between control or dusted leukocytes with any of the dusts although all populations were stimulable by treatment with PMA *in vitro*.

Cells from animals dusted with quartz, coalmine dust, titanium dioxide or chrysotile all showed reduced ability to chemotact towards serum, whether normal, zymosan-treated or dust-treated (chemotaxis studies were not undertaken with amosite asbestos). This reduction was least marked with titanium dioxide and most marked with quartz and it increased with exposure time particularly for studies using the 10mg/m^3 dose level (Table 1). Spontaneous production of chemotaxins by the different leukocyte populations showed no pattern consistent with levels of cell recruitment found in the lungs. The only consistent evidence for any triggering of chemotaxin release was zymosan-stimulated release of chemotaxins for control bronchoalveolar cells. This stimulation was evident in all the lavaged cell populations regardless of dust treatment.

Discussion

A combination of *in vivo* and *in vitro* techniques has permitted an examination of the early inflammatory effects produced by the inhalation of a range of mineral dusts. As might have been predicted, quartz a known toxic dust caused a severe pulmonary inflammation which continued to increase after the end of dust exposure. Coalmine dusts of both high and low rank, while less active, also produced significant inflammation as indicated by the presence of neutrophils and LDH in lavage fluids but this tended to subside after the end of dusting. Even titanium dioxide, a supposedly innocuous dust, caused an influx of neutrophils with LDH release after 50 days of dusting at a high dose. Amosite asbestos caused significant levels of pulmonary inflammation but no higher than coalmine dusts at the same dose levels and this is in keeping with low levels of *in vitro* toxicity previously reported (Gormley *et al.*, 1985). The findings for chrysotile asbestos appear somewhat anomalous. *In vitro* studies show chrysotile to be a highly toxic dust and this is probably reflected in the high levels of LDH found in lavage fluid. At the same time neutrophil levels, while raised, were less than with some coalmine dusts and numbers of lavageable macrophages were only above control levels between 8 and 52 days. After the recovery period following 32 days exposure, macrophage levels were below control figures. A possible explanation of these findings is that macrophages containing asbestos fibres, which accumulate around the respiratory and terminal bronchioles, tend to become bound together in agglomerates which sometimes fuse to form foreign body giant cells. It may well be that macrophages containing asbestos fibres are much less easily dislodged by the lavage process than macrophages containing particulate dust so that the number of cells obtained underestimates the inflammatory response in the lung. Neutrophils may well become involved in the same cellular aggregates as macrophages with their numbers in lavage fluid correspondingly reduced.

While previous *in vitro* studies have demonstrated that at least some of the dust types considered here were able to kill macrophages within a few hours (Gormley *et al.*, 1980,1985) the present study found that >90% of lavaged macrophages were still viable at all time points while a similar proportion of cells contained visible dust. This indicates that under physiological conditions of exposure macrophage killing by dusts may occur but a great excess of the leukocytes are alive and have altered phenotypes from the resting population. The altered activity of these cells is likely to be expressed as increased release of cytokines and mediators of great relevance to disease production. Dusted macrophages were indeed shown to be more activated than controls and activation is associated with increased levels of secretion. At the same time, the dusted cells showed reduced ability to chemotact which may reduce clearance rates and retain the cells in the lung tissue for long periods during which time these products can affect surrounding cells. This ability of macrophages and other inflammatory cells to orchestrate disease development in lung tissue is a key field of research which can best be undertaken using cell populations derived from *in vivo* exposure to harmful materials. We have already examined some aspects of leukocyte response that could be important in pulmonary disease (Brown *et al.*, 1988; Donaldson *et al.*, 1988[a,b]) but much remains to be done before we can fully understand the mechanisms by which inhaled mineral particles are able to initiate and control the disease process.

TABLE 1

CHEMOTAXIS OF BRONCHOALVEOLAR LEUKOCYTES FOLLOWING DUST EXPOSURE AT 10mg/m^3

DUST TYPES	SERUM TREATMENT	CONTROL	DUST-EXPOSED	CONTROL	DUST-EXPOSED
	-	13.5 (0.7)	8.5 (3.5)	17.0 (5.7)	14.0 (8.5)
TiO_2	Zymosan	80.0 (4.2)	55.5 (2.1)	73.5 (3.5)	66.0 (8.5)
	TiO_2	17.5 (4.9)	8.0 (2.8)	10.5 (0.7)	18.5 (2.1)
	-	26.5 (0.7)	11.0 (0.0)	54.5 (6.4)	14.0 (0.0)
Quartz	Zymosan	69.5 (3.5)	37.0 (2.8)	50.0 (17.0)	38.0 (2.8)
	Quartz	29.5 (3.5)	10.5 (2.1)	57.5 (6.4)	16.0 (1.4)
	-	26.0 (2.8)	9.5 (2.1)	15.5 (3.5)	4.0 (1.4)
H	Zymosan	89.0 (9.9)	58.0 (14.1)	60.0 (9.9)	38.5 (2.1)
	H	27.0 (7.1)	12.5 (4.9)	13.0 (0.0)	3.0 (0.0)
	-	35.0 (1.4)	21.0 (2.8)	39.5 (4.9)	14.5 (0.7)
A	Zymosan	67.0 (0.0)	48.0 (14.1)	70.0 (8.5)	35.5 (2.1)
	A	42.5 (9.2)	26.5 (3.5)	30.0 (1.4)	7.0 (0.0)
	-	16.5 (2.1)	9.5 (3.5)	4.0 (1.4)	6.0 (0.0)
L	Zymosan	69.5 (6.4)	34.0 (1.4)	38.5 (4.9)	8.5 (0.7)
	L	37.0 (0.0)	14.5 (2.1)	2.5 (0.7)	8.5 (0.7)
	-	23.5 (4.9)	6.5 (0.7)	30.0 (4.2)	9.0 (1.4)
Chrysotile	Zymosan	63.0 (4.2)	20.0 (1.4)	62.5 (0.7)	20.0 (4.2)
	Chrysotile	32.0 (1.4)	7.0 (0.0)	39.5 (4.9)	7.0 (0.0)

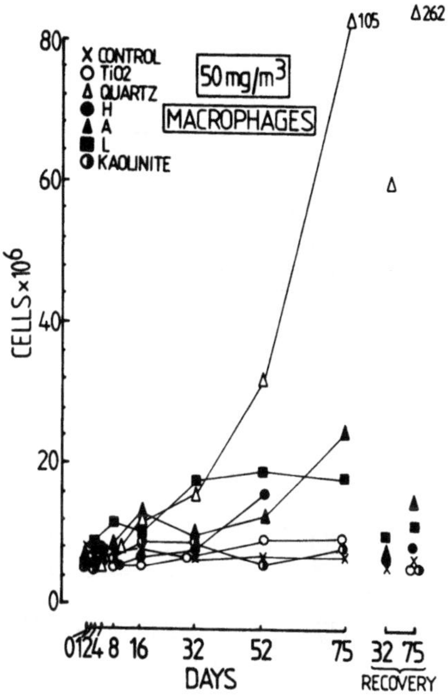

Figure 1

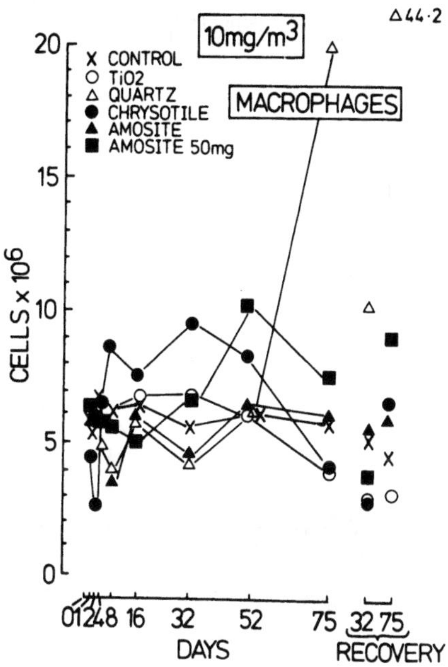

Figure 2

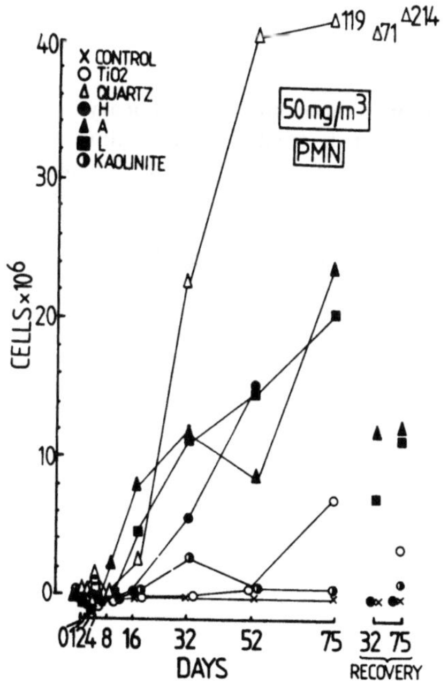

Figure 3

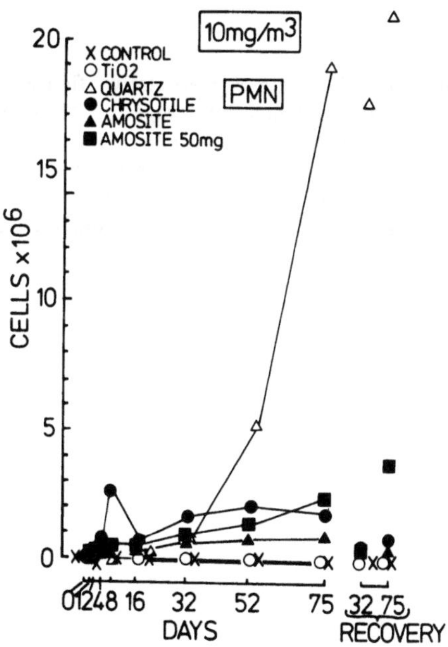

Figure 4

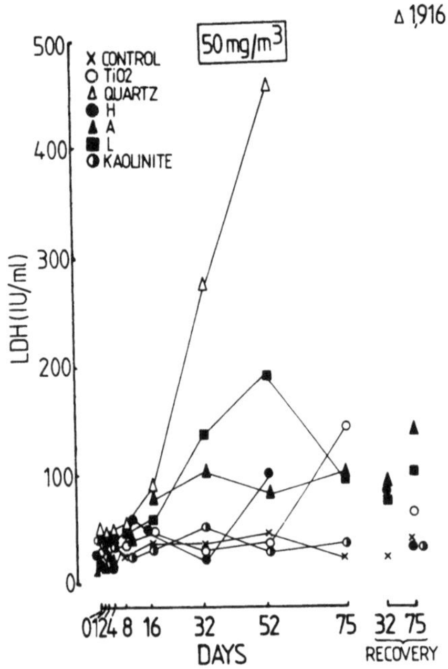

Figure 5

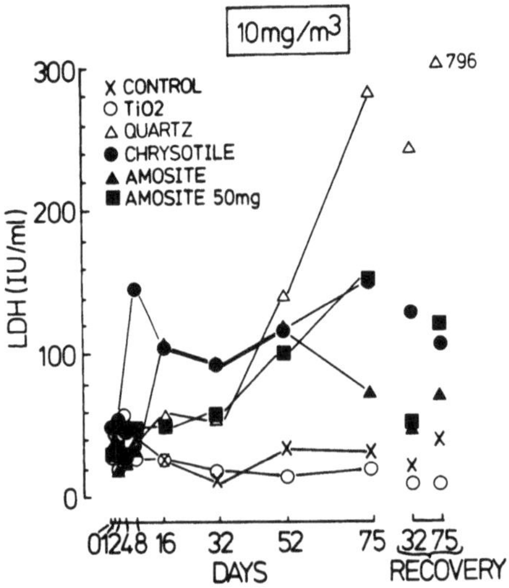

Figure 6

Figure Captions

Figures 1–6 illustrate representative figures for different components of pulmonary inflammation obtained by lavage. These were from animals treated by inhalation to a variety of mineral dusts at either 10mg/m^3 or 50mg/m^3 for periods of up to 75 days with some animals allowed to survive during a recovery period of 64 days. Figures 1 and 2 represent macrophage recovery from lungs. Figures 3 and 4 illustrate levels of PMN present. Figures 5 and 6 demonstrate the LDH content of lavage fluid.

References

Brown GM, Donaldson K (1988) Degradation of connective tissue components by lung-derived leukocytes *in vitro*: role of proteases and oxidants. Thorax 43:132–139

Donaldson K, Bolton RE, Brown DM, Douglas A (1984) An improved macrophage spreading assay. Immunol Comm 13:229–244

Donaldson K, Slight J, Brown GM, Bolton RE (1988[a]) The ability of inflammatory bronchoalveolar leukocyte populations elicited with microbes or mineral dust to injure alveolar epithelial cells and degrade extracellular matrix *in vitro*. Br J Exp Path 69:327–338

Donaldson K, Slight J, Bolton RE (1988[b]) Oxidant production by control and inflammatory bronchoalveolar leukocyte populations treated with mineral dusts *in vitro*. Inflammation 12:231–243

Gormley IP, Brown GM, Collings P, Davis JMG, Ottery J (1980) The cytotoxicity of respirable dusts from collieries. In: Brown RC, Gormley IP, Chamberlain M, Davies R (eds) The *in vitro* effects of mineral dusts, Academic Press, London, pp 19–24

Gormley IP, Brown GM, Cowie H, Wright A, Davis JMG (1985) The effects of fibre length on the *in vitro* cytotoxicity of asbestos samples in three different assay systems. In: Beck EG, Bignon J (eds) The *in vitro* effects of mineral dusts, NATO A51 series, Ecological Sciences, vol G3. Springer, Berlin, pp 397–404

Johnstone RB (1981) Secretion of superoxide anion. In: Adams DO, Edelson PJ, Koren HS (eds) Methods for studying mononuclear phagocytes, Academic Press, London, pp 489–497

Pick E, Keisari Y (1980) A simple colorimetric method for the measurement of hydrogen peroxide produced by cells in culture. J Immunol Methods 38:161–170

Wroblewski F, Ladue JS (1955) Lactic dehydrogenase activity in blood. Proceedings of the Society for Experimental Biology and Medicine 90:210–213

PREDICTIVE VALUE OF *IN VITRO* PULMONARY MACROPHAGE FUNCTIONAL ASSAYS TO ASSESS *IN VIVO* CLEARANCE OF INHALED PARTICLES

David B. Warheit, Arnold R. Brody*, and Mark A. Hartsky

E.I. du Pont de Nemours and Co. Inc.
Haskell Laboratory for Toxicology and Industrial Medicine
P.O. Box 50, Elkton Rd.
Newark, DE 19714

and

*Laboratory of Pulmonary Pathobiology
National Institute of Environmental Health Sciences
Research Triangle Park, NC 27709

INTRODUCTION

The development of *in vitro* methods in biology has provided important tools for carrying out mechanistic studies. These techniques offer the advantage of studying biological responses under rigorously controlled experimental conditions and are particularly useful for the study of readily accessible cells (e.g., pulmonary macrophages). In addition, as more effective methodologies evolve, the numbers of experimental animals used for toxicological studies may be reduced. However, *in vitro* experimental systems do not reflect realistic exposure conditions and cannot simulate cell-cell interactions and corresponding microenvironments which are integral to a number of physiologic and pathologic events.

Pulmonary macrophages (PM) form the first line of defense in the clearance of inhaled particulates from the alveolar regions of the lung (Brain 1985). Macrophage chemotactic and phagocytic responses are essential for efficient particle clearance (Warheit *et al*. 1988a). In earlier studies we reported that rat pulmonary macrophages accumulate selectively at sites of asbestos fiber deposition and phagocytize fibers following inhalation exposure (Warheit *et al*. 1984a). In attempting to elucidate the mechanism(s) of macrophage migration, we showed that asbestos fibers activate complement *in vitro* in both serum and in lavaged proteins leading to the generation of chemoattractants (Warheit *et al*. 1985,1986). A variety of

NATO ASI Series, Vol. H30
Effects of Mineral Dusts on Cells
Edited by B.T. Mossman and R.O. Bégin

inorganic dusts have also been shown to activate complement *in vitro* (Hill *et al*. 1982, Olenchock *et al*. 1978, Saint-Remy and Cole 1980, Warheit *et al*. 1984c, Wilson *et al*. 1977, Yano *et al*. 1984). In addition, in a recent comparative study we reported that rat PM demonstrated a significantly greater chemotactic response to complement-dependent chemotactic factors when compared to mouse, hamster, or guinea pig PM (Warheit *et al*. 1988b).

In the present study we have postulated that *in vitro* PM functional assays can be used to predict macrophage clearance responses of inhaled particles *in vivo*. Particle recognition by pulmonary macrophages is a prerequisite for dust clearance from the alveolar regions of the lung. To test this hypothesis, we have carried out 2 sets of experiments to compare PM chemotactic responses *in vitro* with PM accumulation at sites of particle deposition following brief inhalation exposures. First, using a variety of particulates we showed that a correlation exists between complement activation by inorganic particles *in vitro* and rat PM migration to sites of particle deposition *in vivo*. In the second set of comparative experiments using rodent PM, we attempted to show that rat PM, which are the most responsive to complement-derived chemoattractants *in vitro*, migrate to sites of particle deposition *in vivo* at greater rates than similarly exposed hamsters, guinea pigs, or mice.

METHODS

Animals and Pulmonary Lavage

CD® rats were used in the first set of studies. In the comparative rodent study, one strain each of rat (strain Crl:CD®BR), mouse (strain Crl:CD®-1 BR), syrian golden hamster (strain Lak:LVG(SYR), and Hartley guinea pig (strain Crl:(HA)BR VAF/Plus), all males and 8 weeks of age were utilized. All animals were virus-antibody free and no evidence of viral or other infectious disease was discovered during regular health screenings.

Prior to the lavage procedure, all animals were anesthetized by intraperitoneal injections of sodium pentobarbital and the tracheas were isolated and clamped. An incision was made 2 rings below the larynx and a butterfly catheter was secured into place. The catheter was connected to a reservoir located 10 cm above the thorax of the animal. The lungs were lavaged with a warmed phosphate-buffered-saline solution (Ca++- Mg++ free) which flowed into the lungs by gravity. This procedure was repeated a minimum of 5 times per animal (Warheit et al. 1984b).

Inorganic Particulates

UICC crocidolite asbestos fibers were obtained from NIEHS and have been previously characterized (Warheit et al. 1984b). The glass fibers (Code 100) were provided by Dr. James Leineweber, Manville Corp., Denver CO. The frequency distribution of lengths and diameters of Code 100 glass fibers have been reported previously (Hesterberg and Barrett 1984). Carbonyl iron particles were purchased from GAF Corp (New York, NY). Volcanic ash from the Mt. St. Helen's volcano eruption in May 1980 were obtained from the Environmental Protection Agency.

Serum Activation and Chemotaxis

Pulmonary macrophage chemotaxis was used as a bioassay for serum complement activation as described previously (Warheit et al. 1986). To test for serum complement activation, fresh rat serum was incubated with either crocidolite asbestos fibers, fiberglass, carbonyl iron particles, or Mt. St. Helens ash particles at concentrations of 25 mg/ml. In the comparative rodent studies, zymosan particles were incubated with autologous sera (i.e., rat, hamster, guinea pig, or mouse serum) at concentrations of 10 mg/ml. The serum-particle suspensions were incubated in conical tubes for 45 minutes in a shaking water bath at 37°C. The "activated" sera was then incubated for an additional 30 minutes at 56°C to inhibit any further activation from occurring, and then centrifuged to remove the particles (250 x g for 10 minutes). Normal heated serum (37°C for 40 minutes) and decomplemented sera (56°C for 30 minutes, then incubated with fiberglass after decomplementation) were used as additional negative controls. The activated sera was diluted to concentrations of 1, 5 and 10% (v/v) in an RPMI 1640 buffer solution containing 10 mM Hepes and 1% bovine serum albumin.

The chemotactic responses of pulmonary macrophages were assessed using polycarbonate filters inserted into blind well chambers (Snyderman 1982). The cells were plated at concentrations of 1.5 $x10^6$/ml into the upper portion of the chamber and incubated for 3.5 hours. Following termination of the assay, the filters (pore size = 5 μm) were fixed with ethanol and stained for light microscopy. To quantitate macrophage migration, the numbers of PM which had migrated from the top to the opposite side of polycarbonate filters were counted in 20 predetermined high power fields (Warheit et al. 1985).

Inhalation Studies

In the first set of experiments, rats were exposed to aerosols of crocidolite asbestos fibers, glass fibers, carbonyl iron (CI) particles, or Mt. St. Helens ash particles at concentrations ranging from 10-20 mg/m^3 total mass for 1, 3 or 5 hours in whole body inhalation chambers. Control rats were similarly sham-exposed to room air. In the second set of experiments, rats, mice, guinea pigs, and hamsters were exposed to aerosols of carbonyl iron particles for 6 hr in nose-only holders at concentrations of 100 mg/m^3.

Twenty-four or 48 hr after exposure to the dust aerosol, the animals were anesthetized with a lethal dose of pentobarbital and the lungs were infused with Karnovsky's fixative (i.e., 1% paraformaldehyde, 1% glutaraldehyde) through the trachea at pressures of 15 cm of water. The fixed lung tissues were then dissected and prepared for scanning electron microscopy (SEM) as previously described (Warheit et al. 1984a). Bifurcation surfaces revealed by random dissection of critical-point-dried lung tissue were evaluated by SEM. The numbers of macrophages which had migrated to sites of particle deposition (i.e., alveolar duct bifurcations) were recorded in both dust-exposed and sham-exposed rats.

RESULTS AND DISCUSSION

In Vitro Chemotaxis

Pulmonary macrophage chemotaxis was utilized as a bioassay for complement activation. Addition of crocidolite asbestos fibers, fiberglass or carbonyl iron particles to rat sera generated increased chemotactic responses in comparison to rat serum incubated with Mt. St. Helens ash particles, normal heated sera, or decomplemented controls (Table 1).

In a comparative study of rodent PM chemotactic responses to complement-derived chemoattractants, CD® rat PM were the most responsive species to zymosan-activated sera ($p<0.01$). The responses of CD®-1 mice PM were intermediate, and hamster and guinea pig PM were unresponsive to complement-dependent chemotactic factors (Table 2).

Macrophage Accumulation In Vivo

Scanning electron microscopic techniques were used to quantify pulmonary macrophage accumulation at sites of particle/fiber deposition following brief exposures. As previously described, the inhaled particulates deposited preferentially on alveolar duct bifurcations (Brody *et al*. 1981, Brody and Roe 1983). Within 48 hours after exposure, pulmonary macrophages migrated to sites of particle/fiber deposition in the lungs of rats exposed to crocidolite asbestos, fiberglass, and carbonyl iron particles ($p<0.05$ compared to sham controls) (Table 3). In contrast, macrophages did not accumulate at sites of particle deposition in animals exposed to aerosols of Mt. St. Helens ash (Table 3), and this was consistent with the inability of the volcanic ash to activate complement in serum (Table 1).

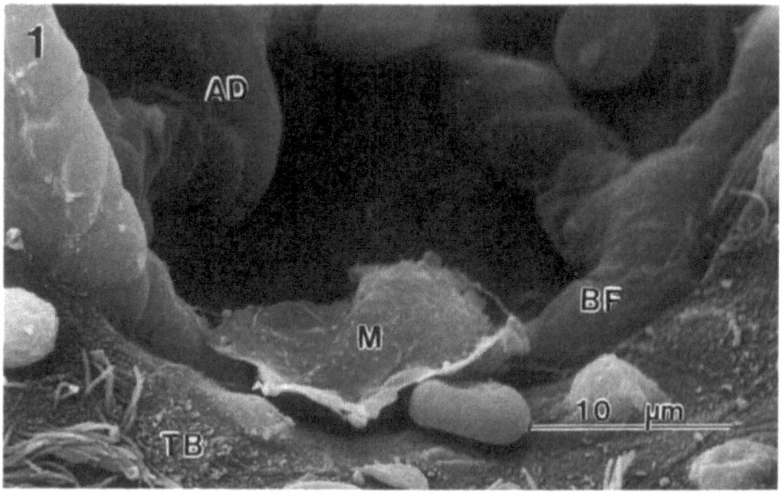

Fig. 1. Scanning electron micrograph of a first alveolar duct bifurcation (BF) located at the junction of the terminal bronchiole (TB) and alveolar duct(s) (AD). A rat pulmonary macrophage (M) has accumulated at the bifurcation within 24 hr following carbonyl iron aerosol exposure.

In a comparative rodent study, quantitative data on PM accumulation *in vivo* demonstrated that increased numbers of CD® rat PM migrated to sites of carbonyl iron particle deposition compared to the other 3 species (Fig. 1, Table 4). These data are consistent with recently reported time course studies which showed that increased percentages of rat PM phagocytized CI particles *in vivo* at 24 and 48 hours postexposure (Warheit *et al*. 1988c). In addition, the finding of increased accumulation of rat PM on bifurcations correlated well with the *in vitro* chemotactic results presented in Table 2, wherein rat PM were the most chemotactically active cells responding to complement-dependent chemotactic factors.

Pulmonary macrophages (PM) play an important role in clearing inhaled particulates from the lung (Brain 1985). Previously we demonstrated that inhaled asbestos fibers activate complement-derived chemoattractants to recruit PM to sites of particle deposition. In the current study we tested the hypothesis that *in vitro* PM functional assays can be used to assess macrophage clearance of inhaled particles *in vivo*.

First we showed that a correlation exists between complement activation by inorganic particles *in vitro* and PM migration to sites of particle deposition *in vivo*. In this regard, crocidolite asbestos fibers, fiberglass, and carbonyl iron (CI) particles generated chemotactic activity in rat serum and concentrated lavaged proteins. In contrast, Mt. St. Helens ash particles did not induce any chemotactic activity in either serum or lavaged proteins (Warheit *et al*. 1988a). Subsequently, rats were exposed to brief aerosols of each of the particulates listed above. With the exception of Mt. St. Helens ash, all of the particulates induced significant recruitment of PM to sites of particle/fiber deposition within 48 hr post-exposure.

In an earlier set of experiments, 2 strains of rats, 3 strains of mice and 1 strain each of hamster and guinea pig were lavaged for *in vitro* PM morphologic and functional studies (Warheit *et al*. 1988b). Hamster and rat PM demonstrated the highest phagocytic activity while mouse PM phagocytosis was depressed. Chemotaxis studies showed that both strains of rat PM migrated best to particle-generated chemotactic factors while the response to n-formyl peptides was enhanced in hamsters. In the current study we have confirmed these *in vitro* chemotactic data in rodents using 1 strain each of rat, mouse, hamster and guinea pig. To test the predictive value of *in vitro* PM functional assays to assess *in vivo* clearance of CI particles, all species were exposed to aerosols of CI particles for 6 hr at 100 mg/m3. Using scanning electron microscopic analysis of PM accumulation at sites of particle deposition, we determined that increased numbers of rat PM migrated to sites of CI deposition ($p<0.01$)(Table 4) and phagocytized particles (Warheit *et al*. 1988c) compared to the other 3 species. Taken together with our previous findings, the results from these studies suggest that the rat is the most efficient species for clearing inhaled CI particles. In addition, it seems clear that our *in vitro* data correlated well with the results of inhalation studies and thus, we conclude that *in vitro* PM functional assays are useful in assessing pulmonary macrophage clearance responses to inhaled particles and fibers.

TABLE 1.
RAT PULMONARY MACROPHAGE CHEMOTACTIC RESPONSES TO DUST-ACTIVATED SERA

	Numbers of Migrating Macrophages/20 HPF		
	1%	5%	10% DAS
Normal Heated Sera	30 ± 5	62 ± 15	109 ± 15
Decomplemented Sera + Fiberglass	26 ± 15	38 ± 3	53 ± 14
Mount St. Helens Ash	37 ± 11	64 ± 15	115 ± 24
Carbonyl Iron	69 ± 18	112 ± 20*	138 ± 15*
Crocidolite Asbestos	78 ± 16	127 ± 15^	227 ± 18^
Fiberglass	67 ± 8	159 ± 40^	282 ± 17^

DAS = Dust-activated sera
N= 9 for each particle tested
* $p < 0.05$ when compared to Mt. St. Helens or normal heated sera
^ $p < 0.01$ when compared to Mt. St. Helens or normal heated sera

Table 2.
SPECIES COMPARISONS OF PULMONARY MACROPHAGE CHEMOTACTIC RESPONSES TO COMPLEMENT-DERIVED CHEMOTACTIC FACTORS

	Numbers of Migrating Macrophages/20 HPF		
SPECIES	1%	5%	10% ZAS
GUINEA PIG	51 ± 13	37 ± 9	28 ± 3
HAMSTER	55 ± 10	50 ± 17	35 ± 6
CD®-1 MOUSE	85 ± 27	150 ± 44	97 ± 24
CD® RAT	468 ± 97	693 ± 80	487 ± 76

ZAS = Zymosan-activated sera
HPF = High power fields

TABLE 3.
PULMONARY MACROPHAGE CHEMOTAXIS TO ALVEOLAR DUCT BIFURCATIONS 48 HR FOLLOWING PARTICULATE EXPOSURE.

EXPOSURE	# RATS	# BIFURCATIONS EXAMINED	% RESPONSE OF SHAM-EXPOSED CONTROLS
SHAM	6	43	100 ± 150%
MT. ST. HELENS VOLCANIC ASH	3	36	97 ± 120%
CARBONYL IRON	4	25	500 ± 254%*
CROCIDOLITE ASBESTOS	3	15	433 ± 320%*
FIBERGLASS	2	6	715 ± 320*

Values are % response of accumulated pulmonary macrophages over sham controls.
* $p < 0.01$ when compared to sham or Mt. St. Helens ash by Student's t-test.
Mt. St. Helens = 3 hr exposure and 48 hr recovery period.
Carbonyl iron = 3 hr exposure and 48 hr recovery period.
Crocidolite asbestos = 1 hr exposure and a 48 hr recovery period.
Fiberglass = 1 hr exposure and a 48 hr recovery period.

TABLE 4.
SPECIES COMPARISONS OF PM ACCUMULATION AT ALVEOLAR DUCT BIFURCATIONS 24 OR 48 HR FOLLOWING CARBONYL IRON PARTICLE EXPOSURE.

SPECIES	# OF BIFURCATIONS EXAMINED	# PM AFTER 24 HRS	# PM AFTER 48 HRS
GUINEA PIG	31	2.4 ± 1.3	3.0 ± 1.6
HAMSTER	24	3.4 ± 1.8	2.9 ± 1.4
CD®-1 MOUSE	41	3.4 ± 1.5	3.9 ± 1.5
CD® RAT	18	4.8 ± 1.8*	4.5 ± 1.1*

* $p < 0.01$ compared to guinea pig, hamster, and mouse.

MANUSCRIPT REFERENCES

Brain JD (1985) Physiology and pathophysiology of pulmonary macrophages. In: Reichard SM, Filkins J (eds) The Reticuloendothelial System, vol 7B. Plenum, New York, pp. 315-337.

Brody AR, Hill LH, Adkins B Jr., O'Connor RW (1981) Chrysotile asbestos inhalation in rats: Deposition pattern and reaction of alveolar epithelium and pulmonary macrophages. Am Rev Respir Dis 123:670-679.

Brody AR and Roe MW (1983) Deposition pattern of inorganic particles at the alveolar level in the lungs of rats and mice. Am Rev Respir Dis 128:724-729.

Hesterberg TW and Barrett JC (1984) Dependence of asbestos and mineral dust-induced transformation of mammalian cells in culture on fiber dimension. Cancer Res 44:2170-2180.

Hill JO, Rothenberg SJ, Kanapilly GM, Hanson RL, and Scott BR (1982) Activation of immune complement by fly ash particles from coal combustion. Environ Res 28:113-122.

Olenchock SA, Mull JC, Major PC, Peach MJ III, Gladish MD, and Taylor G (1978) In vitro activation of the alternative pathway of complement by settled grain dust. J Allergy Clin Immunol 62:295-302.

Saint-Remy JMR, Cole P (1980) Interactions of chrysotile asbestos fibres with the complement system. Immunology 41:431-436.

Snyderman R (1982) Chemotaxis of human and murine mononuclear phagocytes. In: Adams DO, Koren H, Edelson P (eds) Methods for Studying Mononuclear Phagocytes. Academic Press, New York, p 535.

Warheit DB, Chang LY, Hill LH, Hook GER, Crapo JD, and Brody AR (1984a) Pulmonary macrophage accumulation and asbestos-induced lesions at sites of fiber deposition. Am Rev Respir Dis 129: 301-310.

Warheit DB, Hill LH, and Brody AR (1984b) Surface morphology and correlated phagocytic capacity of pulmonary macrophages lavaged from the lungs of rats. Exp Lung Res 6: 71-82.

Warheit DB, Hill LH, and Brody AR (1984c) In vitro effects of crocidolite asbestos and wollastonite on pulmonary macrophages and serum complement. Scanning Electron Microscopy II:919-926.

Warheit DB, George G, Hill LH, Snyderman R and Brody AR (1985) Inhaled asbestos activates a complement-dependent chemoattractant for macrophages. Lab Invest 52: 505-514.

Warheit DB, Hill LH, George G and Brody AR (1986) Time course of chemotactic factor generation and the corresponding macrophage response to asbestos inhalation. Am Rev Respir Dis 134: 128-133.

Warheit DB, Overby LH, George G, and Brody AR (1988a) Pulmonary macrophages are attracted to inhaled particles through complement activation. Exp Lung Res 14: 51-66.

Warheit DB, Hartsky MA and Stefaniak M (1988b) Comparative physiology of rodent pulmonary macrophages: In vitro functional responses. J Appl Physiol. 64: 1953-1959.

Warheit DB, Brody AR, and Hartsky MA (1988c) Lung clearance of inhaled particulates: Mechanism of macrophage recruitment and phagocytosis at sites of particle/fiber deposition. In: Wehner AP (ed) Biological Interactions of Inhaled Mineral Fibers and Cigarette Smoke, in press, 1988.

Wilson MR, Gaumer HR, Salvaggio JR (1977) Activation of the alternative complement pathway and generation of chemotactic factors by asbestos. J Allergy Clin Immunol 60:218-225.

Yano E, Takeuch A, Yukiyamo Y, and Brown RC (1984) Chemotactic factor generation by asbestos: Fibre type differences and the effect of leaching. Br J Exp Pathol 65:223-230.

Leukotriene B_4 and Tumor Necrosis Factor Production after <u>in vitro</u> Exposure of Rat Alveolar Macrophages to Mineral Dust: Potential Role in Fibrogenesis

C. Dubois, E. Bissonnette and M. Rola-Pleszczynski
Immunology Division, Faculty of Medicine
University of Sherbrooke, Sherbrooke, Québec, Canada, J1H 5N4.

INTRODUCTION

Prolonged inhalation of asbestos fibers and silica particles causes chronic inflammatory reactions and interstitial fibrosis. Although the exact mechanism by which these particles can induce lung diseases remains poorly understood, increasing evidence suggests that alveolar macrophages (AM) play a crucial role in the onset and development of asbestosis and silicosis through their ability to release potent inflammatory and fibrogenic mediators (Thomas and Hunninghake, 1987). Activated macrophages produce growth stimulatory activities toward fibroblasts and well characterized cytokines such as tumor necrosis factor (TNF) (Lemaire and al. 1986; Sayers and al. 1987). This cytokine, initially recognized because of its cytotoxic and anti-tumor properties, was shown in recent years to exert a role in systemic and lung inflammation (Moldawer and al. 1987; Bagby and Summer 1988), and in fibroblast growth regulation (Sugarman and al. 1985). Pulmonary macrophages also represent equally a particularly rich source of leukotriene B_4 (LTB_4). This lipoxygenase product may amplify the inflammatory process partly by its ability to modulate the production of immune and inflammatory cytokines, such as interleukin 1, interleukin 2 and gamma-interferon (Rola-Pleszczynski and Lemaire 1985; Rola-Pleszczynski and al. 1986). In this report, we investigated the role of LTB_4 as a potential mediator of lung disease by examining the effect of endogenous and exogenous LTB_4 on the regulation of mineral dust-induced TNF release by rat AM. The contribution of TNF in macrophage-derived fibrogenic activity was also examined.

NATO ASI Series, Vol. H30
Effects of Mineral Dusts on Cells
Edited by B. T. Mossman and R. O. Bégin

MATERIALS AND METHODS

Rats: Male Wistar rats weighing 200 to 250 g were purchased from Charles River Canada Inc. (St-Constant, Québec).

Chemical reagents and mineral dusts: LTB_4 was kindly supplied by Drs A.W. Ford-Hutchinson and J. Rokash, Merck-Frosst Laboratories (Dorval, Canada); nordihydroguaiaretic acid (NDGA) was purchased from the Sigma Chemical Co. (St-Louis, MO) and AA861 was generously provided by Dr Masao Nishikawa, Takeda Chemical Industries (Osaka, Japan). All reagents were dissolved in ethanol and diluted with RPMI 1640 medium giving final ethanol concentrations of less than .03%. Chrysotile A asbestos fibers (UICC standard reference preparation) and min-U-Sil-5 particles (Pennsylvania Glass Sand Co. Pittsburg, PA) were suspended in sterile saline, sonicated and used in AM cultures as indicated. Murine anti TNF-α was purchased from Genzyme Corporation (Boston MA). One milliliter of antiserum neutralized approximatively 10^6 units of TNF.

Preparation and assay of AM supernates: Alveolar macrophages were obtained by lavaging the lung with a total volume of 50 ml PBS in 6-8 ml aliquots. Cells were counted in a hemocytometer chamber and viability was determined by trypan blue exclusion. For production of TNF supernates, 0.5 X 10^6 AM were incubated in RPMI 1640 medium supplemented with 5% FBS. After 24 h incubation, the cells were centrifuged and the supernatant collected. TNF activity was estimated by the L-929 cytotoxic assay as described by Ruff and Gifford (18). One unit of TNF activity was defined as the amount required to lyse 50% of L-929 target cells. Each assay was standardized with a human recombinant TNF preparation containing 9.6 X 10^6 U/mg protein (BIOGEN Research Corp. Cambridge, MA). The LTB_4 content of AM supernatants was mesured by radioimmunoassay (RIA, Seragen, Boston MA). The antiserum had less than 0.1% of crossreactivity with either LTC_4, LTD_4, LTE_4, 5-HETE, 12-HETE, 15-HETE and cyclooxygenase metabolites. For determination of fibroblast growth factor production, AM supernatants were added to human embryonic lung fibroblasts IMR-90 and fibroblast proliferation was assessed by thymidine incorporation (Lemaire and al. 1986) and the absorbance (590nM) of cultures stained with crystal violet.

Statistical analysis: Data were analysed for statistical significance using two-tailed Student's t test and Wilcoxon sign rank test. Level of significance was $p < 0.05$.

RESULTS

Graded concentrations of silica and chrysotile A particles were added to AM and TNF production was measured following a 24 h incubation. Table 1 shows that both silica and asbestos augmented TNF production in a concentration-dependent fashion. From the same AM preparation, LTB_4 production was measured. Data indicate that LTB_4 production paralleled TNF production in AM. Indeed, we found a positive correlation between LTB_4 and TNF production when AM are stimula ed with chrysotile A fibers ($r = 0.80$ $p < 0.001$) or silica particles ($r = 0.72$ $p < 0.01$). Maximal TNF and LTB_4 production was seen at the highest particle concentration (100 ug/ml) used in our experiments.

TABLE 1 Concentration-dependent stimulation of TNF and LTB_4 production by mineral dust.[a]

Particles	TNF (U/ml)	LTB_4 (pg/.5 X 10^6 cells)
-	16 ± 4	125 ± 27
Chrysotile (1 µg)	22 ± 4	202 ± 45
Chrysotile (20 µg)	104 ± 20	1254 ± 289
Chrysotile (50 µg)	140 ± 34	2808 ± 533
Chrysotile (100 µg)	166 ± 45	2947 ± 504
Silica (1 µg)	22 ± 4	125 ± 33
Silica (20 µg)	31 ± 6	383 ± 80
Silica (50 µg)	87 ± 30	1264 ± 290
Silica (100 µg)	157 ± 41	2960 ± 588

[a] Data represent mean ± SEM of 5 separate experiments.

To investigate the relationship between TNF and LTB_4 production by mineral dust stimulated AM, we studied the effect of leukotriene synthesis inhibitors on TNF production. For these experiments, NDGA, a nonspecific lipoxygenase inhibitor (Osaki and al. 1986) and AA861, a specific 5-lipoxygenase inhibitor with little activity on cyclooxygenase or on 12- or 15-lipoxygenase (Yoshimoto and al. 1982), were employed. AM were preincubated for 1 h at 37°C with the indicated inhibitors and further incubated for 24 h in the presence of 50 ug/ml asbestos or silica. Data shown in Fig 1 indicate that both inhibitors reduced asbestos or silica-induced TNF production in a concentration-dependent fashion. NDGA and AA861 were concomitantly able to abolish TNF production by AM at 10 ug/ml and 50 ug/ml respectively.

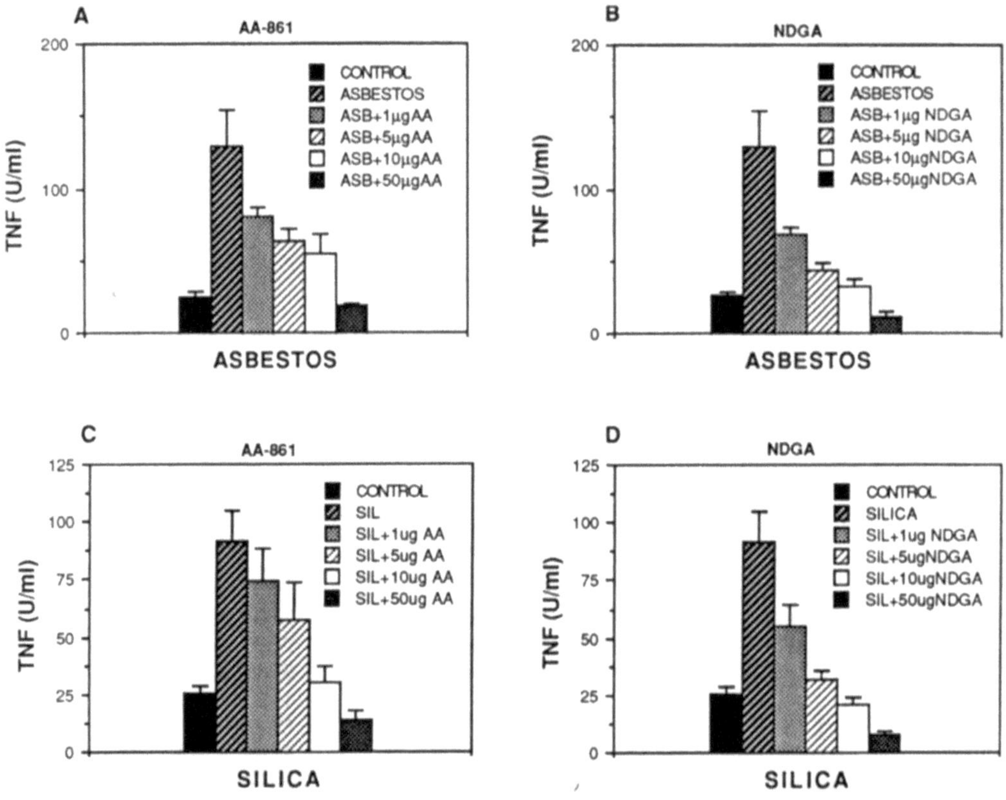

Fig. 1 Effect of leukotriene inhibitors on mineral dust-stimulated TNF production. AM were incubated 24 h with (A-B) chrysotile A (50 ug/ml); (C-D) silica (50 ug/ml) and the indicated concentration of leukotriene inhibitors. Data represent means ± SEM of 6 experiments.

We next tested whether the addition of exogenous LTB_4 to chrysotile or silica-stimulated AM containing the 5-LO inhibitor AA861 could restore TNF production. Results shown in Fig 2 A-B indicate that exogenous addition of LTB_4 at concentrations ranging from 10^{-14} to 10^{-8}M restored TNF production due to silica or asbestos stimulation. Maximal restoration was observed at 10^{-10}M concentration of LTB_4 ($p < 0.05$) giving respectively 73% and 86% recovery of original asbestos and silica stimulation of TNF production.

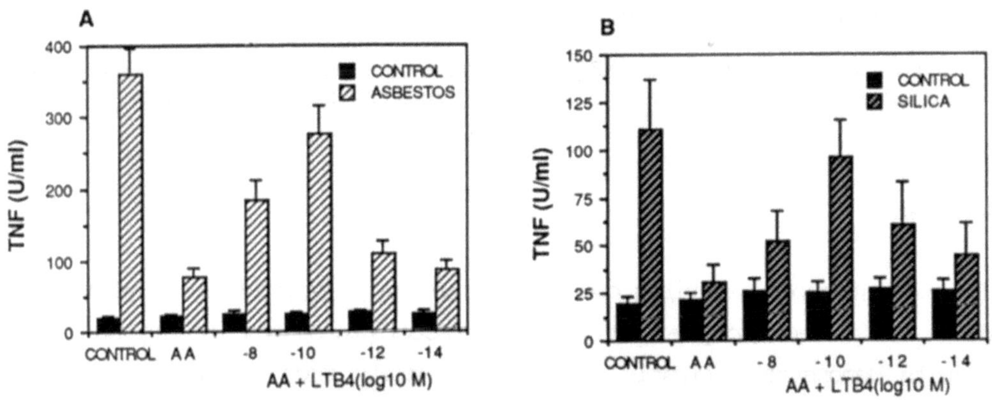

Fig. 2 Reversal of inhibitory effect of AA 861 on mineral dust-induced TNF production by leukotriene B_4. AM were stimulated with (A) chrysotile A (50 ug/ml) or (B) silica (50 ug/ml) and the indicated concentrations of leukotriene B_4. AA861 (10 ug/ml) was added 1 h prior to LTB_4 addition. Data represent means $\pm$ SEM of 5 different experiments.

In order to determine the contribution of TNF in fibroblast growth activity stimulated by mineral dust, AM supernatants were preincubated 1 h in the presence of murine anti-TNFα and TNF and fibroblast growth activity were mesured. As shown in table 2, supernatants from AM exposed to asbestos or silica induced an increase in fibroblast growth and TNF activity which were reduced concomitantly by the addition of anti-TNF antisera. The concentration of antisera used in this study was not toxic to any target cells as demonstrated by cell viability (data not shown).

Table 11 Effect of TNF-specific antisera on mineral dust-stimulated TNF and fibroblast growth activities.

Culture conditions[a]	anti-TNFα[b]	TNF (U/ml)	FB proliferation (index)	
			^{3}H-TdR	O.D.(590nM)
SN-AM	-	29	2.11	1.53
SN-AM	+	<1	1.11	1.15
SN-AM+ASBESTOS	-	116	2.95	2.10
SN-AM+ASBESTOS	+	<1	1.96	1.32
SN-AM+SILICA	-	61	2.64	1.86
SN-AM+SILICA	+	<1	1.37	1.18

[a] Supernatant (SN) were generated from AM cultured 24 h with or without mineral dust used at 50 μg/ml.

[b] Polyclonal rabbit anti-TNF α antisera was added to the macrophages supernatant at dilution 1:100 (10,000 neutralizing units/ml).

DISCUSSION

The present paper describes our studies on the in vitro effects of mineral dusts on inflammatory and fibrogenic factors produced by rat AM. Our results demonstrate firstly that mineral dusts such as silica particles and asbestos fibers can enhanced TNF production in a concentration-dependent fashion. A recent study indicated that recombinant TNF exerts potent growth factor activity for many types of human fibroblasts with maximal effect around 50-500 U/ml (Vilcek and al. 1986). Interestingly, this corresponds to the amount of TNF released by AM stimulated with high doses of chrysotile A or silica. Moreover, the use of TNFα-specific antisera indicated that TNF represent most of fibroblast growth activity generated by mineral dust-stimulated AM. These findings further support the hypothesis that AM may play an important role in mineral dust-induced pulmonary fibrosis through the increased release of potent fibroblast growth factors such as TNF, but the mechanism underlying this process remains unknown.

The inflammatory properties and the ability of lipoxygenase products, mainly LTB_4, to modulate the production of different cytokines has been reported previously (Rola-Pleszczynski and Lemaire 1985; Rola-Pleszczynski and al. 1986). In our study, we found a positive correlation between endogenous LTB_4 and TNF production and a concentration-dependent effect of lipoxygenase inhibitors on the generation of TNF activity. Moreover, the addition of exogenous LTB_4 (10^{-8}- 10^{-12}M) to AM cultures, in the presence of lipoxygenase inhibitors, restored to a large extent the level of mineral dust- stimulated TNF production. Collectively, these data suggest that endogenous leukotrienes produced by AM play a role in the cellular events which lead to enhanced TNF activity in response to asbestos and silica particles. Such LTB_4 effects could be mediated by its known ability to affect the influx of external calcium and the intracellular calcium concentration as well as other intracellular activation messengers (White and al. 1983).

The enhancement of TNF production by LTB_4 can be of crucial importance in the process of lung inflammation and fibrogenic disease such as found in silicosis and asbestosis. 1- TNF can amplify the inflammatory response by stimulating neutrophil emigration, mobilisation and activation at the site of inflammatory response (Klebanoff and al. 1986; Tsujimoto and al. 1986); 2- Inflammatory cells, like neutrophils, found at sites of inflammation, (Schoenberger and al. 1982) can provide an exogenous supply of LTB_4. Finally, TNF has been found to potentiate LTB_4 production by neutrophils stimulated with calcium ionophore (A23187) (Roubin and al. 1987). These interactions may establish a powerful amplification loop of inflammatory and fibrogenic cytokine production. In this context, the possibility of inhibiting TNF activity by lipoxygenase inhibitors could represent a rational basis for the development of specific agents to interrupt the process of chronic inflammation and fibrosis.

REFERENCES

Bagby G, Summer W (1988) Intratracheal lipopolysaccharide elicits tumor necrosis factor activity in bronchoalveolar lavage luid. Am Rev Respir Dis 137:43

Klebanoff SJ, Vadas MA, Harlan JM, Sparks LH, Gamble JR, Agosti JM, Watersdorph AM (1986) Stimulation of neutrophils by tumor necrosis factor. J Immunol 136:4220-4225

Lemaire I, Beaudoin H, Dubois C (1986) Cytokine regulation of lung fibroblast proliferation. Am Rev Respir Dis 134:653-658

Moldawer LL, Gelin J, Schersten T, Lundholm G (1987) Circulating interleukin 1 and tumor necrosis factor during inflammation. Am J Physiol 253:R922-R928
Osaki Y, Ohashi T, Niwa Y (1986) A comparative study on the effects of inhibitors of the lipoxygenase pathway on neutrophil function: inhibitory effects on neutrophil function may not be attributed to inhibition of the lipoxygenase pathway. Biochem Pharmacol 35:3481-3488
Rola-Pleszczynski M, Lemaire I (1985) Leukotrienes augment interleukin 1 production by human monocytes. J Immunol 135:3958-3961
Rola-Pleszczynski M, Chavaillaz P-A, Lemaire I (1986) Stimulation of interleukin 2 and interferon-gamma production by leukotriene B_4 and human lymphocyte culture . Prostagl Leukotrienes and Med 23:207-210
Roubin R, Elsas PP, Fiers W, Dessein AJ (1987) Recombinant human tumor necrosis factor (rTNF) enhances leukotriene biosynthesis in neutrophils and eosinophils stimulated with the CA^{2+} ionophore A23187. Clin Exp Immunol 70:484-490
Ruff MR, Gifford GE (1981) Tumor necrosis factor in Lymphokine Reports, vol 11. Academic Press, NY, 235-272
Sayer TJ, Macher I, Chung J, Kugler E (1987) The production of tumor necrosis factor by mouse bone marrow-derived macrophages in response to bacterial lipopolysaccharide and a chemical synthesized monosaccharide precursor. J Immunol 138:2935-2949
Schoenberger CI, Hunninghake GW, Kawanami O, Ferrans VJ, Crystal RG (1982) Role of alveolar macrophages in asbestosis: Modulation of neutrophil migration to the lung after acute asbestos exposure. Thorax 37:803-809
Sugarman BJ, Aggarwal BB, Hass PE, Figari IS, Palladino MA, Shepard HM (1985) Recombinant human tumor necrosis factor alpha: effects on proliferation of normal and transformed cells in vitro. Science 230:943-945
Thomas PD, Hunninghake GW (1987) Current concepts of the pathogenesis of sarcoidosis. Am Rev Respir Dis 135:747-760
Tsujimoto M, Yokota S, Vilcek J, Weissman N (1986) Tumor necrosis factor provokes superoxide anion generation from neutrophils. Biochem Biophys Res Commun 137:1094-1100
Vilcek J, Palombella VJ, Henriksen-De Stefano D, Swenson C, Feinman R, Hirai M, Tsujimoto M (1986) Activity of tumor necrosis factor and its relationship to other polypeptide growth factors. J Exp Med 163:632-643
White JR, Naccache PH, Molski TFP, Borgeat P, Sha'afi RI (1983) Direct demonstration of increased intracellular concentration of free calcium in rabbit and human neutrophils following stimulation by chemotactic factors. Biochem Biophys Res Comm 113:44-58
Yoshimoto T, Yokoyama C, Ochi K, Yamamoto S, Maki Y, Ashida Y, Terao S, Shiraishi M (1982) 2,3,5-trimethyl6-(12-hydroxy-5, 10-dodedecadiynyl)-1, 4-benzoquinone (AA861), A selective inhibitor of the 5-lipoxygenase reaction and the biosynthesis of slow-reacting substance of anaphylaxis. Biochem Biophys Acta 713:470-473

Modulation of Macrophage-Mediated Cytotoxicity by Toxic Particles

E. Bissonnette, B. Carré, C. Dubois and M. Rola-Pleszczynski
Immunology Division, Faculty of Medicine,
University of Sherbrooke, Sherbrooke, Québec, Canada, J1H 5N4.

Introduction

Prolonged inhalation of toxic environmental particles, like asbestos fibers and silica particles, causes immune disorders (Kagan and al., 1977a,b). The effect of these particles on mouse splenocytes and human lymphocytes proliferation is well documented (Miller and Zarkower, 1974; Scheuchenzuber and al., 1984; Lange and al., 1978). But, the first cell population which is in contact with the dusts is the alveolar macrophage (AM). These cells play a pivotal role in the protection of the lung. They represent the best defense against air pollutants and bacterial and viral infections (Hocking and Golde, 1979). It is also conceivable that AM play an important role in the initial lesions and subsequent pathologic processes of asbestosis and silicosis (Reiser and Last, 1986; Bowden, 1987). Asbestos fibers can stimulate AM to release interleukin-1 (IL-1), tumor necrosis factor (TNF), lysosomal enzymes, oxygen radicals and prostaglandins (PGs) (Pernis and Vigliani 1982; Hartmann and al., 1984; Bissonnette and Rola-Pleszczynski, 1988; Davis and al., 1974; Fantone and Ward, 1982; Sirois and al., 1982), but little is known concerning the effects of inhaled dust particles on their functions. In addition, macrophages and monocytes have been shown to be tumoricidal against various tumor target cells (Wood and Gillespie, 1975; Miller and Feldman, 1976; Fidler, 1985), and may also represent an important defense mechanism against neoplasia. We report here the effects of asbestos fibers and silica particles on rat AM and human monocyte cytotoxicity.

Materials and Methods

Rats: Male Wistar rats weighing 200 to 250 g were purchased from Charles River Canada Inc. (St-Constant, Québec).

Cell lines: P815 mastocytoma cells were maintained as suspension culture in RPMI 1640 medium supplemented with 10% heat-inactivated foetal bovine serum

NATO ASI Series, Vol. H30
Effects of Mineral Dusts on Cells
Edited by B.T. Mossman and R.O. Bégin

(complete medium). The murine L929 fibroblast cell line was grown as adherent cell monolayers in Earle's complete medium and was passaged by trypsinization.

Reagents: Chrysotile A asbestos fibers (UICC standard reference preparation) and min-U-Sil-5 particles (Pennsylvania Glass Scand. Co. Pittsburg, PA) were sterilized in autoclave, suspended in sterile phosphate-buffered saline, sonicated, and used at different concentrations as indicated. Indomethacin, a cyclooxygenase inhibitor, was purchased from Sigma Chemical Co. (St-Louis, MI).

Bronchoalveolar lavage: Animals were anesthetized by intramuscular injection (0.5 ml/kg) of a 6.5:1 mixture of ketamine (100 mg/ml) and xylazine (2 mg/ml; rogar/STB, Mtl, P.Q. and Haver-Lokhart, Mississauga, Ont.) and bronchoalveolar lavage was performed as described elsewhere (Lemaire, 1986).

Preparation of adherent cells: Venous blood from healthy volunteers was sedimented on Ficoll-Hypaque gradients, and the lymphocyte-monocyte layer was removed, washed, and suspended in Earle's complete medium. The cells were then allowed to adhere to the surface of plastic petri dishes previously coated with baby hamster kidney (BHK) cell microexsudate (Douglas et al., 1981). This was effective in enriching the adherent population to > 90% monocytes.

Cytotoxic assay: 5×10^5 monocytes/macrophages were distributed into microtiter wells and incubated for 1 hr to adhere. At the end of this incubation, different concentrations of toxic particles were added and the cells were incubated at 37^oC for two hours. Cytotoxic activity of monocytes/macrophages was measured by using a ^{51}Cr-release microassay.

Statistical analysis: Differences between groups of experiments were analysed for statistical significance by using the Student's T test.

Results and Discussion

Effect of toxic particles on monocyte/macrophage cytotoxicity against TNF-resistant target cells.

Different concentrations (20, 50 and 100 µg/ml) of asbestos fibers were added to rat AM with or without indomethacin. Figure 1 illustrates the cytotoxic assay against P815 target cells. AM-mediated cytotoxicity was significantly inhibited. This inhibition became non-significant, at all concentrations used, in the presence of indomethacin. Similar results were observed when we used human monocytes (fig. 2). At high asbestos concentrations (50 and 100 µg/ml), the inhibition of human monocyte-mediated cytotoxicity was more marked than that of AM-mediated cytotoxicity. However, when PGs synthesis was inhibited, the impairment of cytotoxicity was prevented or not significant.

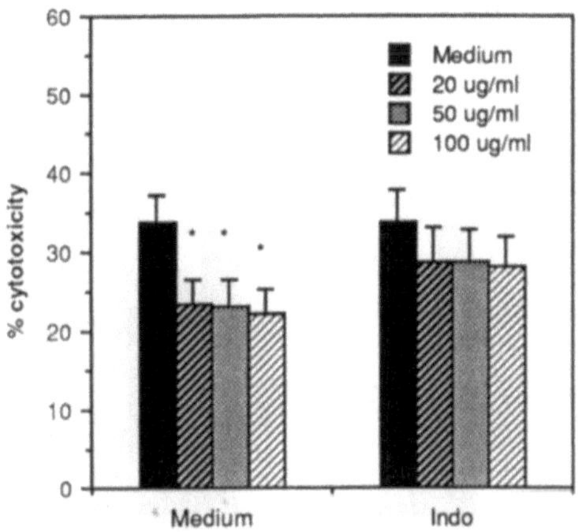

Fig. 1. Effect of asbestos fibers on rat AM cytotoxicity against P815 target cells (* $p<0.05$).

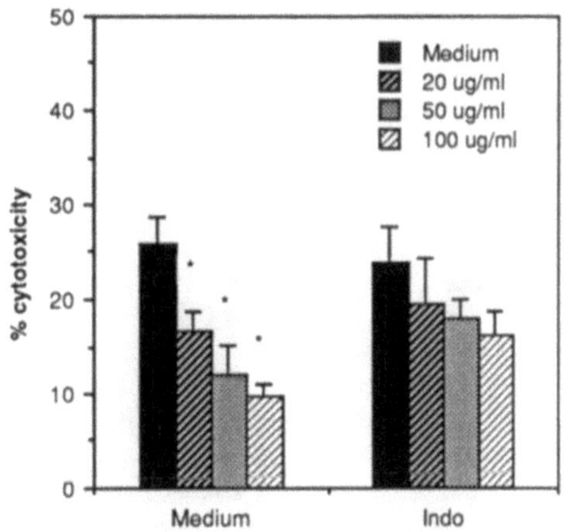

Fig. 2. Dose-dependent inhibitory effect of asbestos on human monocyte cytotoxicity against P815 target cells (* $p<0.05$).

The same experiments were performed with silica particles (fig. 3), but silica had no significant effect, at any concentration used, on AM-mediated cytotoxicity. Similar results were observed with human monocyte-mediated cytotoxicity (data not shown).

We found an inhibition of AM cytotoxicity by asbestos fibers but not by silica particles. This inhibition could be prevented by the addition of indomethacin. Prostaglandin production seems to be a major cause of inhibition of AM cytotoxicity by asbestos. We previously demonstrated that asbestos fibers stimulate AM to produce PGs (Sirois and al., 1982), and the present results suggest that the inhibition of AM cytotoxic activity is mediated by PG production. Human monocytes respond in the same way as rat AM.

Silica particles do not stimulate PGs production, nor do they inhibit monocytes/AM cytotoxicity. This difference could be relevant, at least partially, to the low incidence of pulmonary cancers in silicosis as compared to asbestosis.

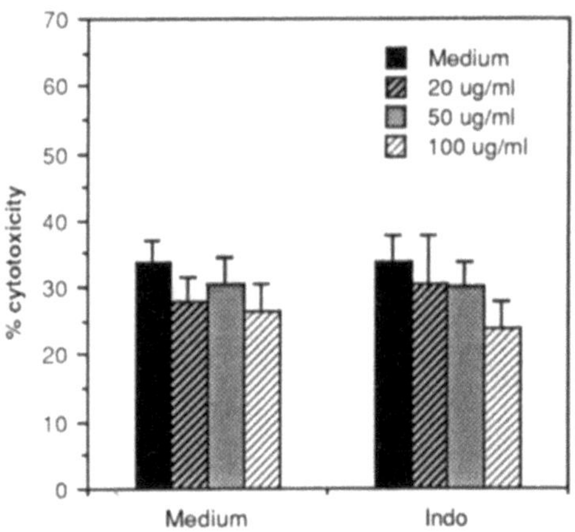

Fig. 3. Effect of silica particules on rat AM cytotoxicity against P815 cells.

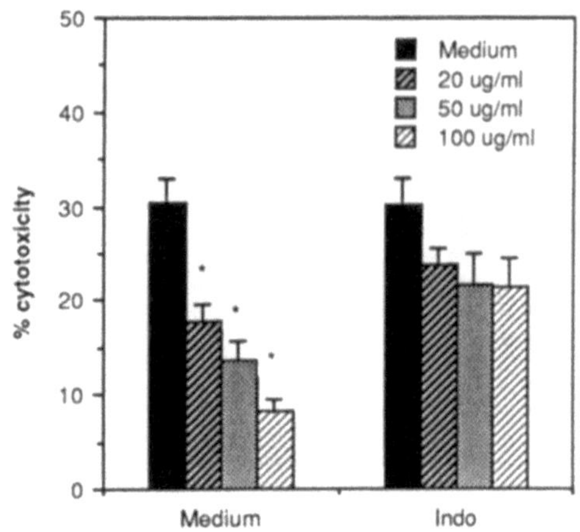

Fig. 4. Dose-dependent inhibitory effect of asbestos fibers on AM cytotoxicity against the TNF-sensitive cells line, L929 (* $p<0.01$).

<u>Effect of asbestos and silica on monocyte/macrophage cytotoxicity against TNF-sensitive target cells.</u>

We know that asbestos fibers and silica particles can stimulate AM to produce TNF (Bissonnette et Rola-Pleszczynski, 1988). Therefore, we studied the effect of toxic particles on cytotoxicity against TNF-sensitive target cells (L929) (fig. 4). Asbestos fibers caused a dose-dependent inhibition of

AM-mediated cytotoxicity which was reversed by the addition of indomethacin. Similar results were obtained with human monocytes and asbestos (data not shown). On the other hand, no inhibition of cytotoxicity was observed when silica particles were added to AM or human monocytes (data not shown). The effect of asbestos may be mediated by more than one mechanism, including TNF production.

The modulation of TNF production by asbestos does not affect AM cytotoxicity. Asbestos inhibits monocytes and AM cytotoxicity against the two types of target cells used (TNF-resistant and TNF-sensitive). This inhibition can be reversed by the addition of indomethacin in both cases. The stimulation of PGs production by asbestos seems to be more effective than the stimulation of TNF production. This could explain asbestos inhibition rather than asbestos stimulation of AM cytotoxicity against L929 cell. Another possibility is that asbestos fibers could modulate indirectly the production of PGs by fibroblasts. It is well known that asbestos stimulates IL-1 production by AM (Pernis and Vigliani, 1982; Hartmann and al., 1984). Recently, Newton and Carrington (1987) showed that IL-1 stimulates fibroblasts to produce PGE. The target cells L929 are fibroblasts. They can be indirectly stimulated by asbestos fibers to produce PGs which will inhibit AM cytotoxicity against themselves. This inhibition seemed to be more effective than the stimulation of TNF production by asbestos. Silica particles also stimulate IL-1 and TNF production (Gery and al., 1981; Oghiso and Kubota, 1986; Bissonnette and Rola-Pleszczynski, 1988), but these two stimulations resulted in no effect on L929 target cell cytotoxicity.

Effect of alveolar macrophage and P815 target cells supernatants

PGs production seems to be important in the inhibition of monocyte/macrophages cytotoxicity by asbestos. It is known that inert particle cause inhibition of natural cytotoxicity (Sibbitt and al., 1986). We wanted to verify if the presence of asbestos fibers during the cytotoxic assay could cause the observed inhibition or if the inhibition was mediated by soluble mediators. AM were incubated at 37°C with different concentrations of asbestos fibers during 2 hr. Then, cells were centrifuged, cell-free supernatants, containing no fibers, removed and added to other AM for cytotoxic assay against

P815 cells. In this way, we would see the effect of mediators released by AM in presence of asbestos. Figure 5 shows the results: inhibition of cytotoxicity was also observed with the supernatants of AM incubated with high asbestos concentrations (50 µg/ml and 100µg/ml). This inhibition became non significant when indomethacin was added with toxic particles. The diminution of AM cytotoxicity seemed to be due to PGs production by macrophages.

Finally, we verified the effect of asbestos fibers on target cells (P815). The mastocytoma P815 cells were incubated with asbestos during 2 hr. Cell-free supernatants were added to AM for the cytotoxic assay against P815 cells. Figure 6 shows the results: a strong inhibition of cytotoxicity was observed at all asbestos concentrations used. AM cytotoxicity was not affected however when PGs production was inhibited by indomethacin. In the light of these results, we can deduct that asbestos fibers can stimulate other cells types, beside macrophages, to produce PGs which will inhibit AM cytotoxicity. Silica particles did not affect the AM cytotoxicity when added to P815 target cells.

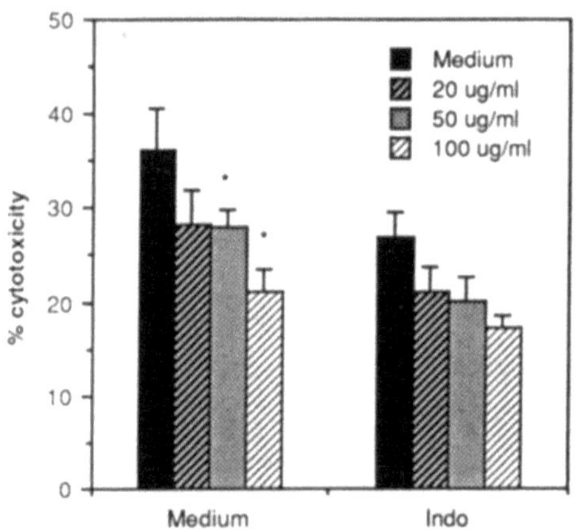

Fig. 5. Effect of supernatants from asbestos-treated AM on fresh AM cytotoxicity against P815 cells (* $p<0.05$).

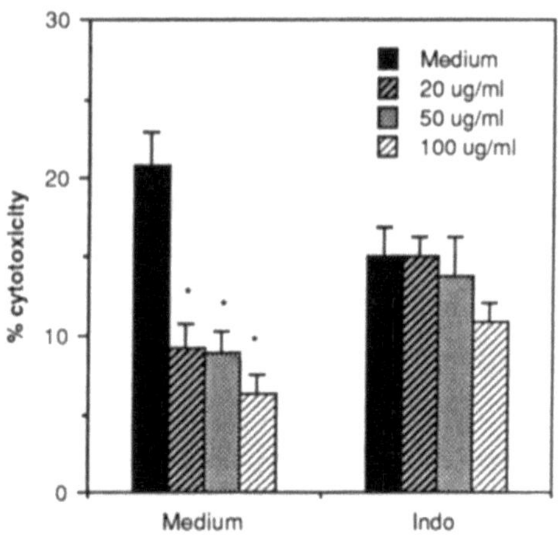

Fig. 6. Inhibitory effect of supernatants from asbestos-pretreated mastocytoma P815 on AM cytotoxicity (* $p<0.01$).

Another important point which emerged from our studies is the modulation by asbestos fibers of PGs production by other cells. We saw a stimulation, by asbestos, of PGs production by the mastocytoma P815. It is known that stimulated mast cells secrete large amounts of arachidonic acid metabolites (Lee and al., 1986). In addition, mast cell hyperplasia is observed in chronic inflammation and fibrosis (Gupta, 1970; Watanabe and al., 1974; Kawanami and al., 1979), but the role of these cells is not elucidated. It is possible that asbestos stimulates mast cells to release PGs which will inhibit AM cytotoxicity against tumor cells. Further experiments with pulmonary mast cells would be necessary to confirm this hypothesis.

In conclusion, asbestos fibers inhibit human monocyte and rat AM cytotoxicity by stimulation of PGs production by different cell populations (monocytes, alveolar macrophages, mast cells, fibroblasts). Silica particles do not affect human monocytes or rat AM cytotoxicity. These two inhalated toxic particles have different effects on AM cytotoxicity, a finding which may be relevant to the different incidence of pulmonary cancers between asbestosis and silicosis.

REREFENCES

Bissonnette, E and Rola-Pleszczynski, M (1988) Pulmonary inflammation and fribrosis in a murine model of asbestosis and silicosis: possible role of tumor necrosis factor, Inflammation in press.

Davies, P, Allison, AC, Ackerman, J, Butterfield, A and Williams, S (1974) Asbestos induces selective release of lysosomal enzymes from mononuclear phagocytes. Nature 251: 423-425.

Douglas, SD, Zuckerman, SH and Ackerman, SK (1981) Obtaining and cutturing human monocytes. In Methods for studying mononuclear phagocytes. DO Adams,PS Edelson and H. Koren) Academic Press, New York pp 33-42.

Fantone, JC and Ward, PA (1982) Role of oxygen-derived free radicals and metabolites in leukocyte-dependent inflammatory reactions. Am J Pathol 107:397-418.

Fidler, IJ (1985) Macrophages and metastasis - a biologic approach to cancer therapy. Cancer Res 45: 4714-4726.

Gery, I, Davies, P, Derr, J, Krett, M and Barranger, JA (1981) Relationship between production and release of lymphocyte activating factor (interleukin 1) by murine macrophages. I. Effects of various agents. Cell Immunol 64: 293-303.

Gupta, RK (1970) Mast cell variations in prostate and urinary bladder. Arch Pathol 89: 302-305.

Hahn, H and Kaufmann, SHE (1981) The role of cell-mediated immunity in bacterial infections. Rev Infect Dis 3: 1221-1250.

Hartmann, DP, Georgian, MM, Oghiso, Y and Kagan, E (1984) Enhanced interleukin activity following asbestos inhalation. Clin Exp Immunol 55: 643-650.

Hocking, WG and Golde, DW (1979) The pulmonary alveolar macrohages. N Engl J Med 301: 580-587.

Kagan, E, Solomon, A, Cochrane, JC, Beissner, EI, Gluckman, J, Rocks, P.H. and Webster, I (1977a) Immunological studies of patients with asbestosis. I.Studies of cell-mediated immunity. Clin Exp Immunol 28: 261-267.

Kagan, E, Solomon, A, Cochrane, JC, Kuba, P, Rocks, PH and Webster, I (1977b) Immunological studies of patients with asbestosis. II. Studies of circulating lymphoid cell numbers and humoral immunity, Clin Exp Immunol 28: 268-275.

Kawanami, O, Ferrans, VJ, Fulmer, JD and Crystal, RG (1979) Ultrastructure of pulmonary mast cells in patients with fibrotic lung disorders. Lab Invest 40: 717-734.

Lange, A, Smolik, R, Chmielarczyk, W, Garncarek, D and Gielgier, Z (1978) Cellular immunity in asbestosis. Arch Immunol Ther Exp 26: 899-903.

Lee, TDG, Swieter, M and Befus, AD (1986) Mast cell responses to Helminth infection. Parasitology Today 2: 186-191.

Lemaire, I (1985) Characterization of the bronchoalveolar cellular response in experimental asbestosis. Different reactions depending on the fibrogenic potential, Am Rev Respir Dis 131: 144-149.

Miller, CA and Feldman, JD (1976) Genetic role of rat macrophage cytotoxicity against tumor. Int J Cancer 18: 168-175.

Miller, SD and Zarkower, A (1976) Silica-induced alterations of murine lymphocyte immunocompetence and suppression of B lymphocyte immunocompetence: a possible mechanism. J Reticuloendothel Soc 19: 47-61.

Newton, RC and Carrington, M (1987) The activation of human fibroblast prostaglandin E production by interleukin 1. Cell Immunol 110: 338-349.

Oghiso, Y and Kubota, Y (1986) Enhanced interleukine 1 production by alveolar macrophages and increase in la-positive lung cells in silica-exposed rats. Microbiol Immunol 30: 1189-1198.

Pernis, B and Vigliani, EC (1982) The role of macrophages and immunocytes in the pathogenesis of pulmonary diseases due to mineral dusts, Am J Ind Med 3: 133-137.

Reiser, KM and Last, JA (1986) Early cellular events in pulmonary fibrosis, Exp Lung Res 10: 331-355.

Scheuchenzuber, WJ, Eskew, ML and Zarkower, A (1985) Effects of prolonged inhalation of silica and olivine dusts on immune functions in the mouse. Env Res 38: 389-399.

Sibbitt, WL, Imir, T and Bankhurst, AD (1986) Inert particles inhibit natural killer cell function in vitro. Cell Immunol 97: 146-154.

Sirois, P, Cadieux, A, Rola-Pleszczynski, M and Bégin, R (1982) Perifused alveolar macrophages. A technique to study the effects of toxicants on prostaglandin release. Experimentia 38: 1125-1127.

Watanabe, S, Watanabe, K, Ohishi, T, Aiba, M and Kageyana, K (1974) Mast cells in the rat alveolar septa undergoing fibrosis after ionizing irradiation ultrastructural and histochemical studies. Lab Invcest 31: 555-567.

Wood, GW and Gillespie, GY (1975) Studies on the role of macrophages in regulation of growth and metastasis of murine chemically-induced fibrosarcomas. Int J Cancer 16: 1022-1029.

EFFECTS OF ASBESTOS FIBRES ON SOME ASPECTS OF THE CELL SECOND MESSENGER SYSTEM

R.C.Brown, J.A.Hoskins, K.J.Cole, C.E.Evans & E.A.Sara, MRC Toxicology Unit, Woodmansterne Road, Carshalton, Surrey UK.

Several cell types have been treated with asbestos fibres and diacylglycerol accumulation, protein phosphorylation and gap junction integrity have been studied. The results obtained are consistent with the fibres causing the activation of phospholipases and the resulting release of a second messenger being responsible for the tumour promoting activity of asbestos.

Introduction

While exposure to asbestos alone can result in the development of mesothelioma, in lung cancer asbestos exposure seems to act as a promoter with the initiator coming from tobacco smoke.

There have been several efforts to model the interaction between smoking and asbestos exposure in both animals (eg Topping & Nettesheim, 1980; Harrison & Heath, 1986) and in vitro (eg Brown et al., 1983, Hei et al., 1985, Mossman et al.,1980). The mechanism by which fibres exert one effect of promoters (the induction of ornithine decarboxylase (ODC)) has been linked to the activity of protein kinase C (PKC) and the production of diacylglycerols (DAG's) (Marsh & Mossman, 1988).

The classical tumour promoter 12-O-tetradecanoylphorbol 13-acetate (TPA) is known to exert some or all of its effects through its ability to replace DAG's as an activator of PKC. It is therefore postulated that the prolonged production of DAG's could have a similar effect to treatment with TPA. We have investigated the effect of fibre treatment on DAG production and some related activities using a variety of cell types chosen for their response in these systems rather than for any resemblance to the target cells in fibre carcinogenesis.

Materials and Methods

Materials: The UICC standard samples of asbestos were used. Dusts were sterilised by autoclaving (15 p.s.i. for 15 min) and then suspended in complete medium and added to the cultures to give the desired concentrations. All suspensions were prepared immediately before use. Foetal calf serum and other tissue culture media and reagents were obtained from Flow Laboratories, Irvine, Scotland. Lipids, TPA and other chemicals were from Sigma Chemical Co., Poole, Dorset unless otherwise stated.

Cell culture: C3H10T1/2, V79-4 and P388D1 cells were cultured as described previously (Chamberlain & Brown, 1978; Brown et al., 1983; Brown & Poole, 1984). BL9L cells (Skilleter et al., 1983) were grown to confluence in Williams' Medium E supplemented with 5% heat inactivated foetal calf serum in an atmosphere of 5% CO_2.

Extraction and measurement of DAG's: DAG's were estimated by two methods:- 1. Labelling the cells with [^{14}C]-glycerol, extracting the lipid by a modified Bligh and Dyer technique and separation on TLC. 2. By acetylation of extracted cold DAG's with [^{14}C]-acetic anhydride prepared from sodium [^{14}C]-acetate (Du Pont, Stevenage, Herts. U.K.). This method is being published elsewhere; briefly

NATO ASI Series, Vol. H30
Effects of Mineral Dusts on Cells
Edited by B.T. Mossman and R.O. Bégin

cells were treated with 150 µl of 10% w/v trichloroacetic acid (TCA) plus physiological saline (1 ml). [^{3}H]-triolein was added to assess the extraction efficiency and lipids were extracted with chloroform/methanol (2:1). The organic layer was removed and the solvent evaporated. The residue was acetylated and dissolved in chloroform/methanol (50 µl; 9:1) and separated by TLC on silica gel. In both methods the bands of interest were cut out and counted in a liquid scintillation counter.

Protein Phosphorylation: Confluent cultures of C3H10T1/2 cells in 50mm Petri dishes were treated overnight with asbestos or TPA. After treatment the monolayer was washed twice with phosphate free medium and labelled with [^{32}P]-phosphate (60 µCi/ml) for two hours. Either acetone or TPA in acetone was added to the medium and 30 minutes later the monolayer was scraped off in ice cold 5% TCA. The pellet was washed three times with TCA and twice with phosphate buffered saline (PBS), blotted and 50 µl of sample buffer (63mM Tris with 2% SDS, 10% glycerol, 5% 2-mercaptoethanol, and 0.1% bromophenol blue) added. The protein was dissolved and heated to 100°C for 15 minutes; any fibres remaining were removed by centrifugation. Protein content was measured using the Bio-Rad protein assay method total radioactivity counted. The proteins were separated on SDS-polyacrylamide gels (10%), samples were applied such that each lane received the same number of counts. Gels were fixed, stained, dried and autoradiographed at room temperature.

Integrity of cell communication: A method based on that of Oh et al. (1988) was used: essentially confluent cultures of BL9L cells were treated with fibres or TPA in medium containing 1% serum. After exposure the cultures were treated with TPA for 10 minutes, washed several times with PBS and scrape lines made with a specially constructed comb through a layer of 0.05% Lucifer yellow CH. After 3 minutes at room temperature, the monolayers were again washed to remove excess dye. The scratches and surrounding cell layers were examined immediately with a fluorescence inverted microscope. The width of the fluorescent band (in number of cells) was estimated at at least 10 randomly selected sites along the scratches.

Results

DAG levels: Confluent cultures of C3H10T1/2 cells were labelled overnight with [^{14}C]-glycerol and treated with varying concentrations of crocidolite. Lipid was extracted and the DAG content measured. The results in table 1 show a dose related increase in the levels of DAG.

Table 1

Accumulation of Diacylglycerols in C3H10T1/2 cells

Crocidolite	Mean DAG content (dpm/cm^2)
0 (control)	59.7
20 μg/cm²	74.1
40 μg/cm²	84.5
80 μg/cm²	86.0*
120 μg/cm²	101.8*

*significantly different from control at the 5% level
C3H10T1/2 cells were treated in 25cm² flasks and were pre-labelled with [^{14}C]-glycerol, DAG's were estimated as described above. The results given are the means for 4 cultures at each point.

Similar results have been found with other cell types and following treatments with other fibres. However the results were variable; with [^{3}H]-glycerol the problem was worse, since the label distributed into many compounds. For this reason we have developed a method using unlabelled cells and post-labelling the DAG's with [^{14}C]-acetate.

Table 2

Accumulation of Diacylglycerols in P388D1 cells

Chrysotile	Mean DAG content (dpm/cm^2)	Ratio of $^{14}C/^{3}H$ in triglyceride band
0 (control)	261.4	7.1
4 μg/cm²	371.3**	9.3
8 μg/cm²	445.7*	10.0
16 μg/cm²	421.2	9.4

*significantly different from control at the 5% level
**significantly different from control at the 1% level
P388D1 cells were grown in 6 well dishes, treated with Canadian chrysotile for 18 hours and the lipid extracted and acetylated before separation by TLC. Tritiated triolein was added to correct for differences in extraction and DAG's estimated as described above. Both the total ^{14}C-DAG counts and the increase in the $^{14}C/^{3}H$ ratio confirm the increase in DAG's in treated cultures. The results given are the mean for three cultures at each point. The highest dose of asbestos produced more variable results which just fail to attain significance at the 5% level ($p=0.058$).

Protein phosphorylation: There was an increase in the total incorporation of ^{32}P into the protein of asbestos exposed cultures. This effect was observed in many types of cell cultures and is apparent at non-toxic doses of fibre. Typical results are shown in Table 3.

Table 3

Phosphorylation of protein in C3H10T1/2 cells

Fibre treatment	Dust alone (cpm/µg protein)	With TPA (100nM)
0 (control)	558 ± 29	901 ± 49
5 µg/cm²	648 ± 41	1118 ± 38
20 µg/cm²	890 ± 38	1196 ± 63
50 µg/cm²	1013 ± 188	1806 ± 64

Confluent cultures of C3H10T1/2 cells were exposed to amosite asbestos for 18h then treated as described above. Each point is the mean and standard error for 4 replicate cultures. Both TPA and amosite increased the labelling and these effects seem to be roughly additive.

The activation of PKC is reported to stimulate particularly the phosphorylation of an 80KD protein (Blackshear et al., 1986). This effect was observed in both C3H10T1/2 cells and primary fibroblasts. Other cell lines lacked this target protein. Fibre treatment did not appear to increase the specific phosphorylation of this protein though such an effect might be expected from the release of DAG's reported above. We have observed, however, that while there was an overall increase in protein phosphorylation, fibre treatment caused a diminution in the TPA stimulated phosphorylation of this target protein.

Integrity of intercellular communication: The effect of TPA and fibres on the integrity of gap junction formation was estimated by the ability to transfer Lucifer yellow across a band of cells bordering a scratch made in a confluent monolayer through a layer of dye. This "band width" was reduced by TPA and to a lesser degree by treatment with fibre. Most significantly the effect of TPA was much less in cells which had been pre-treated with fibre. Typical results are shown in table 4.

Table 4.

Lucifer yellow transfer in cells treated with asbestos and TPA

[TPA] (nM)	Pre-treatment none	amosite	chrysotile
0	9.9 ± 0.62	7.4 ± 0.48	10.9 ± 0.72
5	12.7 ± 0.80	9.2 ± 0.66	12.1 ± 0.71
25	5.2 ± 0.44	7.1 ± 0.54	10.0 ± 0.78
100	3.1 ± 0.43	6.7 ± 0.63	6.5 ± 0.6

BL9L cells were treated overnight with fibre and for 10 minutes with TPA. Then the ability of the cells to transfer Lucifer yellow through gap junctions was estimated as described above. The figures are the mean number of fluorescent cells normal to the scratch ± 1 standard deviation. The fibres were used at 25µg/cm². At low concentrations of TPA there was an increased band width. Treatment with either fibre reduced the subsequent sensitivity of these cells to higher concentrations of TPA.

Discussion

The promoting effect of mineral fibres could have several explanations: the retention of carcinogens at sensitive sites, their enhanced transport into cells, or inhibition of detoxification have all been suggested as possible mechanisms. One activity of fibres that has been extensively studied is the increased release of phospholipid derived inflammatory mediators from exposed cells (eg Brown & Poole,1984). This has been thought to be related to tumour promotion only through some indirect effect on cell proliferation. The realisation that the best known class of tumour promoter, the phorbol esters, act by mimicking one product of phospholipid breakdown suggests a more direct link between the activation of phospholipases and tumour promotion.

The results in this paper and those from several other authors in this same volume confirm that treatment with fibres can result in phospholipid hydrolysis and the activation of a number of pathways of the so-called second messenger systems. We suggest that cell membrane/fibre interactions result in the slow onset of phospholipid hydrolysis and thus a prolonged release of second messengers. One reason for the dynamics of this process differing from the more common "burst" of second messengers is that short term treatment with fibres and other particulates is impossible. The fibres take some minutes to settle onto the monolayer and usually several hours to exert their effects. This response to fibres differs from the rapid self-limiting effect of treatment with ligands known to bind to specific cell surface receptors and from that of materials such as TPA which mimic the messengers themselves.

Some fibres with a positive charge in suspension, chrysotile for example, can interact with negatively charged cells much more rapidly and may produce a more rapid response.

The effects of fibres should be compared with longer term

effects of exposure to other promoters. Oh et al. (1988) found that the effect of TPA on gap junction formation and on PKC was maximal after 10 minutes. This response was abolished at 6 hours when the activity of PKC was minimal In the study reported here short term treatment with TPA dramatically reduced intercellular communication. Fibre reduced the subsequent effect of TPA. Overnight treatment with TPA also produced this "down regulation" and this coincided with a reduction in phorbol ester binding (unpublished observations). Similarly, long-term treatment with TPA blocked the effect of subsequent short-term treatment on phosphorylation of the 80KD protein. Fibre pre-treatment duplicated this effect even though the effect of fibre and TPA on total protein phosphorylation was additive.

One longer term effect of TPA is its ability to block metabolic cooperation as measured by the reduction in the cloning efficiency of 8-azaguanine resistant cells in the presence of this compound and wild type cells. Fibres are unable to mimic this effect of TPA (Chamberlain, 1982). It is not clear why this effect of TPA is not subject to the same "down regulation" seen in the systems described above.

Treatment with cell-permeable DAG's has also failed to duplicate the effect of fibres or prolonged TPA treatment. This might be due to the short life of these compounds in cells. Blocking one route of DAG metabolism with an inhibitor of DAG-kinase has resulted in results similar to those seen with fibres (unpublished observations). Further studies are underway on the levels and distribution of PKC and on the number and distribution of phorbol ester binding sites. There remain some questions such as the source of the DAG's and the type of phospholipases activated during fibre exposure but there seems to be every reason to believe that these activities are central to the tumour promoting activities of mineral fibres.

References

Blackshear P.J., Wen L., Glynn B.P. and Witters L.A. (1986) Protein kinase C stimulated phosphorylation in vitro of a M_r 80,000 protein phosphorylated in response to phorbol esters and growth factors in intact fibroblasts. J.Biol.Chem. 261: 1459-1469

Brown R.C., Poole A., and Fleming G.A., (1983) The influence of asbestos dust on the oncogenic transformation of C3H10T1/2 cells. Cancer Letters 18: 221-227

Brown R.C. and Poole, A.(1984) Arachidonic acid release and prostaglandin synthesis in a macrophage like cell line exposed to asbestos. Agents and Actions 15: 336-340

Chamberlain, M. (1982) The influence of mineral dusts on metabolic co-operation between mammalian cells in tissue culture. Carcinogenesis 3: 337-339.

Chamberlain M. and Brown R.C. (1978) The cytotoxic effects of asbestos and other mineral dust in tissue culture cell lines. Br.J.Exp.Path. 59: 183-189

Harrison P.T.C. and Heath J.C. (1986) Apparent synergy in lung carcinogensis: interactions between N-nitroso-heptamethyleneimine, particulate cadmium and crocidolite asbestos fibres in rats. Carcinogenesis 7: 1903-1909

Hei T.K., Geard C.R., Osmak R.S. and Travisano M. (1985) Correlation in vitro genotoxicity and oncogenicity induced by radiation and asbestos fibres. B.J.Cancer 52: 591-597

Marsh J.P. and Mossman B.T. (1988) Mechanisms of induction of ornithine decarboxylase activity in tracheal epithelial cells by asbestiform minerals. Cancer Res. 48: 709-714

Mossman B.T., Landesman J.M. and Craighead J.E. (1980) Asbestos stimulates ornithine decarboxylase (ODC) activity in hamster tracheal epithelial cells. J. Cell Biol. 87: 307-312

Oh S.Y., Madhukar B.V. and Trosko J.E. (1988) Inhibition of gap junctional blockage by palmitoyl carnitine and TMB-8 in a rat liver epithelial cell line. Carcinogenesis 9: 135-139

Skilleter D.N., Price R.J. and Legg R.F. (1983) Specific G_1-S phase cell cycle block by beryllium as demonstrated by cytofluorometric analysis. Biochem. J. 216: 773-776

Topping D.C. and Netteshem P. (1980) Two stage carcinogenesis studies with asbestos in Fischer 344 rats. J. Natl. Cancer Inst. 65: 627-680

A MODEL FOR THE INFLUENCE OF INHALED MINERAL FIBERS ON THE CELLULAR BURDEN OF POLYCYCLIC AROMATIC HYDROCARBONS IN THE LUNG

P. Gerde and P. Scholander
Department of Chemical Engineering,
The Royal Institute of Technology,
S-100 44 Stockholm,
Sweden

INTRODUCTION

Ever since it was discovered that tobacco smoking and asbestos exposure have a synergistic interaction on the risk of lung cancer (Selikoff *et al*, 1968), there has been a search for a plausible mechanism for this effect. One early proposal was that carcinogens from tobacco smoke such as polycyclic aromatic hydrocarbons (PAHs) are adsorbed onto the mineral fiber surfaces (Harington, 1973). This should increase both the retention and the cellular burden of such hydrocarbons in the lung. Bronchial cancer is the most prevalent neoplastic manifestation of both asbestos exposure and tobacco smoking. A rapid initial mucociliary clearance from the lung of both mineral fibers and tobacco tar indicates a dense density of deposition of these agents in the bronchial lining layer (Morgan *et al*, 1975: Lippmann and Schlesinger, 1984). This paper presents a model for how inhaled mineral fibers may influence the penetration of PAHs through the bronchial lining layer.

CELLULAR BURDEN OF PAHs IN A BRONCHIAL EPITHELIUM UNDISTURBED BY FIBERS

Both the exact structure of the bronchial lining layer and its degree of coverage over the surface of the bronchial tree are still matters of debate. From a mechanistic model, Van As (1977) argues that an entirely covering mucus layer in the airways cannot exist. Electron microscopy preparations, however, points in the direction of a mostly covering layer of the airways (Luchtel, 1978: Ueda *et al*, 1984).

The main composition of the bronchial lining layer is one major aqueous phase, including both the mucus layer and the liquid surrounding the cilia, and one minor lipid phase constituting some 0.3-0.5% surfactant lipids (Lopez-Vidriero and Reid, 1980). Most of the lipid fraction seems to be located at the air interface on top of the mucus layer (Ueda et al, 1984).

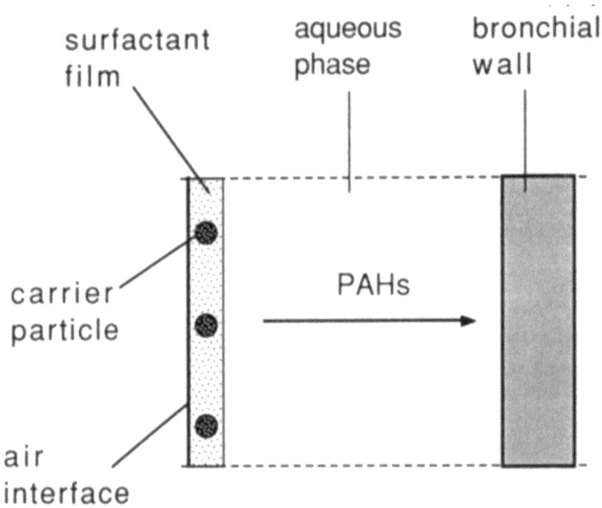

Figure 1. A schematic model of the bronchial lining layer.

In a theoretical model we have simulated the penetration of a typical PAH from the air interface across the lipid and aqueous layers of the bronchial lining layer and into the epithelial cells below (Gerde and Scholander, 1987a). The model barrier concentrates on the lipid-aqueous nature of the bronchial lining layer. The dominating aqueous fraction is in contact with the lipophilic cellular membranes at the bronchial wall and the lipophilic surfactant phase is covering the aqueous phase at the air interface, see fig 1.

Particle associated PAHs can be assumed to be deposited mainly at the air interface in contact with the surfactant lipids. The overall transport of PAHs from their carrier particles to the bronchial epithelium can then be described in four steps.

First is the release of the hydrocarbons from their carrier particles. Second is their dispersion within the surfactant layer at the air interface. Third is the penetration of the aqueous phase, and the final fourth step is the absorption of PAHs by the epithelial cellular membranes.

In the model only the slowest step, the penetration of the aqueous phase, will be considered. The other steps are assumed to be instantaneous. Results show that the thin layer of surfactant lipids at the air interface gives a drastic decrease in the rate of penetration of lipophilic substances like PAHs, compared to if this had been a purely aqueous barrier, see fig 2. The model PAH has a half-life within the bronchial lining layer model of about two hours. If this is true *in vivo*, this enables even dissolved PAHs to be removed from the upper

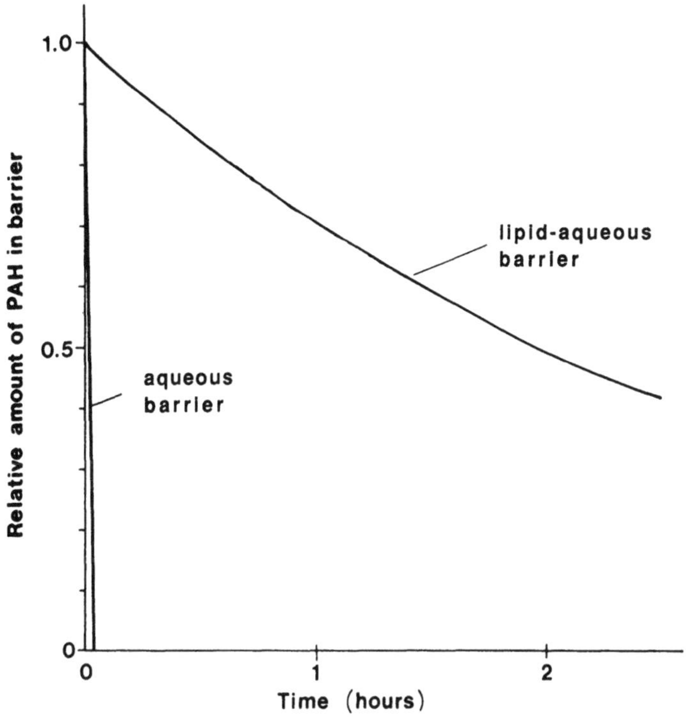

Figure 2. The amount of a typical PAH retained by the model barrier as a function of time. The lipid-aqueous barrier is compared with a purely aqueous barrier just for reference.

bronchial tree through mucociliary clearance before they have reached into the epithelial cells below. Another interesting protective property of the lining layer model is that a load of PAHs from a single carrier particle is dispersed over a large area within the lipid layer before it penetrates into the bronchial epithelium. The cellular doses of PAHs below such a barrier will thus be evened out so that peak doses on few cells close to the particles instead will be shared among a greater number of cells. This very characteristic transport property of a lipid-aqueous model barrier can be demonstrated quite easily in non-biological experiments (Gerde, 1987).

CELLULAR BURDEN OF PAHs IN A BRONCHIAL EPITHELIUM PENETRATED BY FIBERS

A necessary condition for the low permeability of PAHs in the bronchial lining layer is the distribution of its lipid and aqueous phases in different layers. At the same time, during the first hour or so after deposition, a slow penetration of PAHs in the bronchial lining layer is likely to give very high concentrations of such substances there. Any mechanism that may create a continuous lipid connection, a 'lipid-link', across the aqueous phase of the lining layer thus can drasticly increase the penetration of lipophilic substances. Fibers then assume importance as being the only inhalable particles likely to have one dimension exceeding the thickness of the bronchial lining layer.

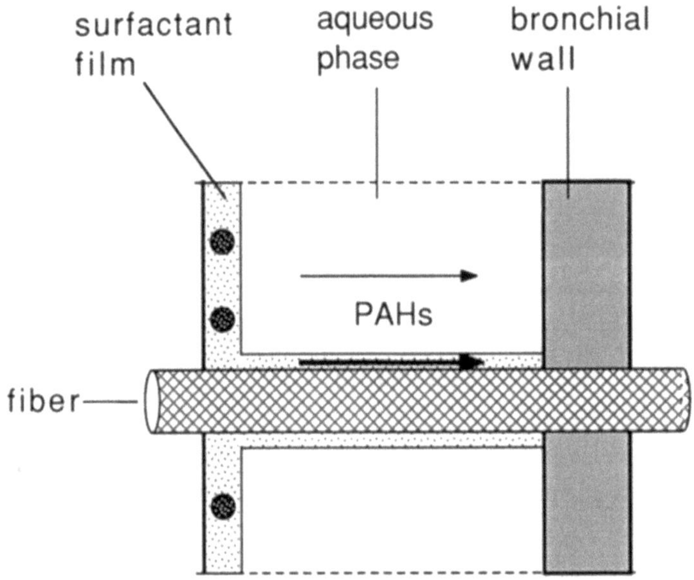

Figure 3. The bronchial lining layer traversed by an inhaled fiber, which has an adsorbed layer of surfactant phospholipids. The two transport routes of PAHs into the bronchial epithelium are indicated: one through the aqueous phase and the other within the adsorbed lipid layer.

Although asbestos fibers probably adsorb PAHs poorly in the lung (Gerde and Scholander, 1988) they are good adsorbents of phospholipids from lung surfactant (Light and Wei, 1977). It can also be demonstrated that during this process PAHs will follow the phospholipids onto the fibers and remain dissolved or solubilized within the adsorbed lipid layers (Gerde and Scholander, 1988). If such a phospholipid-covered fiber is traversing the bronchial lining layer and is penetrating an epithelial cell, then the necessary conditions are fulfilled for a short-circuiting. A pathway is created through which lipophilic substances contaminating the surfactant lipid layer at the air interface can rush into the epithelial cellular membrane (Gerde and Scholander, 1987b), see fig 3.

Calculations show that cells penetrated by such a fiber may receive on the order of 100 times as much PAHs as cells covered by the undisturbed lining layer. This hypothetical effect should give a strong correlation between inhalation of lipid-adsorbing fibers and tobacco smoking for the induction of very local but extremely high cellular doses of PAHs in the lung. Consequently, this should increase the risk for lung cancer.

One prerequisite for such an effect is that the fiber is long enough to reach across the bronchial lining layer, a risk that should increase with length. At the same time the fiber has to be thin enough at all to reach into the tracheo-bronchial region. This points in the direction of an increased risk with a long and thin fiber. One key surface property is a strong adsorption of phospholipids whereas the direct adsorption of PAHs is rather irrelevant. One further important mechanical property is resistance against crosswise cleavage.

CONCLUSIONS

The presented model points out two new important findings. First, it predicts that if the bronchial lining layer of the airways consists of separate lipid and aqueous phases this will give a drastic decrease in the rate of penetration of lipophilic substances, compared with a lining layer consisting of a homogeneous phase. This result should have a significant bearing on the mechanism behind the uptake of e.g. PAHs by the human lung.

Secondly, a fiber that is traversing the bronchial lining layer and is penetrating an epithelial cell may adsorb lipid corridors along its surface. These lipid connections may constitute an extraordinary short-circuit which can break the isolating property of the undisturbed bronchial lining layer. Hereby the cellular uptake of PAHs can be increased on the order of 100 times per penetrated cell.

The most crucial prerequisite for these mathematical model simulations is the assumption that the bronchial lining layer consists of separate lipid and aqueous phases. Further biological experiments are needed to verify the relevance of both this assumed structure and the two significant consequences hereof.

ACKNOWLEDGMENT

This work was supported by the Swedish Work Environment Fund.

REFERENCES

Gerde P (1987) A model for the uptake of polycyclic aromatic hydrocarbons by the human lung - influence of inhaled mineral fibers. PhD Thesis, The Royal Institute of Technology

Gerde P, Scholander P (1987a) A mathematical model of the penetration of polycyclic aromatic hydrocarbons through the bronchial lining layer. Environ Res 44:321-334

Gerde P, Scholander P (1987b) A hypothesis concerning asbestos carcinogenicity: the migration of lipophilic carcinogens in adsorbed lipid bilayers. Ann Occup Hyg 31:395-400

Gerde P, Scholander P (1988) Adsorption of benzo(a)pyrene onto asbestos and man-made mineral fibers in an aqueous solution and in a biological model solution. Br J Ind Med 45:682-688

Harington JS (1973) Chemical factors (including trace elements) as etiological mechanisms. In: Bogovski P, Gilson JC, Timbrell V, Wagner JC (ed). Biological effects of asbestos. IARC Scientific publications No. 8, p304-311

Light WG, Wei ET (1977) Surface charge and hemolytic activity of asbestos. Environ Res 13:135-145

Lippmann M, Schlesinger RB (1984) Interspecies comparisons of particle deposition and mucociliary clearance in tracheobronchial airways. J Toxicol Environ Health 3:441-469

Lopez-Vidriero MT, Reid L (1980) Respiratory tract fluid - chemical and physical properties of airway mucus. Eur J Respir Dis 61(suppl 110):21-26

Luchtel DL (1978) The mucous layer of the trachea and major bronchi in the rat. Scanning Electron Microscopy 11:1089-1098

Morgan A, Evans JC, Evans RJ, Hounam RF, Holmes A, Doyle SG (1975) Studies on the deposition of inhaled fibrous material in the respiratory tract of the rat and its subsequent clearance using radioactive tracer techniques. Environ Res 10:196-207

Selikoff IJ, Hammond EC, Churg J (1968) Asbestos exposure, smoking and neoplasia. J Amer Med Assoc 204:106-112

Ueda S, Kawamura K, Ishii N, Matsumoto S, Hayashi K, Okayasu M, Saito M, Sakurai I (1984). Ultrastructural studies on surface lining layer (SLL) of the lungs: Part III. J Jap Med Soc Biol Interface 15:67-88

Van As A (1977) Pulmonary airway clearance mechanisms: a reappraisal. Am Rev Resp Dis 115:721-726

ONCOGENIC TRANSFORMATION BY ASBESTOS FIBERS AND RADON-SIMULATED ALPHA PARTICLES

Tom K. Hei
Radiological Research Laboratory
College of Physicians and Surgeons
Columbia University, New York, N.Y. 10032
USA

The association between exposure to asbestos fibers and the development of lung cancer and mesothelioma have been well established in both man and experimental animals (Barum et al., 1958; Wagner and Berry, 1969). Futhermore, cigarette smoking can enhance the incidence of lung cancers in asbestos workers in a synergistic fashion (Selikoff et al., 1968). It is not known for certain whether specific chemicals or the high Linear Energy Transfer (LET) α-irradiation from deposited polonium is the important factor in carcinogenesis by cigarettes and its enhancement by asbestos. The importance of these high LET α-particles (radon daughters) has recently been re-emphasized with the realization that environmental radon represents the largest source of natural background radiation to the U.S. population (NCRP Report 78).

Studies designed to identify a link between lung cancer and radon in the home, by seeking a correlation between lung cancer incidence and variations in regional radon levels, have produced results that are equivocal. In epidemiological studies, uranium miners exposed for many years to high radon levels in the mines show an enhanced incidence of lung cancer (Lundin et al., 1969; Whittemore and McMillan, 1983). The perceived risk to the public from breathing lower levels of radon in the home are based on an extrapolation from the high concentrations experienced by the miners. However, extrapolating from the uranium miner data involves complications because most miners studied were also heavy cigarette smokers. The number of non-smoking miners is too small to separate the effects of radon from those of the cigarette smoke.

The magnitude of the problem, and the complexity of interacting factors, emphasizes the urgent need for basic research in the biological effects of radon. The present studies were undertaken to compare the

NATO ASI Series, Vol. H30
Effects of Mineral Dusts on Cells
Edited by B. T. Mossman and R. O. Bégin

oncogenic transforming potential of radon-simulated high LET Helium-3 ions and X-rays, and to investigate the interaction of this densely ionizing radiation with asbestos fibers. Furthermore, the possible protective effects of superoxide dismutase (SOD) and catalase in the transformation incidence induced by asbestos fibers and γ-rays were examined in an attempt to elucidate the role of oxygen free-radicals in such interaction.

Materials and Methods

C3H 10T1/2 mouse embryo fibroblasts isolated and characterized by Reznikoff et al. (Reznikoff et al., 1973) were used. These cells exhibit contact inhibition of growth and anchorage dependence under normal culture conditions. When treated with physical or chemical carcinogens, isolated foci consisting of morphologically altered cells appear. These cells are anchorage independent and produce fibrosarcoma upon inoculation into syngeneic animals. Cells were maintained in Eagle's Basal medium supplemented with 10% heat-inactivated fetal bovine serum (Hyclone Laboratories) and 25 μg/ml gentamycin (U.S. Biochemicals). Only cells from passage 8-13 were used for these studies.

Helium-3 particles were accelerated in the 4-MeV van de Graaff accelerator at the Radiological Research Accelerator Facility. These ions are virtually identical to the α-particles emitted by radon daughter products. A beam of helium-3 ions was directed vertically through the Mylar bottom of specially constructed irradiation dishes. Cells were grown as monolayers attached to the Mylar which had an average thickness of 6 μm. The details of the physical arrangement for track-segment irradiation has been published previously (Hei et al., 1988). Briefly, selected LETs were obtained by adjusting the accelerator energy such that charged particles of the appropriate energies were directed onto the irradiated cells. The specimen wheel had a capacity for 20 dishes and was remotely rotated by a stepping motor so that an individual dish was passed over a slot-shaped beam at a rate determined by both the dose to be delivered and the instantaneous dose rate of the beam. For low LET radiations, either X-rays generated by a Westinghouse X-ray machine delivered at a dose rate of 125 rads/min or γ-rays from a ^{137}Cesium irradiator at an absorbed dose rate of 136 rads/min were used.

For the transformation assay, exponentially growing C3H 10T1/2 cells on Mylar dishes were irradiated with graded doses of helium-3 particles or X-rays. Following irradiation, cells were trypsinized, counted and plated in 100 mm diameter petri dishes at a density such that approximately 300-400 viable cells would survive the treatments. In studies where the interaction between asbestos fibers and radiation were to be examined, the cultures were pretreated wih 5 µg/ml crocidolite fibers for 24 hours followed by exposure to X-rays or helium-3 ions. All treated and control cultures were maintained for a total of 6 weeks, with the medium changed every 10 days. The cultures were then fixed, stained and type II and III foci scored as transformants as described previously (Hei et al., 1984, 1985). All dishes were coded and scored independently twice. In experiments involving either SOD or catalase, the drug (Sigma Chemical), at 25 units/ml was added to the culture medium concurrently with the fibers. After the 24 hrs. treatment time, the cultures were washed and replated with fresh drug as above.

Cell survival after high vs. low LET radiations

The helium-3 ions used in the studies have a LET value of 120 keV/µm and

are well in range of the energy spectrum of radon α-paraticles. Figure 1 shows the survival curve for C3H 10T1/2 cells irradiated with either X-rays or helium-3 ions. The survival curve for X-rays has a characteristic initial shoulder region before becoming steeper and approaching an exponential function of dose at higher doses. By contrast, the cell-survival curve for the high LET helium ions, approximates closely to an exponential function of dose, indicating single-hit kinetics - i.e., the cell is killed by the traversal of a single particle. However, at a dose corresponding to a surviving fraction of 37%, about 14 particles traverse the nucleus of each cell killed (Lloyd et al., 1979). While one particle kills the cell, on average about 13 particles traverse the cell without killing it for every one that does! This may be a clue to the high efficiency with which high LET-particles induce transformed foci.

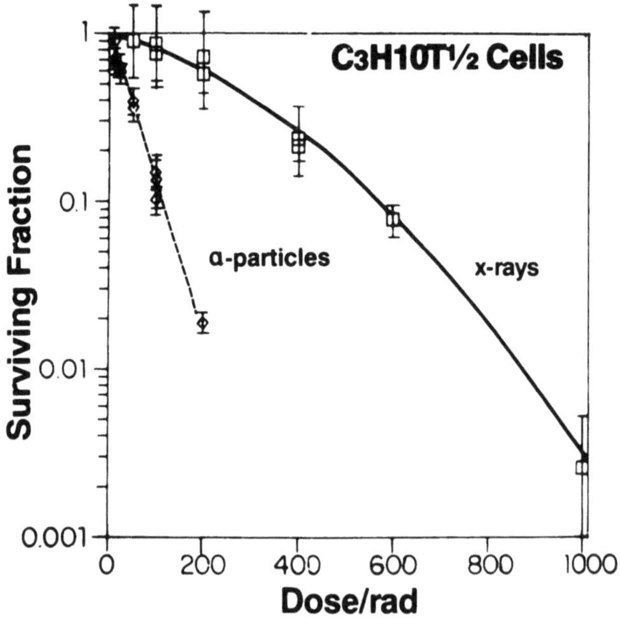

Figure 1: Survival curves for C3H/10T-1/2 cells exposed to graded doses of either X-rays or helium-3 ions (From Hall and Hei, 1987).

Transformation frequencies by high vs. low LET radiations

Transformation data are most commonly plotted in terms of transformation incidence per surviving cell as a functin of dose. This is convenient since this is the way in which various transformation data are compared. However, in applying these in vitro data to whole organisms, a more relevant quantity is transformation incidence per initial cell at risk. This is shown for X-rays and radon in Figure 2. The overall shape of the curves reflects the combined effects of transformation incidence and cell lethality. The interesting point is that at a dose of about 50 rad, the transformation incidence per initial cell at risk for helium-3 ions is five-fold higher than can be achieved by any dose of X-rays. This is presumably a consequence of a differing balance between the induction of transformed cells and the killing of transformed cells. High LET particles are much more effective at cell killing than are X-rays, but they are even more efficient at inducing transforming events.

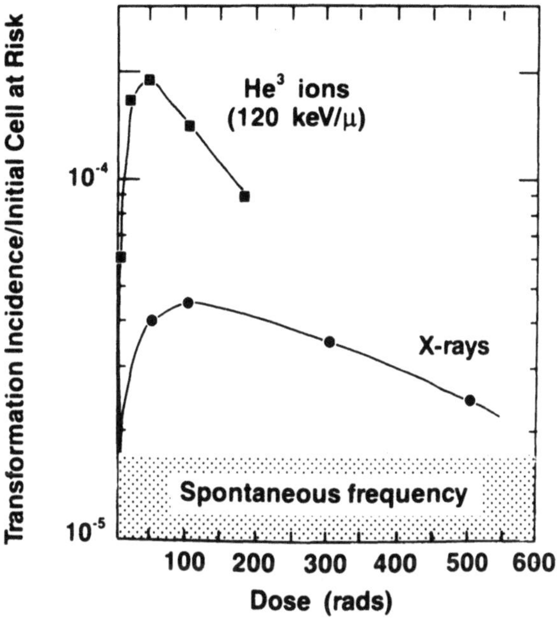

Figure 2: Transformation incidence expressed per initial cell at risk for graded doses of X-rays and helium-3 ions (From Hall and Hei, 1987).

Transformation induction by asbestos and radon-simulated α-particles

The effects of asbestos fibers together with densely ionizing particles on oncogenic transforming incidence have also been examined. Figure 3 shows transformation data for C3H 10T1/2 cells resulting from a 24 hr. pretreatment with 5 μg/ml crocidolite fibers; or dose of 66 rad helium-3 ions. As shown previously, crocidolite fibers, at concentrations that resulted in only moderate cell killing, induced transformation frequencies indistinguishable from the spontaneous rate. However, cells pretreated with crocidolite fibers for 24 hours and subsequently irradiated with α-particles exhibited a significantly higher transformation incidence than after radiation alone. The combination of asbestos fibers and high-LET α-particles produces a transformation incidence higher than the simple sum of the two agents alone. This apparent supra-additivity has implications for the possible interaction of radon and asbestos in the environment, and possibly for combinations of other environmental pollutants.

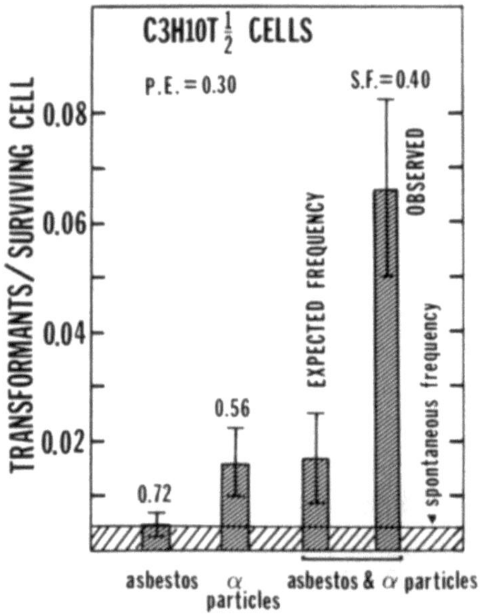

Figure 3: Transformation incidence in C3H/10T-1/2 cells for asbestos fibers and helium-3 ions, delivered alone or in combination. Transformation appears to be supra-additive when asbestos and high LET particles are delivered together.

Effects of SOD and catalase on transformation incidence

Previous studies from this laboratory have demonstrated that asbestos and low LET radiation such as γ-rays can interact in a supra-additive fashion in transformation induction (Hei et al., 1984, 1985). Figure 4 shows the transformants per surviving cell resulting from such a treatment either in the presence or absence of oxygen scavenging enzymes. As reported previously, crocidolite fibers alone, at concentrations used, induced transformation frequencies that do not differ significantly from the spontaneous rate. Addition of SOD or catalase to the culture medium has little or no effect on the cytotoxicity of the fibers as determined by the clonogenicity of treated cells. However, cells pretreated with fibers for 24 hours and subsequently irradiated with 4 Gy of γ-rays produced significantly higher transformation frequencies than radiation alone (Hei et al., 1987). Addition of SOD obliterated the supra-additive response characteristic of the combination of radiation and asbestos and resulted in a transformation frequency similar to that due to radiation by treatment alone. In the present studies, addition of SOD to the irradiated cells had no protective effect either in terms of transformation yield or cytotoxicity. The addition of catalase to the culture medium, however, suppressed the expression of the oncogenic transforming phenotypes due to radiation.

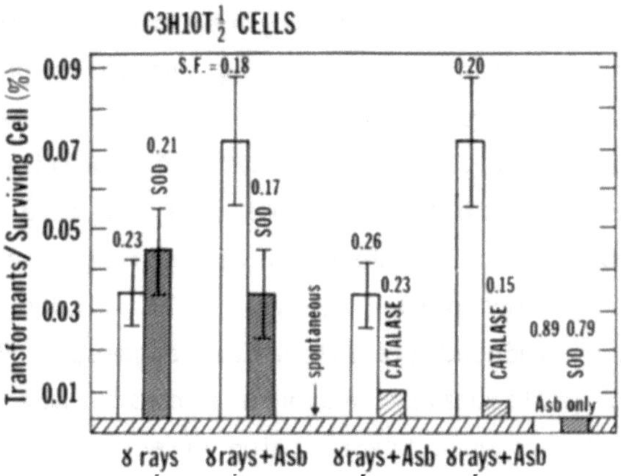

Figure 4: Transformation incidence in C3H 10T 1/2 cells produced by the various treatments either in the presence (closed bar) or absence (open bar) of either SOD or catalase. Results are pooled data from 3 experiments. Bars represent ± S.E. of means. S.F. = surviving fraction.

Summary

In vitro assays for oncogenic transformation allow a rapid and relative inexpensive comparisons to be made of the oncogenic transforming potential of a variety of physical and chemical agents. These *in vitro* systems, based principally on cells of rodent origin, and free of host-mediated homeostatic influences, afford the opportunity to evaluate both the quantitive and qualitative aspects of oncogenic transformation. Furthermore, these systems make it possible to study, under defined laboratory conditions, the cellular and molecular mechanisms involved in radiation or chemically induced transformation. It is evident that high LET, densely ionizing radiation such as the alpha-emitting radon-daughter particles are more oncogenic than low LET radiations even at doses that produce little or no cell killing. The finding that asbestos fibers and α-particles interact in a supra-additive fashion in inducing oncogenic transformation further suggests the importance of interacting factors in human carcinogenesis.

The cellular and molecular basis of the apparent supra-additivity between radiation and asbestos fibers in the induction of transformation is not known. The correlation between fiber dimension and carcinogenicity, however, suggests the importance of fiber-cell interactions (Stanten et al., 1977). The involvement of superoxide amions in asbestos-induced cytotoxicity has been suggested (Hansen and Mossman, 1987) and the radical scavenging enzyme, superoxide dismutase has been shown to protect against the toxicity of asbestos in vitro (Goodglick and Kane, 1987). The fact that SOD significantly reduced the transformation incidence in asbestos and γ-ray treated C3H 10T1/2 cells strongly suggested that the mechanisms of such synergistic interaction may be mediated by superoxide anions. The findings that catalase suppressed radiation induced transformation suggested that free radical and its products, particularly hydrogen peroxide may be an important intermediate in the oncogenic transforming effects of radiation.

Acknowledgement

The authors would like to thank Mr. Steve Kushner for his technical assistance and to Mrs. Jeanne Kramer for typing the manuscript. This work was suppored by Grant CA-12536 from the National Cancer Institute.

References

Barum, DC and Truan, RD (1958) An epidemiological study of lung cancer in asbestos miners. Arch Ind Hyg Occup Med 17:634-637

Goodglick, LA and Kane, AB (1987) Role of reactive oxygen metabolites in crocidolite asbestos toxicities to mouse macrophages. Cancer Res 46:5558-5566

Hall, EJ and Hei, TK (1987) Radon Induced Transformation. In: Cerutti, P, Nygaard, O and Simic, M (eds) Anticarcinogenesis and Radiation Protection. Plenum Press, New York, NY, pp 329-334

Hansen, K and Mossman, BT (1987) Generation of superoxide from alveolar macrophages exposed to asbestiforms and non-fibrous particles. Cancer Res 47:1681-1686

Hei, TK, Hall, EJ and Osmak, R (1984) Asbestos, radiation and oncogenic transformation. Br J Cancer 50:717-720

Hei, TK, Geard, CR, Osmak, RS and Travisano, M (1985) Correlation of in vitro genotoxicity and oncogenicity induced by radiation and asbestos fibers. Br J Cancer 52:591-597

Hei, TK and Kushner, S (1987) Radiation and asbestos fibers: Interaction and possible mechanism. In: Cerutti, P, Nygaard, O and Simic, M (eds) Anticarcinogenesis and Radiation Protection. Plenum Press, New York, NY, pp 345-348

Hei, TK, Komatsu, K, Hall, EJ, and Zaider, M (1988) Oncogenic transformation by charged particles of defined LET. Carcinogenesis 9:747-750
Lloyd, EL, Giemmell, MA, Henning, CB, Giemell, DS and Zabransky, BJ (1979) Int J Radiat Biol 35:23-31
Lundin, FE, Lloyd, JW, Smith, EM, Archer, VE and Holaday, DA (1969) Mortality of uranium miners in relatin to radiation exposure, hard-rock mining and cigarette smoking, 1950 through September 1967. Health Phys 16:571-578
NCRP Report No 79, (1984) Evaluation of occupational and environmental exposures to radon and radon daughters in the US. National Council on Radiation Protection and Measurements, Washington, DC
Reznikoff, CA, Bertram, JS, Brankow, DW and Heidelberger, C (1973) Quantitative and qualitative studies on chemical transformation of cloned C3H mouse embryo cells sensitive to post confluence inhibition of cell division. Cancer Res 33:3239-3249
Stanton, MF, Layard, M and Tegeris, A (1977) Carcinogenicity of fibrous glass: pleural response in the rat in relation to fiber dimension. J Natl Cancer Inst 58:587-604
Selikoff, IJ, Hammond, EC, and Churg, J (1968) Asbestos exposure, smoking and neoplasia. J Am Med Assoc 204:104-110
Wagner, JC and Berry, G (1969) Mesothelioma in rats following inoculation with asbestos. Br J Cancer 23:567-581
Whittemore, AS and McMillan, A (1983) Lung cancer mortality among US uranium miners; A reappraisal. J Natl Cancer Inst 71:489-493

STUDIES ON CELL CYCLE OF RAT PLEURAL MESOTHELIAL CELLS SUBMITTED TO DIRECT OR INDIRECT HIT BY CHRYSOTILE FIBRES.

F. Levy*, A. Van der Meeren**, A. Renier**, A. Katz***, J. Bignon**, and M.C. Jaurand**.

* Laboratoire de Biologie Cellulaire. Faculté de Pharmacie, 4 ave de l'Observatoire. 75006 Paris, France

** Equipe Pathologie Cellulaire de l'Environnement.INSERM U 139. CHU Henri-Mondor. 94010. Créteil Cedex, France

*** Faculté de Médecine. CHU Henri-Mondor. 94010. Créteil France

INTRODUCTION

The cytotoxicity of mineral dusts, specially asbestos, has been studied in vitro with cell types from different origins and species. In our laboratory, we use rat pleural mesothelial cells (RPMC). From all data reported, there is a general agreement to conclude that treatment of cell cultures with chrysotile fibres induces an inhibition of cell growth (Wade et al. 1976, Neugut et al. 1978, Brown et al. 1978, Reiss et al. 1980, Jaurand et al. 1983, Mossman et al. 1983, Hesterberg et al. 1984, Hei et al. 1985, Lechner et al. 1985, Lu et al. 1988). This inhibition has been shown either from a reduction in cloning efficiency or from an increase in the mean population doubling time. With RPMC we have observed a dose dependent decrease in cell proliferation (Jaurand et al. 1983). This effect might be related to a cell killing and/or to an increase in the duration of the cell cycle. To test this second hypothesis we studied the cell cycle of normal RPMC and examined the effects of chrysotile on its regulation, either directly after treatment with chrysotile or indirectly after treatment with conditioned medium obtained from mesothelial cells transformed by chrysotile (Van der Meeren et al. 1988). In addition to the direct effects of asbestos fibres, treatment of cell cultures, including RPMC, with asbestos fibres results in a cell

NATO ASI Series, Vol. H30
Effects of Mineral Dusts on Cells
Edited by B. T. Mossman and R. O. Bégin

transformation and, in vivo, intrapleural inoculation or inhalation of chrysotile in rats results in occurrence of pleural mesothelioma (Wagner et al. 1969, Monchaux et al. 1981). It is known that transformed cells exhibit a different regulation of their proliferation when compared to their normal counterparts. An autocrine stimulation of cell growth is often observed, indicating a production of growth factors. We therefore studied if chrysotile-transformed RPMC produced external factor(s) which might modify cycle of normal cells.

MATERIALS AND METHODS

Cells and culture conditions. Rat pleural mesothelial cells were obtained from the parietal pleura as reported elsewhere (Jaurand et al. 1981). Cells were routinely cultured in Ham F 10 medium supplemented with 10 % fetal calf serum (FCS), 50 µg/ml streptomycin, 100 U/ml penicillin, 1 mM ascorbic acid, 2.7 mM cystein, 2 mM glutamine and 10 mM HEPES.

One mesothelioma cell line was obtained from a pleural mesothelioma induced in vivo by chrysotile fibres (Jaurand et al 1986). Mesothelioma cells were routinely cultured in RPMI 1640 medium supplemented with 10% FCS and 50 µg/ml streptomycin, 100 U/ml penicillin, 2 mM glutamin and 10 mM HEPES.

Mineral fibres. Attapulgite was obtained from Mormoiron (France); mean fibre length was 0.77 µm. Rhodesian chrysotile was provided by the Union Internationale contre le Cancer. Fibres were dispersed in culture medium by sonication for 5 min. (Vibra Cell TM, power 5 watts)

Production of conditioned medium. Mesothelioma cells were cultured until confluence in Nunc 75 cm^2 flasks as described above. Medium was removed and the cells were washed with phosphate buffered saline (PBS pH 7.35) and incubated in serum-free medium for 48 h. Media were collected after 48, 96 and 144 h, according to a method derived from Todaro et al. (1980) and concentrated 50 fold. The YM5 membranes used for the concentration procedure were purchased from Amicon. The resulting supernatants were stored at -80°C until use.

Flow cytometry.

Direct treatment of RPMC with chrysotile fibres. Ten millions of normal cells were plated per 25 cm^2 flask. Twenty four hours later medium was replaced with Ham F10 supplemented with 10 % FCS with or without 2 $\mu g/cm^2$ of chrysotile fibres (10 µg/ml) for 24 or 48 hours. After the 48 hours treatment, medium was replaced with medium without fibres.

Treatment of RPMC with conditioned medium from asbestos-transformed RPMC. Normal cells were plated as described above, and incubated in Ham F10 supplemented with 5 % FCS and either 100 µl of 50× concentrated conditioned medium or no conditioned medium. Cells were treated during 24 or 48 hours.

After treatment, cells were fixed with 70° ethanol, stored at -20°C and further stained with 100 µg/ml mythramycin solution containing 15 mM $MgCl_2$ according to a method derived from Tobey and Crissman (1975); pH was adjusted at 7.4 with NaOH. Fifty thousand cells were analyzed with a flow cytometer ATC 3000 (Odam) equipped with an Argon ion laser operating at 457 nm. In these conditions, fluorescence intensity of each nucleus is proportional to DNA content ("2C" for a diploid cell).

When cells were treated with chrysotile, a sorting of "4C" population was carried out. Sorted cells were examined by microscopy to discriminate between binucleated and G2 cells.

RESULTS AND DISCUSSION.

Cell repartition in the different phases of the cell cycle following treatment with mineral fibres.

Table 1 reports data obtained after treatment of RPMC with chrysotile fibres. It can be observed a time dependent increase in the percentage of "4C" population in the treated cultures. Concerning cell population with a DNA content between "2C-4C" (S cells), there was also a higher percentage of cells in the treated cultures. In parallel, percentage of cells with a DNA content of "2C" was decreased. When the same experiment was performed with attapulgite-treated RPMC, there was no difference in the cell distribution when compared to untreated cells (data not shown).

Table 1 - Flow cytometry : DNA distribution of the RPMC population after treatment with 10 µg/ml chrysotile (Mean values from 3 experiments ± standard deviation)

Treatment	Duration of Treatment	2C %	2C< ... <4C %	4C %
Untreated	24h	75.3 ± 3.2	17.0 ± 2.0	7.7 ± 2.1
Chrysotile		60.8 ± 6.8	19.5 ± 4.9	19.7 ± 5.3
Untreated	48h	81.2 ± 2.7	13.7 ± 1.2	5.2 ± 1.5
Chrysotile		51.3 ± 6.8	20.3 ± 4.8	28.3 ± 4.0
Untreated	72h	85.0 ± 3.5	10.5 ± 2.8	4.5 ± 0.7
Chrysotile		48.2 ± 7.4	18.2 ± 1.8	33.5 ± 9.2

DNA distribution of RPMC treated with attapulgite was identical to untreated cells.

Table 2 - RPMC Distribution in the various phases of the cell cycle taking into consideration percentage of binucleated cells (Mean values from 3 experiments ± standard deviation)

Treatment	Duration of Treatment	G0-G1 %	S %	G2-M %
Untreated	24h	75.5 ± 3.2	17.0 ± 2.0	7.5 ± 2.0
Chrysotile		74.2 ± 10.1	16.7 ± 1.8	9.0 ± 2.8
Untreated	48h	81.3 ± 2.7	13.7 ± 1.2	5.0 ± 1.4
Chrysotile		75.1 ± 3.6	20.3 ± 4.8	4.6 ± 1.4
Untreated	72h	85.1 ± 3.5	10.5 ± 2.8	4.4 ± 0.7
Chrysotile		76.8 ± 8.5	18.2 ± 1.7	4.9 ± 0.7

Distribution of RPMC after sorting are shown in Table 2. The percentage of binucleated cells was subtracted from the "4C" population and added to G1 population. From the results it might be concluded that there was no difference in the percentage of G2 cells between treated and untreated cells. A concomitent decrease in the repartition of the cells in G1 was observed in the treated cells. In contrast, the percentage of S cells remained different between treated and untreated cells from 48 hours.

The increase in percentage of S cells might be due either to recruitment of S cells or to an increased duration of S phase. To test this second hypothesis we determine the thymidine incorporation after a pulse of ^{3}H thymidine. We observed a decreased thymidine incorporation in the treated cultures when compared to untreated cultures (data not shown). Therefore this result does not sustain a recruitment of S cells and might indicate an increased duration of S phase.

Cell repartition in the different phases of the cell cycle following treatment with conditioned medium.

Figure 1 shows histograms obtained with untreated cells or with cells treated with conditioned medium.

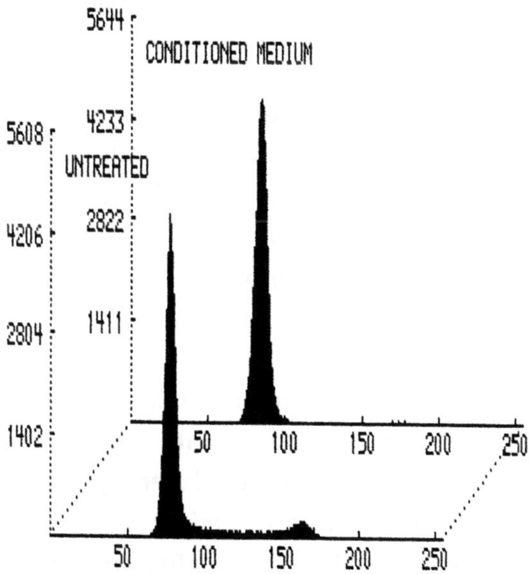

Fig. 1. Experimental DNA histograms with their fitting curves (the mean DNA content expressed in arbitrary units is on the x axis, while the number of the cells is on the y axis).

Cell repartition in the different phases of the cell cycle following treatment with conditioned medium is reported in Table 3. There is a strong decrease in the percentage of S and G2 cells and a concommitent increase in the percentage of cells in G1. These results are in agreement with a blocking of normal cells in G1 by conditioned medium.

It has been demonstrated elsewhere that conditioned medium from these transformed cells also inhibited thymidine incorporation and cell proliferation (Van der Meeren et al. 1988).

Table 3 - DNA distribution of the RPMC population after treatment with conditioned medium (100 μl/ml) obtained from in vivo transformed cells with chrysotile.

Treatment	Duration of Treatment	G0-G1 %	S %	G2-M %
Untreated	24h	77.9 ± 6.4	15.5 ± 4.7	6.1 ± 2.4
Conditioned medium	24h	93.7 ± 2.5	4.4 ± 2.6	1.4 ± 0.6
Untreated	48h	83.7± 7.2	11.2 ± 5.4	5.7 ± 1.8
Conditioned medium	48h	91.4 ± 2.6	6.4 ± 2.3	2.1 ± 0.7

From these results it might be suggested that chrysotile fibres impair cell cycle of normal rat pleural mesothelial cells in culture and that chrysotile transformed cells produce factors inhibiting growth of these normal cells. It is not known if direct effects of chrysotile result from a direct interaction with DNA or if it is a consequence of a modification of the cell response to growth factors; further experiments are necessary to understand the mechanisms of cell growth reduction.

Inhibitory effects of conditioned medium on cell proliferation has been also observed with RPMC in vitro transformed with chrysotile fibres (data not shown). This likely results from the presence of inhibitory factors produced by transformed cells. It would be of interest to determine if this effect might play a role in tumor growth in vivo and if these factors are produced soon after treatment with chrysotile fibres.

References

Brown RC, Chamberlain M, Griffiths DM and Timbrell V (1978) The effect of fibre size on the in vitro biological activity of three types of amphibole asbestos. Int J Cancer 22:721-727

Hei TK, Geard CR, Osmak RS and Travisano M (1985) Correlation of in vitro genotoxicity and oncogenicity induced by radiation and abestos fibres. Br J Cancer 52:591-597

Hesterberg TW, Barrett JC (1984) Dependence of asbestos and mineral dust-induced transformation of mammalian cells in culture on fiber dimension. Cancer Research 44:2170-280

Jaurand MC, Bernaudin JF, Renier A, Kaplan H and Bignon J (1981) Rat pleural mesothelial cells in culture. In Vitro, 17:98-106

Jaurand MC, Bastie-Sigeac I, Bignon I and Stoebner P (1983) Effect of chrysotile and crocidolite on the morphology and growth of rat pleural mesothelial cells. Environ Res 30:255-269

Jaurand MC, Renier A, Van der Meeren A, Magne L, Pinchon MCC, Bignon J (1986) In vitro growth characteristics of rat mesothelioma cells in culture. Biol Cell 57:249-256

Lechner JF, Tokiwa T, La Veck M, Benedict WF, Banks-Schlegel S, Yeager H, Barnerjee JA and Harris CC (1985) Asbestos-associated chromosomal changes in human mesothelial cells. PNAS 82:3884-3888

Lu YP, Lasne C, Lowy R and Chouroulinkov I (1988) Use of orthogonal design method to study the synergistic effects of asbestos fibres and 12-o-tetradecanoylphorbol-13-acetate (TPA) in the BALB/3T3 cell transformation. Mutagenesis 3:355-362

Mossman BT and Landesman JM (1983) Importance of free radicals in asbestos-induced injury to airway epithelial cells. Chest, 83s:50s-51s

Monchaux G, Bignon J Jaurand MC, Lafuma J, Sebastien P, Masse R, Hirsch A and Goni J (1981) Mesotheliomas in rats following inoculation with acid-leached chrysotile asbestos and other mineral fibres. Carcinogenesis 2:229-236

Neugut AI, Eisenberg D, Silverstein M, Pulkrabek P and Weinstein IB (1978) Effects of asbestos on epithelioid cell lines. Environ Res 17:256-265

Reiss B, Solomon S, Weisburger JH and Williams GM (1980) Comparative toxicities of different forms of asbestos in a cell culture assay. Environ. Res. 22:109-129

Tobey RA and Crissman HA (1975) Unique technique for cell cycles analysis utilizing mithramycin and flow microfluorometry. Exp Cell Res 93:235-239

Todaro GJ, Fryling C and De Larco JE (1980) Transforming growth factors produced by certain human tumor cells: polypeptides that interact with epidermal growth factor receptors. Proc Natl Acad Sci USA 77:5258-5262

Van der Meeren A, Levy F, Bignon J and Jaurand MC (1988) Growth of normal and neoplastic rat pleural mesothelial cells in the presence of conditioned medium from neoplastic mesothelial cells. Biol Cell 62:293-296

Wade MJ, Lipkin LE and Frank AL (1979) Studies on in vitro asbestos-cell interaction. J Environ Pathol Toxicol 2:1029-1039

Wade MJ, Lipkin LE, Tucker RW and Frank AL (1976) Asbestos cytotoxicity in a long term macrophage-like cell culture. Nature 264:444-446

Wagner JC and Berry G (1969) Mesotheliomas in rats following inoculation with asbestos. Brit J Cancer 23:567-581

EFFECTS OF 3METHYLCHOLANTHRENE-COATED ASBESTOS ON HEMATOLOGICAL STATUS OF F1(NZB*C57B1/6) MICE

B. Pietrzyca, A. Moniewska, A. Lange

Institute of Immunology and Experimental Therapy
53-114 Wrocław
Czerska 12
Poland

Introduction

We have previously shown that asbestos fibres coated with polycyclic aromatic hydrocarbons (PAH) facilitate presentation of PAH to microsomal enzymes. This produces genotoxic effect which was seen in the Salmonella typhimurium/mammalian microsome mutagenicity test (Szyba and Lange 1983). Also we have already documented that 3-methylcholanthrene (3MC)-coated chrysotile induces a high level activity of aryl hydrocarbon hydroxylases (AHH) after intraperitoneal (i.p.) injection to inducible C57Bl/6 and Fl(NZB*C57Bl/6) mice (Dzugaj et al. 1985). The latter mice were chosen to investigate the effect of i.p. injected 3MC coated chrysotile A either any of these compounds alone on the survival rate and the hematological status.

Materials and Methods

Mice. Inbred strains of mice: NZB/Iiw and C57Bl/6JBoyIiw were housed in our animal facility. Fl(NZB * C57Bl/6) hybrids were raised in our institute from parental strains and used for experiments at 8-12 weeks of age.

Chemicals. Rhodesian Chrysotile A (UICC reference sample was obtained from Dr.V.Timbrell), 3MC was purchased from Sigma and Serva Chemicals Co.

NATO ASI Series, Vol. H30
Effects of Mineral Dusts on Cells
Edited by B.T. Mossman and R.O. Bégin

Adsorption of 3MC onto asbestos fibres (3MC/A/PBS). Chrysotile sample (125mg) was mixed with 10 ml solution of 25 mg 3MC in ethyl ether and ethanol (1:2 v/v). The solvent was then evaporated under reduced pressure at 70°C and later 10 ml of phosphate buffered saline (PBS) was added to the residue and the sample was ultrasonicated at 20 kHz for 10 minutes.

Aqueous suspension of Chrysotile A (A/PBS) prepared by ultrasonicating a mixture of 125 mg of asbestos fibres with 10 ml of PBS. This suspension was finally brought up to 50 ml with PBS, and autoclaved.

3MC oil solution (3MC/OIL) was made by dissolving 2.5 mg 3MC in 1 ml of linseed oil.

Antiserum (rabbit) against mouse gamma globulin came from Behring (Marburg,FRG).

Design of experiments. The experiments consisted the following groups of mice which were i.p. injected with: (1) 3MC oil solution (500 ug) (2) chrysotile PBS suspension (2.5 mg) (3) 3MC (500 ug) coated chrysotile (2.5 mg) (4) oil (5) PBS; four days later the injections were repeated in the same groups (1) 3MC oil solution (300 ug females and 30 ug males) (2) chrysotile PBS suspension (1.5 mg females and 0.15 males (3) 3MC (300 ug females, 30 ug males) coated chrysotile (1.5 mg females, 0.15 mg males) (4) and (5) oil and PBS controls, respectively. First injection was given to induce AHH and the second one for the final toxic effect. Mice were observed for 14 weeks. The hematocrit, hemoglobin and reticulocytes were measured and anti-globulin (Coombs') test positivity was estimated at some certain time intervals. Mice which survived were sacrificed at 14 weeks of observation by cervical dislocation.

Hematology. Tail-vein blood samples were collected into heparin coated tubes and examined. One of the tubes was used for direct anti-globulin (Coombs') test, the second was centrifuged for hematocrit. After 14 weeks the mice were bled and serum used for indirect Coombs'test. Hemoglobin level was measured with cyanomethemoglobin method and blood films were made for Cresyl blue staining to estimate reticulocytes as a fraction of at least 1000 erythrocytes counted.

Results

Survival. F1(NZB*C57Bl/6) mice were injected with 3MC-coated chrysotile either each of the compounds alone. In a preliminary experiment the standard initial dose of the appriopriate agents was followed, with a higher dose of chrysotile (1.5 mg) and 3MC (300 ug) given seperatly or after adsorption procedure in both female and male mice. After 8 weeks of injecting the mice with 3MC-coated chrysotile, all the male ones died but none of the females. Therefore, in the following experiments female mice injection regime was not changed but the male mice second dose was ten times lower than that of the females (0.15 mg of chrysotile and 0.03 of 3MC). Figure 1 a,b shows the combined results of 5 different experiments. During the 14 week time limit observation, only 3MC-coated chrysotile shortened the survival of both male and female mice, however, neither of these two compounds injected separately, including 3MC oil solution. The cause of the shortened survival was not apparent from the morbid anatomy. There were no apparent tumors at that time. However, the physical shape of the mice injected with 3MC coated chrysotile was worse than in other groups. The mice were less vigorous, and represented signs of the wasting syndrome.

Hematological observations. Female mice injected with 3MC-coated chrysotile had lower hematocrit values at 8 and 12 weeks after the injection, as compared to other groups
of mice, including those which received 3MC in oil solution The

anemia progresed with time (Fig. 2a). The same was observed in the male mice (Fig.2b). Chrysotile/PBS suspension injected mice did not differ from the PBS control (not shown).

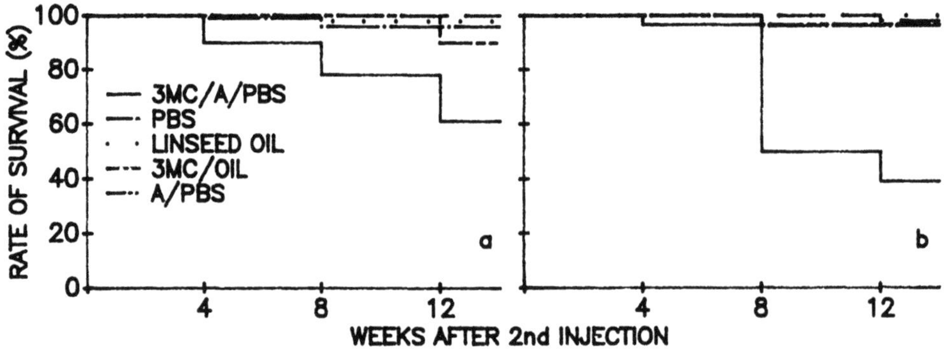

Fig. 1. Survival rates of female (a) and male (b) F1(NZB*C57Bl/6) mice which were injected (i.p.) with different compositions of chrysotile fibres, 3MC and solvents. Mice were injected at 8 - 12 weeks of age two times at four days interval.The first dose to induce AHH activity, and the 2nd dose, to evoke (geno)toxic effect (data from five experiments)

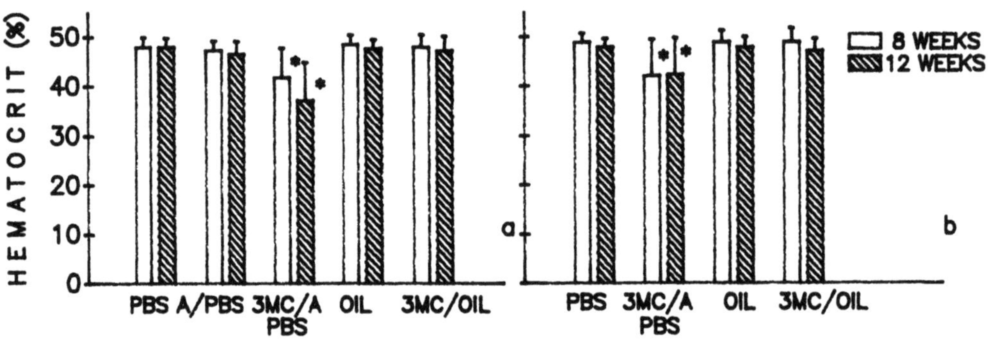

Fig. 2. Hematocrit in (a) female and (b) male F1(NZB*C57Bl/6) mice determined at 8 and 12 weeks after the 2nd injection (data from three experiments).* denotes the statistical significance ($p<0.001$, Student's t-test) as compared separately to the groups injected with PBS and A/PBS

To substaniate the hematocrit results hemoglobin levels were measured in two experiments with female mice. The mice injected with 3MC-coated chrysotile had lower hemoglobin levels as compared to the values of PBS, chrysotile/PBS suspension and 3MC dissolved in linseed oil, respectively (11.39 ± 3.65 vs 15.59 ± 1.35 vs 14.81 ± 1.56 vs 16.09 ± 2.10 g%). Notably, the mice with the 3MC-coated chrysotile induced anemia had higher reticulocyte count than that in other groups (Table 1 and Fig.3). An increase in the reticulocyte count was seen as early as the first hematological work up, meaning before any apparent signs of anemia were registered.

Serological studies. Neither direct nor indirect Coombs' test was positive in two subsequent mice experiments in spite of the anemia and elevation of the reticulocyte count.

Discussion

In our previous works we documented that asbestos fibres can serve as a carrier for benzo(a)pyrene (BP) facilitating genotoxic effect of PAH in the microsomal/Salmonella typhimurium test system (Szyba and Lange 1983). In a subsequent work we showed that this asbestos fibres carrier effect also plays a role in a rodent system. 3MC-coated chrysotile is a more effective inducer of liver AHH than 3MC aqueous microcrystalline suspension and this AHH inducing potential of 3MC-coated asbestos is similar to that exerted by 3MC oil solution (Dzugaj at al. 1985). Therefore, it was shown that PAH, while linked with asbestos, is readily available at least for liver microsomes.

In the present work, F1(NZB*C57Bl/6) mice injected with 3MC-coated chrysotile had considerably shortened life span and the male mice were more susceptibile than female ones. Reason for the poor survival of these mice was not apparent from the present work. Especially, that the observation time (14 weeks) was too short for the tumor induction (Suzuki and Kohyama 1984). The high rate of death in 3MC-coated chrysotile injected mice could be due to the anemia and the deterioration of the

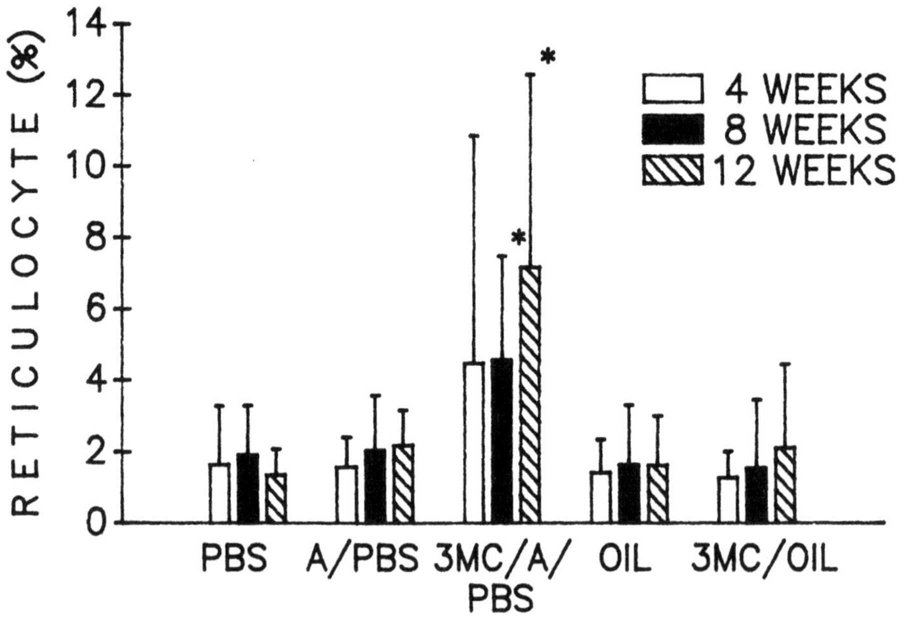

Fig. 3. Reticulocyte counts (%) in female F1(NZB*C57Bl/6) mice determined at different time intervals after the 2nd injection (data from two experiments) * denotes the statistical significance ($p<0.001$, Student's t-test) as compared separately to the groups injected with PBS and A/PBS

Table 1. Reticulocyte counts (%) in male F1(NZB*C57Bl/6) mice determined at 8 and 12 weeks after the 2nd injection (data from two experiments)

	Intraperitoneal exposure			
weeks	PBS	3MC/A/PBS	L.OIL	3MC/OIL
8	1.7±0.8*[a]	2.9±2.7[a]	1.8±1.1	1.3±0.9
	(23)	(25)	(24)	(20)
12	1.3±0.7[b]	2.9±2.3[b]	1.4±0.6	1.9±0.8
	(21)	(16)	(24)	(18)

* mean value ± SD; in brackets number of animals
statistically significant [a] $p<0.05$, [b] $p<0.001$ (Student's t-test)

mice. All other groups of mice including those injected with 3MC oil solution neither had shortened survival nor anemia. Therefore, a presentation of 3MC, as material linked with chrysotile fibres to the body constituents, was crucial for exerting a toxic hit.

Massive i.p., doses of BP and 3MC cause earlier death of mice responsive to AHH induction (Kouri et al. 1980). In the present work we showed that PAH adsorbed onto chrysotile is more toxic than that dissolved in oil. The mechansim of this effect of the combined exposure to PAH and chrysotile is not apparent from the present work. It is known, that asbestos fibres can break to the circulation after exposure (Sebastien et al. 1980). Accumulation of 3MC should therefore rely on the distribution of asbestos fibres which rather accumulate in phagocytic cells rich organs, including the bone marrow. Therefore, the effect upon the bone marrow can be seen at the lower range of doses than that after oil solution injection. The anemia was paraleled with the higher than in other groups reticulocyte count. However, neither anti-globulin test positivity nor spleen enlargement seen in the anemic groups pointed out on the presence of autoimmune hemolytic anemia or hypersplenism in the affected mice. These two latter possibilities should be taken into account because in NZB mice autoimmune hemolytic anemia and spleen enlargement constitute the hallmarks of their autoimmune syndrome (Theofilopoulos and Dixon 1981). This observation rather speaks in favor of the unproper erythropoiesis than simply myelosuppression in our mice. Indeed, it was found, that BP causes a decline in the number of spleen colony forming units (CFU) and granulocyte-macrophages CFU with an overall increase in the cellularity of the bone marrow (Dean et al. 1981). Therefore, it is likely that the anemia caused by 3MC-coated asbestos fibres reflected the damage of the bone marrow stem cells, which results in the unproper differentiation and anemia due to the shorten erythocyte survival. The biological effect of 3MC-coated chrysotile is similar to that exerted by higher doses of PAH injected in an oil solvent. Therefore, we can conclude that asbestos fibres carrying PAH not only facilitate AHH induction (Dzugaj et al. 1985) but also potentiate the advert effect of PAH on F1(NZB*C57Bl/6) mice.

References

Dean JH, Luster MI, Boorman GA, Chae K, Lauer LD, Luebke RW, Lawson LD, Wilson RE (1981) Assessment of immunotoxicity induced by the environmental chemicals 2,3,7,8-tetrachlorodibenzo-p-dioxin,diethylstilbestrol and benzo(a)pyrene. In: Hadden J, Chedid L, Mullen P, Spreafico F (eds) Advances in immunopharmacology. Pergamon Press, New York Toronto Sydney Paris Frankfurt, p 37

Dzugaj A, Szyba K, Lange A (1985) Adsorption of 3-methylcholanthrene onto chrysotile fibres enhances aryl hydrocarbon hydroxylase inducement in mice. In: Beck EG , Bignon J (eds) In vitro effects of mineral dusts. Third International Workshop. NATO ASI Series, Springer-Verlag Berlin Heidelberg New York Tokyo, p 467

Kouri RE, Schechtman LM, Nebert DW (1980) Metabolism of chemical carcinogens. In Kouri RE (ed) genetic differences in chemical carcinogenesis, CRC Press, Boca Raton, p 21

Sebastien P, Masse R, Bignon J (1980) Recovery of ingested asbestos fibers from the gastrointestinal lymph in rats. Env Res 22:201-216

Suzuki Y, Kohyama N (1984) Malignant mesothelioma induced by asbestos and zeolite in the mouse peritoneal cavity. Env Res 35:277-292

Szyba K, Lange A (1983) Presentation of benzo(a) pyrene to microsomal enzymes by asbestos fibres in the Salmonella/mammalian microsome mutagenicity test. Environ Health Perspect 51:337-341

Theofilopoulos AN, Dixon FJ (1981) Etiopathogenesis of murine SLE. Immunological Rev 55:179-216

OCCURRENCE AND MORPHOLOGY OF TUMOURS INDUCED IN NUDE MICE TRANSPLANTED WITH PLEURAL MESOTHELIAL CELLS IN VITRO TREATED WITH CHRYSOTILE OR BENZO 3-4 PYRENE.

L. Saint-Etienne*, M. Nebut**, J. Fleury-Feith***, M. J. Paterour**, L. Kheuang**, J. Bignon**, M. C. Jaurand**.

* LEPI, 44, rue Charles Moureu, 75013 Paris, France

** Equipe Pathologie Cellulaire de l'Environnement, INSERM U 139. CHU Henri Mondor 94010 Créteil Cedex, France

*** Faculté de Médecine, Université Paris XII. 94010 Créteil Cedex, France

Mesothelial cells are a useful model to study carcinogenesis related to mineral fibres (Lechner et al. 1985, Paterour et al. 1985a, Craighead et al. 1987, Mackay et al. 1987).

In our laboratory we have cultured rat pleural mesothelial cells (RPMC) and performed in vitro treatment with chrysotile fibres or with benzo 3–4 pyrene (BP) to study mechanisms of transformation. These treatments resulted in a cell transformation assessed by abnormal morphology in liquid media or by growth in soft agar (Paterour et al. 1985b, Jaurand et al. 1986).

However, relationship between morphological or growth modifications and neoplastic transformation was unknown for these mesothelial cells. Since cell transplantation into nude mice is currently used to assay of cellular tumorigenicity (Freedman and Shin, 1978) it was of great interest to study tumorigenic potency of rat pleural mesothelial cells following treatments with benzo 3-4 pyrene or chrysotile fibres.

Human mesothelioma is sometimes difficult to diagnose. While normal mesothelial cells exhibit characteristics of epithelial cells (junctions, microfilaments, synthesis of basal lamina), tumours often present populations of mixed cells either fusiform or epithelial. Until now, it is not clearly established if fusiform cells are a differentiated state of mesothelial cells. It was therefore

NATO ASI Series, Vol. H30
Effects of Mineral Dusts on Cells
Edited by B. T. Mossman and R. O. Bégin

of interest to study the morphology of tumours formed in nude mice after inoculation of mesothelial cells. It was also useful to compare ultrastructural characteristics of cells from the tumours to that of inoculated cells.

The aim of this work was to determine the tumorigenic potency of treated mesothelial cells and to study ultrastructure of tumours induced in nude mice in comparison with that of inoculated cells.

Materials and methods.

Two series of rat pleural mesothelial cells were used. Cells were cultured using classical methods of culture, based on weekly subculturing 1:3 after trypsinization (Jaurand et al.1981).

a) Mesothelial cells were treated with 1 µg/ml of BP and 0.05% DMSO at passage 10 and subsequently passaged until passage 45. Details are reported elsewhere; cloning was carried out in liquid medium or in soft agar (Paterour et al. 1985a,b). Three cultures were tested: one treated with BP in DMSO as described above, one treated with 0.05 % DMSO, one clone expanded from a colony isolated in liquid medium and showing cell overlapping and growth in multilayer. This clone was subcultured 9 times after colony isolation.

b) Cells were treated with 2 µg/ml of UICC Rhodesian chrysotile at passage 10 and subsequently incubated at each passage with or without chrysotile fibres. A control culture was passaged according to the same process but remained untreated. Repetitively chrysotile-treated cells formed colonies in soft agar after 40 passages (Jaurand et al. 1986).

Before injection into nude mice one tissue culture flask was processed for electron microscopic study of cells. Nude mice were inoculated subcutaneously with 1.5 x 10^6 cells suspended in 0.2 ml phosphate buffered saline, one site per mouse. When tumours or nodules appeared they were allowed to grow for to 2 to 6 weeks. Tumours were dissected after lethal anesthesia and cut into small pieces. Some were kept for culture purposes and others were fixed in 1.2 % glutaraldehyde in 0.045 M sodium cacodylate buffer. Cell cultures were processed for electron microscopic analysis according to standard methods. Fixation was carried out with 2.3 % glutaraldehyde and 1 % osmium tetroxide. Dehydration and Epon embedding were performed according to standard methods.

Results and discussion.

Table 1 summarizes delay between injection and tumour development in nude mice and tumour incidence. It can be observed that tumour occurrence was dependent on treatment and on number of passages. It must be noted that several tumours have been caryotyped and caryotypes confirmed the rat origin of tumours (data not shown).

Table 1: Tumor occurrence in nude mice following inoculation of RPMC.

Series N°	Cell Treatment	Passage number when inoculated	Delay to first tumour	Tumour incidence
1	0.05% DMSO (1 treatment)	39	22	50 %
2	BP (1 treatment)	40	4	90 %
3	as 2, clone	36+9	9	80 %
4	None	75	17	60 %
5	Chrysotile (1 treatment)	75	2	80 %
6	Chrysotile (65 treatment)	75	1	100 %
7	as 6, agar clone	75 + 9	1	100 %
8	as 6, agar clone	75 + 13	2	100 %
9	None	48	-	0 %
10	Chrysotile (1 treatment)	48	-	0 %
11	Chrysotile (36 treatment)	46	13	100 %

From results obtained with series 1 to 3 (Table 1), it appears that there was some transformation of DMSO-treated cells but tumours appeared after more than 5 months and only 50 % of mice injected bore tumour.

After 26 weeks. BP-treated mesothelial cells were strongly more tumorigenic since 90 % of the mice injected developed tumours. In that series, first tumours appeared after 4 weeks. Cloned cells showed a tumour incidence more or less identical.

Electron microscopic studies of rat parietal pleura or of cultures of normal mesothelial cells revealed ultrastructural features of mesothelial cells: microvilli, junctions, basal lamina and bundles of microfilaments (Jaurand et al.1981). Those characteristics were found in tumours recovered after injection of BP-treated cells (Figure 1).

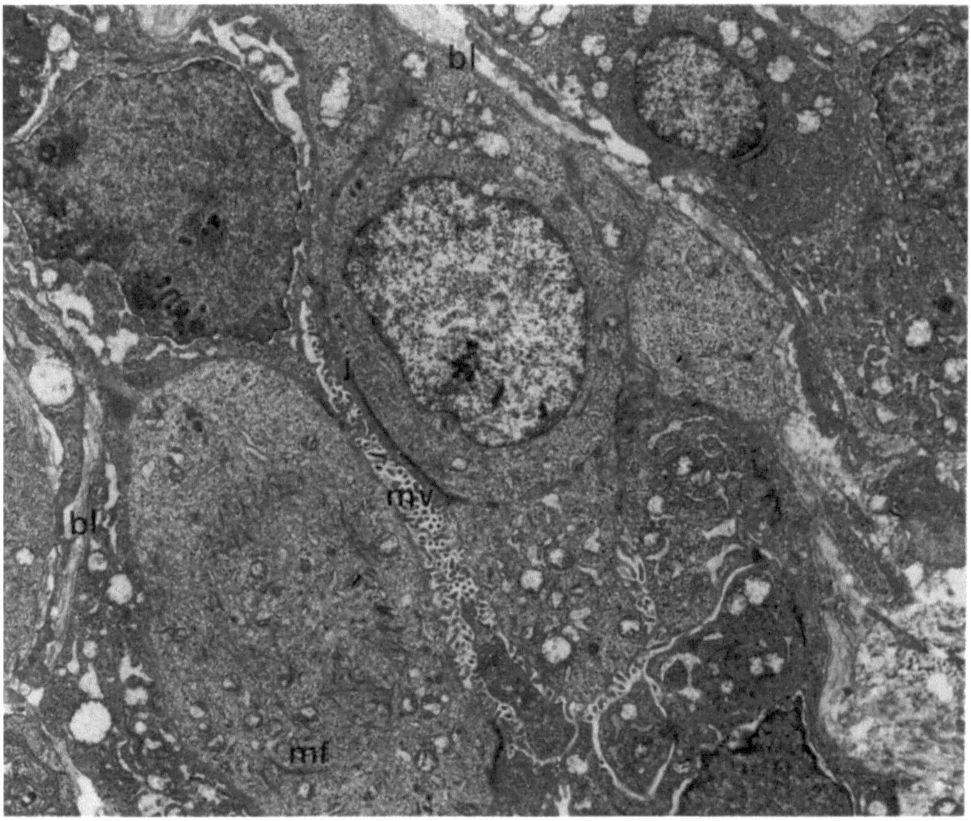

Figure 1 - Transmission electron microscopy of tumour, arising in nude mice 13 weeks following inoculation of BP-treated cells and excised three weeks later. Intracytoplasmic filaments arranged in bundles (m.f.); microvilli (m.v.), basal lamina (b.l.),junctions (j); ×7200.

Untreated cells injected at passage 48 (series 9) were not tumourigenic (Table 1) even after a delay of 40 weeks. In the same way cells treated once with chrysotile (series 10) were not tumourigenic. However tumours appeared after 17 weeks when untreated cells were injected at passage 75 (series 4). Tumours arising after this delay showed microvilli and junctions characteristics of mesothelial cells (Figure 2). Culture of untreated cells was observed in electron microscopy before inoculation (Figure 3). No specific markers was detected. Cells were polygonal and grew as a monolayer. Tumour growth seems likely to result from the proliferation of differentiated cells which might have been present in the culture injected or which evolved with time in host. Adaptation to host environment might also be of importance to explain long delay between injection and tumour occurrence.

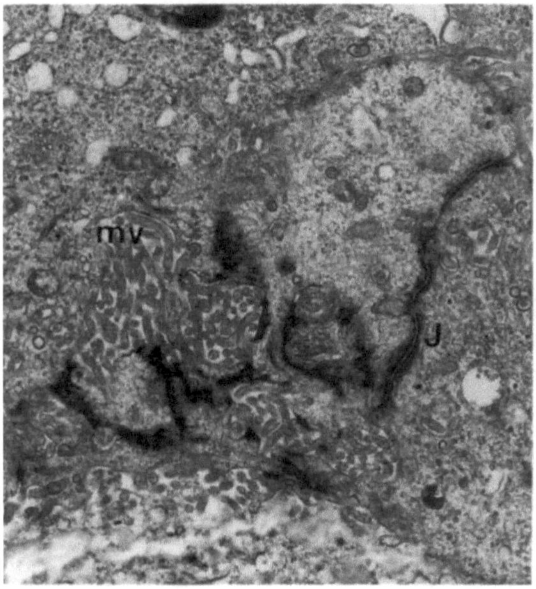

Fig.2- Transmission electron microscopy of tumour arising in nude mice 21 weeks following inoculation of untreated cells and excised one week later. Junctions (j), microvilli(m.v.); ×13000.

When mesothelial cells were treated with chrysotile fibres for a long period (series 6 and 11) or selected by their growth in soft agar (series 7 and 8) a maximal rate of tumour incidence was obtained and tumours appeared as soon as 1 week after cell inoculation. Figure 4 shows that cultures exhibited bundles of microfilaments which were recovered in tumours (Figure 5). Therefore tumours keep the morphological characteristics of cells injected.

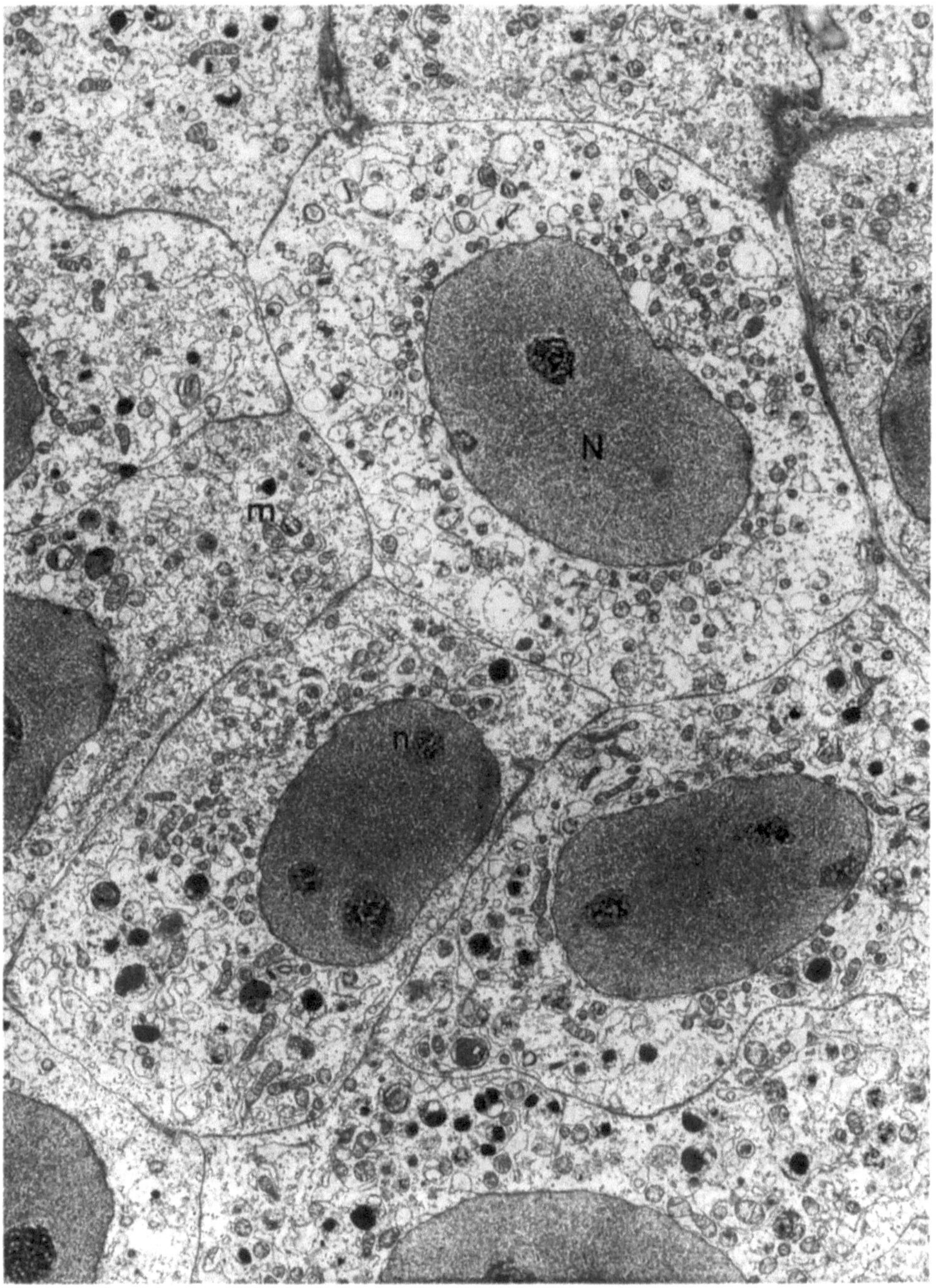

Figure 3 - Transmission electron microscopy of untreated culture at passage 60. Mitochondria (m), nucleoli (n), Nucleus (N), junctions (j); ×4900.

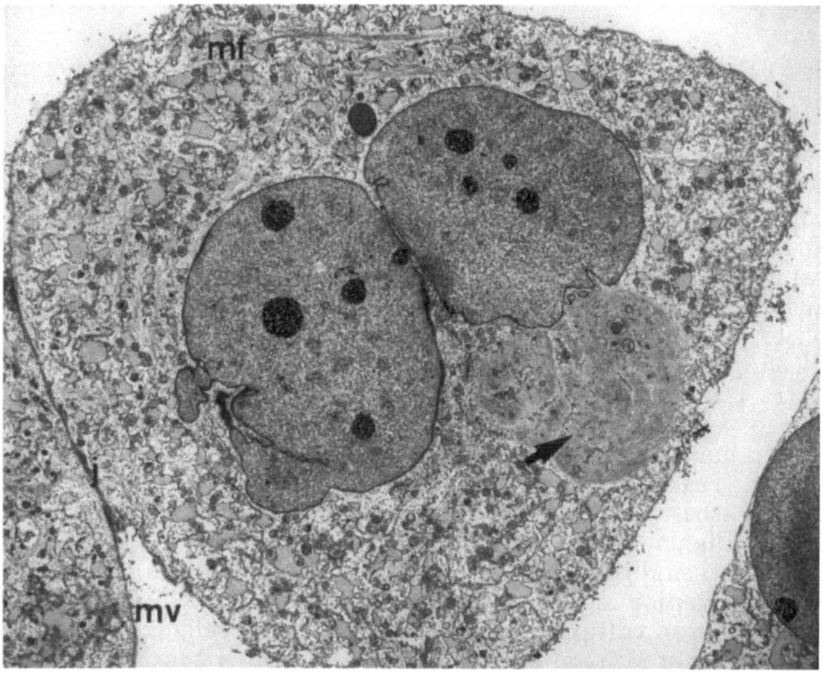

Figure 4 - Transmission electron microscopy of chrysotile-treated culture at passage 58. Microvilli (m.v.), junctions (j), microfilaments (m.f.), arranged in bundles (--->); ×2300.

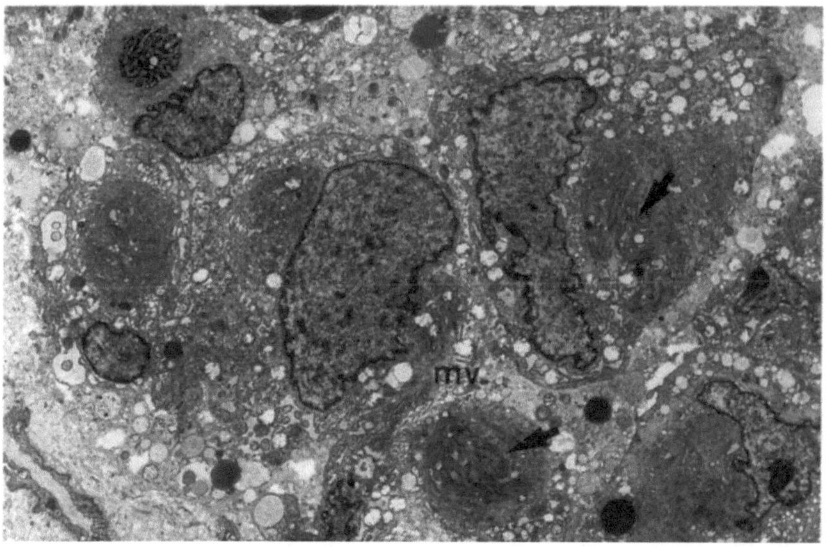

Figure 5 - Transmission electron microscopy of tumour arising in nuce mice one week following inoculation of repetitively chrysotile treated cells and excised 9 weeks later. microvilli (m), bundles of microfilaments (--->); × 4400.

References

Craighead JE, Akley JN, Gould LB and Libbus DL (1987) Characteristics of tumors and tumor cells cultured from experimental asbestos-induced mesotheliomas in rats. Amer J Pathol 129:448-462

Freedman VH and Shin S (1978) Use of nude mice for studies on the tumorigenicity of animal cells. In :The nude mouse in experimental and clinical research. Fogh J, Giovanella BC eds. Acad Press, p 353.

Jaurand MC, Bernaudin JF, Renier A, Kaplan H and Bignon J (1981) Rat pleural mesothelial cells in culture. In Vitro 17:98-106

Jaurand MC, Magne L, Renier A, Kheuang L and Bignon J (1986) In vitro responses of target mesothelial cells to chrysotile asbestos fibres. Eur J Cell Biol 42:35A

Lechner JF, Tokiwa T, La Veck M, Benedict WF, Banks-Schlegel S, Yeager H, Barnerjee JA and Harris CC (1985) Asbestos-associated chromosomal changes in human mesothelial cells. PNAS 82:3884-3888

Mackay AM, Tracy RP and Craighead JE (1987) Intermediate filaments proteins in asbestos-induced mesotheliomas of rat. Cancer Res 47:5461-5468

Paterour MJ, Bignon J and Jaurand MC (1985a) In vitro transformation of rat pleural mesothelial cells by chrysotile fibres and/or benzo(a)pyrene. 6:523-529

Paterour MJ, Renier A, Bignon J and Jaurand MC (1985b) Induction of transformation in cultured rat pleural mesothelial cells by chrysotile fibres. In vitro Effects of Mineral Dusts. Beck EG, Bignon J eds. Nato Series vol G3 Springer Verlag, Berlin, p 209

CYTOTOXIC, HEMOLYTIC AND MUTAGENIC ISSUE CAUSED BY CHRYSOTILE--ASBESTOS IN VITRO

I. Trošić, Đ. Horvat, L. Štilinović, Z. Pišl
Institute for Medical Research
and Occupational Health
Moše Pijade 158
41000 Zagreb
Yugoslavia

Introduction

The development of malignant alterations has been well documented in association with previous occupational or environmental exposure to asbestos dust (Kagan et all. 1977). Interest of action of asbestos has centered largely on the structural and aerodynamic properties of the fibers. However, litlle attention has been paid to the soluble components of asbestos fibers. The biologically harmful effect of asbestos have been extensively investigated but mechanisms by which asbestos fibers cellular and intracellular alteration are not completcly understood.

The aim of our studies was to investigate the cytotoxic, hemolytic and mutagenic effects of ultrasonically disintegrated chrysotile-asbestos (ASB I) and soluble components of disintegrated chrysotile asbestos (ASB II).

Materials and Methods

Asbestos fibers were ultrasonically disintegrated and sterilised. Ultrasonically disintegrated samples of chrysotile-asbestos (ASB I) were used in concentrations of 50, 100 and 200 µg/mL of nutrient medium (MEM).

NATO ASI Series, Vol. H30
Effects of Mineral Dusts on Cells
Edited by B.T. Mossman and R.O. Bégin

The soluble components (ASB II) was obtained by soaking asbestos particles in water. The soluble fractions were filtered (filter Swennex - 25; 0.22) and centrifuged at 1600 g to remove chrysotile fibers from the system.

The cytotoxicity of ASB I and ASB II was examined by the survival and colony forming ability of lung fibroblasts in the chinese hamster - V79 cells, by standard method (Puck et al. 1965). The incubation period lasted to 2,6 and 24 hours.

The direct hemolytic effect of ASB I and ASB II was determined on washed human erythrocytes obtained form healthy non-asbestos exposed subject. After triple washing with PBS, a suspension of packed erythrocytes was prepared. The concentration of erythrocyte suspension in the physiological solution was adjusted to the optical density of lysate 0.7. The experimental procedure used was direct asbestos hemolysis by Harington (Harington et al. 1971). One series of test tubes contained a hemolytic system of asbestos fibers (ASB I), in concentrations of 50, 100 and 200 μg/mL PBS and washed human erythrocytes. The second series of test tubes contained equivalent asbestos fraction (ASB II) and washed erythrocytes. Both systems were incubated for 30, 60, 90, 120 min. and 4 and 8 hours, at 37 ^{o}C, with continouous rotation et 24 rpm. One series of test tubes served as a control and contained packed human erythrocytes and distilled water, so complete hemolysis occured. Relase hemoglobin was determined spectrophotometrically.

The intracellular effect of chrysotile-asbestos was examined by determining the synthesis of macromolecular structures in the V79 cells. The synthesis of deoxyribonucleic acid, ribonucleic acid and proteins were analysed using Miller's method for scientilation counter after incorporation of ^{3}H-thymidine, ^{3}H-uridine and ^{14}C-phenylalanine (Miller et al. 1972). The specific activity of specific precursor of DNA synthesis ^{3}H-thymidine and specific precursor of RNA synthesis ^{3}H-uridine was 1985 GB_q/mMol, respectively. Protein synthesis activity was followed by incorporation of ^{14}C-phenylalanine, specific activity 370 GB_q/mMol. Concentrations of asbestos (ASB I) were 50, 100 and 200 ug/mL nutrient medium. Incubation lasted 2, 6 and 24 hours

The effect of ASB I on the genom of V79 cells was examined by conventional analysis of the structural aberations of chromosomes (Moorcheal et al. 1960) and frequency of sister chromatide exchanges (SCE), (Kato 1974).

Results and Discusion

The harmful effect for both agents, ASB I and ASB II, on the proliferation of V79 was evident. After 24 hours of incubation V79 cells with ASB I, cells survival was 22 %, (Fig. 1.), at the same condition with ASB II cells survival was 24 % (Fig. 2.). The role of soluble components of asbestos (ASB II) in cytotoxicity is obvious.

ASB I and ASB II have hemolytic potency related to the incubation period respectively. Human erythrocytes incubated for 8 hours with highest concentration of ASB I (200 µg/mL) showed hemolysis amounted to 99 % (Fig. 3.). After 8 hours of incubation erythrocytes with ASB II showed hemolytic activity of 43 % (Fig. 4.).

Obviously, the mechanical-irritative properties, like the chemical components of asbestos, plays a significant role in hemolysis.

The relatively rapid effect of asbestos of ^3H-thymidine into DNA in the cell was obvious as shown in Fig. 5. After 6 hours incubation ^3H-thymidine with V79 cells incorporation increased. The reason was an elevated number of cells produced by cell

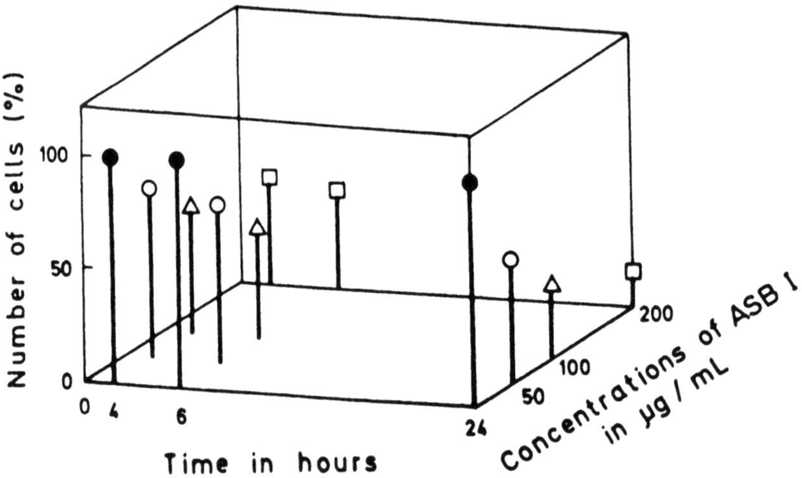

Fig, 1. Survival of V79 cells after treatment with chrysotile--asbestos (ASB I). Average values of six separate samples: control (●), 50 (o), 100 (Δ) and 200 µg/mL (□).

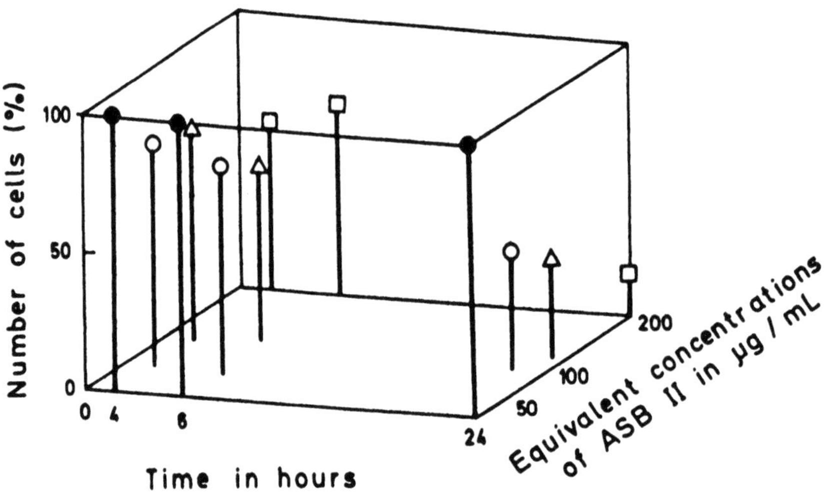

Fig. 2. Survival of V79 cells after treatment with soluble components of chrysotile-asbestos (ASB II). Average of six separate samples: control (●), 50 (o), 100 (Δ) and 200 µg/mL (□).

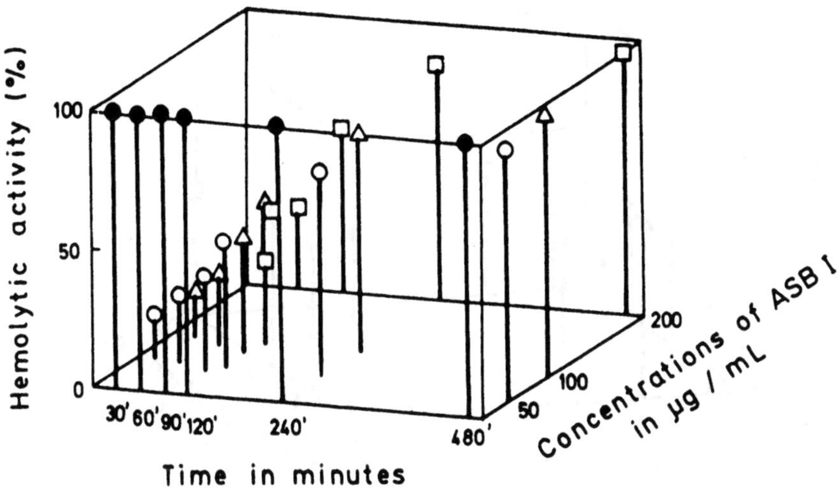

Fig. 3. Hemolytic activity of chrysotile-asbestos (ASB I). Average values of six separate samples: control (●), 50 (o), 100 (△) and 200 µg/mL (□).

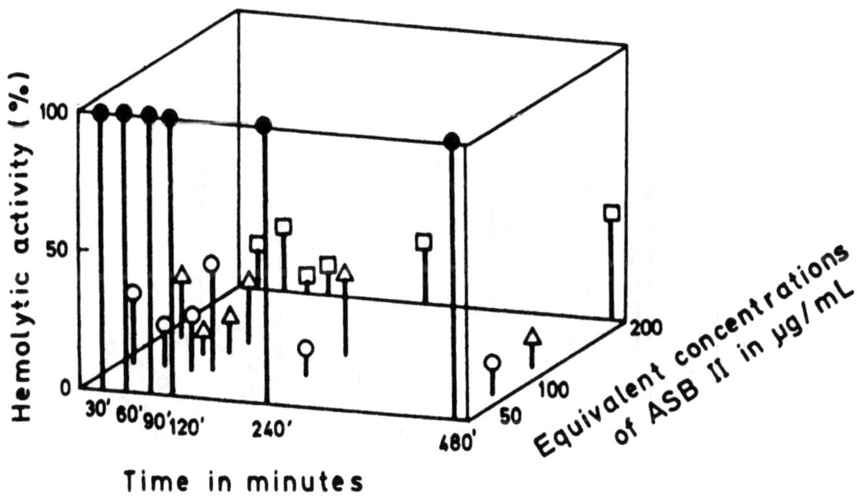

Fig. 4. Hemolytic activity of soluble components of chrysotile--asbestos (ASB II). Average values of six separate samples: control (●), 50 (o), 100 (△) and 200 µg/mL (□).

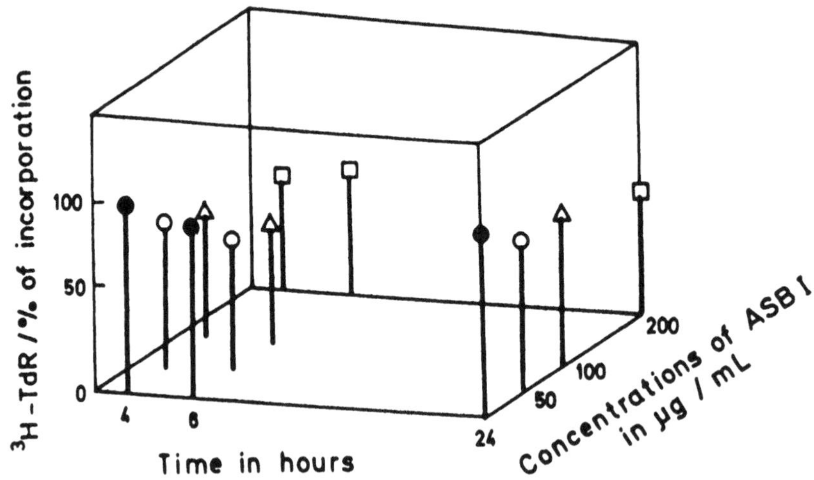

Fig. 5. Incorporation of ^{3}H-thymidine into the DNA of V79 cells treated with chrysotil-asbestos (ASB I). Average values of six separate samples: control (•), 50 (o), 100 (△) and 200 µg/mL (□).

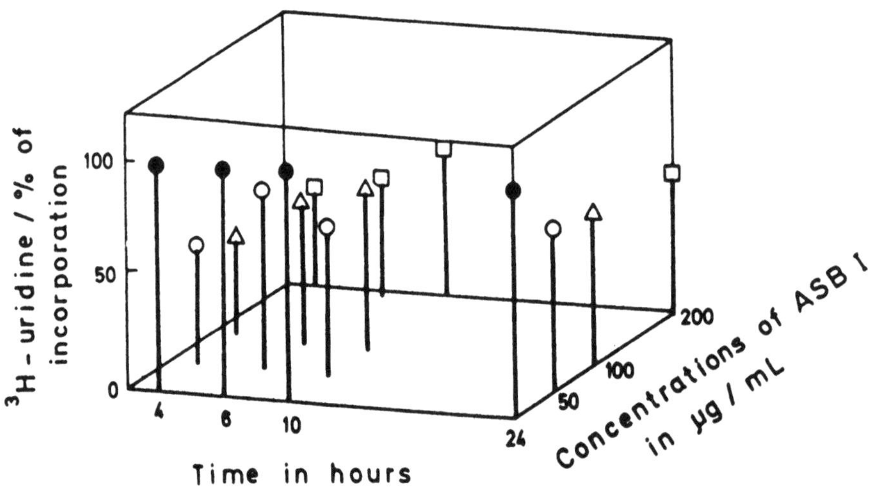

Fig. 6. Incorporation of ^{3}H-uridine into the RNA of V79 cells treated with chrysotile-asbestos (ASB I). Average values of six separate samples: control (•), 50 (o), 100 (△) and 200 µg/mL (□).

division. Increased ^{3}H-tymidine was probably due to pseudo synchronisation of the cells incubated with asbestos and partly also DNA repair.

Continous incorporation of ^{3}H-uridine into the V79 cells was significantly decreased after two-hour of incubation with asbestos. Incorporation of marked precursor was inhibited to about 60 % in treated samples, in relation to the control samples, irrespective of dose. Incorporation of ^{3}H-uridine gradually increased after a two-hour "shock" inhibitory period. The incorporation of ^{3}H-uridine into RNA of V79 cells after incubation for 24 hours was still significantly inhibited, (Fig. 6.).

Incorporation of ^{14}C-phenylalanine into proteins of V79 cells demonstrated that asbestos did not significantly affect this aspect of cell metabolism (Fig. 7.). This surprising result may be due to the method of continued incorporation of the precursor which is not suitable enough to detect possible small changes in the synthesis of proteins. As the results of these experiments relate to a 24-hour period, it is possible that the effect of asbestos on the protein synthesis could be evidenced by apply the method of puls-marcation or other biochemical tests.

Various chromosomal abnormalities was demonstrated in V79 chinese hamster lung cells exposed to chrysotil-asbestos. There were a higher number of chromosome abberations in the cell culture treated with chrycotil-asbestos than untreated samples. In the chrysotil treated samples the significant amount was evident in the dose level of 100 ug/mL of nutrient medium after 24 hours of incubation. The dominant type of chromosome damage were mono and isochromatide breakes and dislocated fragments, dicentric and ring chromosome, and different forms of translocations Fig. 8.

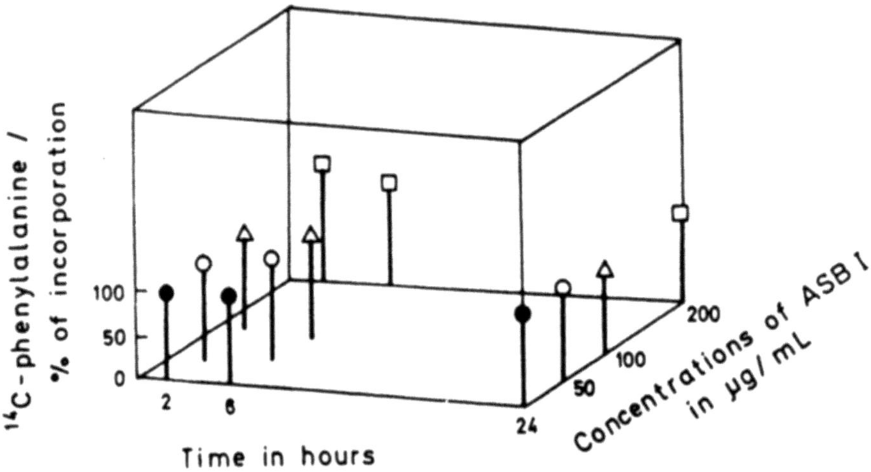

Fig. 7. Incorporation of ^{14}C-phenylalanine into the proteins of V79 cells treated with chrysotil-asbestos (ASB I). Average values of six separate samples: control (•), 50 (o), 100 (Δ) and 200 µg/mL (□).

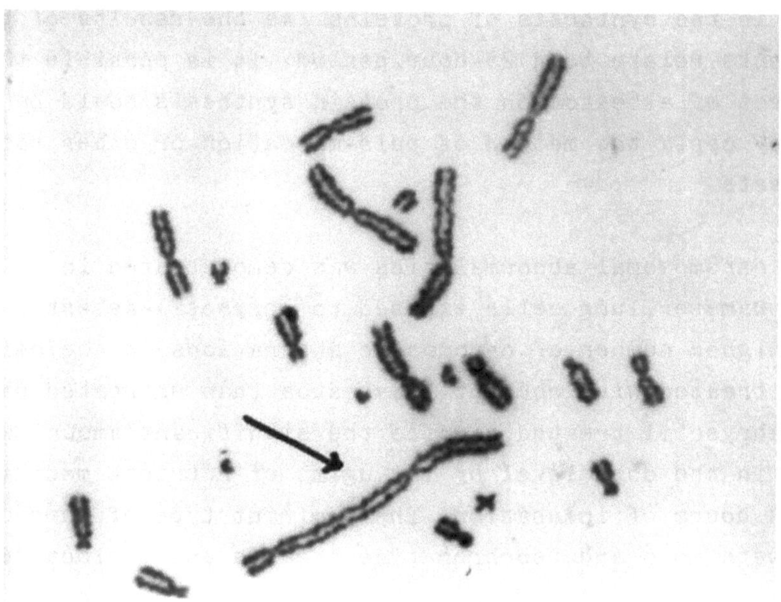

Fig. 8. Chromosome aberation (CA) in the V79 cells treated with chrysotile-asbestos.

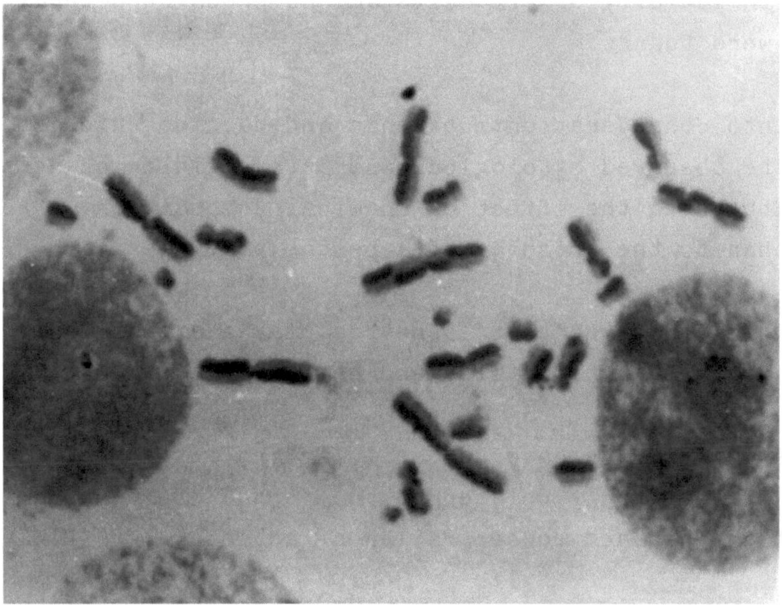

Fig. 9. Sister chromatide exchange (SCE) in V79 cells treated with chrysotile-asbestos.

The incidence of sister chromatide exchanges (SCE) in untreated control samples was 5 $\pm$ 1,5 but in treated samples was 10 $\pm$ 2 SCE per methaphase.

Conclusion

The cytotoxic and hemolytic effect of ultrasonicaly disintregated chrysotil-asbestos (ASB I) and soluble components (ASB II) increased in proportion with asbestos dose and incubation period. Incorporation of ^{3}H-thymidine, and ^{3}H-uridine into DNA and RNA was significantly inhibited after only 2 hours of exposure of V79 cells to ASB I at concentrations of 50, 100 and 200 ug/mL nutritive medium. The same doses did not effect the incorporation of ^{14}C-phenylalanine into proteins of V79 cells.

Changes in ploidity of V79 cells and numerous structuraly damaged were found.

Taking into consideration mutagenic and carcinogenic asbestos action the observed cytological and intracellular changes can be attributed to the effect of chemical composition of asbestos rather than to the presence of asbestos particles.

The soluble constituents of asbestos play a significant role in the proces of cytotoxicity and hemolysis.

In these experiments observed cellular and intracellular changes are significantly related to the period of incubation rather than the used concentration of asbestos.

Acnowledgements. This work was supported by Self-Managed--Community of Interest for Science of SR Croatia, Yugoslavia.

References

Harington JS, Miller K, Macnab G (1971) Hemolysis by asbestos. Environ Res 4: 95-117

Kato H (1974) Spontaneous sister chromatide exchange detect by BUdR-labelling method. Nature 251: 70-81

Miller GG, Walker GWR, Gilbah RE (1972) A rapid method to determine the mamalian cell cycle. Exptl Cell Res 72: 533-548

Moorched PS, Nowel PC, Mallman WL, Bettips DD, Hungerford DA (1960) Chromosome preparations of leucocyte cultures from pherypheral blood. Exptl Cell Res 20: 613-623

Puck TT, Marcus PI, Cieciura SJ (1965) Clonal growth of mammalian cell in vitro. J. Exptl Med 103: 273-275

EFFECTS OF ASBESTOS ON SPECIFIC BINDING OF PHORBOL ESTER TUMOR PROMOTER AND PROTEIN KINASE C ACTIVITY IN HAMSTER TRACHEAL EPITHELIAL (HTE) CELLS

M. Perderiset, J.P. Marsh, B.T. Mossman
Department of Pathology, University of Vermont College of Medicine, Burlington, VT 05405, U.S.A.

Introduction

Crocidolite asbestos causes proliferative changes in tracheobronchial epithelial cells that are similar to those observed after exposure of a number of cell types to phorbol esters, i.e. classical tumor promoters in skin and other organs (reviewed in Mossman et al., 1985). These biological changes, which include increased uptake of ^{3}H-thymidine (Landesman and Mossman, 1982), increased activity of ornithine decarboxylase (ODC), i.e. a rate-limiting enzyme in the biosynthesis of polyamines (Marsh and Mossman, 1988), and the development of squamous metaplasia (Mossman et al., 1980a), may be important in the pathogenesis of bronchogenic carcinoma, a tumor of high risk in asbestos workers who smoke. In an attempt to understand how asbestos initiates alterations in cell proliferation and differentiation, we focused here on whether crocidolite asbestos causes activation of protein kinase C (PKC), a calcium- and phospholipid-dependent enzyme which phosphorylates several intracellular protein substrates and plays a crucial role in growth control and tumor promotion (Nishizuka, 1986). The potent phorbol tumor promoter, 12-0-tetradecanoylphorbol-13-acetate (TPA), and related compounds activate PKC by structurally resembling diacylglycerol.

We have shown previously that accumulation of diacylglycerol precedes crocidolite asbestos-induced proliferation in hamster tracheal epithelial (HTE) cells (Sesko et al., 1988), a cell line developed and characterized in this laboratory (Mossman et al., 1980b). Moreover, induction of ODC by crocidolite in these cells is ameliorated when inhibitors of PKC are added to culture medium (Marsh and Mossman, 1988). These results suggest that PKC is directly involved in asbestos-induced proliferation. In studies here, activation of PKC by crocidolite asbestos was explored using both a ^{3}H-phorbol 12,13 dibutyrate (^{3}H-PDBu) binding assay (Jaken, 1987) in intact cells and a histone phosphorylation assay (Thomas et al., 1987) performed on membrane vs. cytosolic fractions of control and crocidolite asbestos-exposed HTE cells. This brief report describes the highlights of

NATO ASI Series, Vol. H30
Effects of Mineral Dusts on Cells
Edited by B.T. Mossman and R.O. Bégin

these studies whereas a more complete manuscript should be consulted for further details (Perderiset et al., 1989).

Materials and Methods

For the ^{3}[H]PDBu binding assay, HTE cells were grown to confluency in Ham's F12 medium (GIBCO, Grand Island, NY) supplemented with 5% FBS and 5% newborn calf serum. Cells then were exposed to TPA, diluted in dimethyl sulfoxide (DMSO) to achieve a final concentration of 0.1 ug/ml medium for 15 minutes, or crocidolite asbestos (1 ug/cm^2 dish, UICC reference sample) for 30 to 180 minutes. Controls were exposed to DMSO alone. Plates then were washed twice with phosphate-buffered saline (pH 7.4), twice with buffer solution (20 mM Tris-HCl, 0.33 M sucrose, 0.5 mM EGTA, 2 mM EDTA, 2 mM phenylmethylsulfonyl fluoride and 25 ug/ml leupeptin) (Buffer A) and lysed, homogenized and centrifuged in this buffer at 120,000 x g for 1 hour to obtain both cytosolic (supernatant) and membrane (precipitate) fractions (Thomas et al., 1987). After further washing in buffer A without the addition of sucrose or leupeptin (Buffer B), the membrane fraction was homogenized, Triton X-100 was added at a final concentration of 0.5%, and the suspension centrifuged at 12,000 x g for 15 minutes at 4°C. Cytosolic and detergent-solubilized membrane fractions were applied to a DEAE 52 column (Whatman Biosystems Ltd., Maidstone, Kent) and the partially purified enzyme eluted with increasing NaCl concentrations. One ml fractions were assayed for PKC in a standard reaction mixture.

In additional studies, we examined whether crocidolite had an affinity for the phorbol ester receptor using an assay to evaluate binding of ^{3}H-PDBu to HTE cells in the presence and absence of asbestos. In brief, cells were grown to confluency in 12-well dishes and washed twice in Ham's F12 medium containing bovine serum albumin (BSA) (1 mg/ml) and HEPES (10 mM, GIBCO, Grand Island, NY) before addition of buffer (0.14 M NaCl, 5 mM KCl, 1.8 mM $CaCl_2$, 1 mM $MgCl_2$, 10 mM HEPES, 1 mg/ml BSA) at 4°C (Jaken, 1987). For Scatchard analyses, HTE cells were incubated for 90 minutes at 4°C in buffer containing serial dilutions of ^{3}H-PDBu (19.9 Ci/m mol, New England Nuclear, Boston, MA) in DMSO (final concentration $\leq$0.2%). The specific binding was calculated from the difference in cell-associated radioactivity in the presence and absence of unlabeled PDBu (40 uM). Control (untreated) cells and those exposed to crocidolite at 0.5, 1.0 and 1.5 ug/cm^2

dish for 3 and 24 hours were evaluated in these studies.

Results and Discussion

Distribution of PKC activity in the cytosolic and membrane fractions of HTE cells. After application of the cytosolic fraction to a DEAE-cellulose column, a calcium- and phospholipid-dependent peak of PKC activity was eluted at 40 mM NaCl. Addition of TPA to cells for 15 minutes caused redistribution of PKC activity from the cytosolic to the membrane fraction. Exposure to crocidolite at 1 ug/cm^2 dish resulted in an increase in membrane-associated PKC activity which peaked at 60 minutes, and was still elevated at 180 minutes. However, the majority of the PKC activity was still associated with the cytosolic fraction of the cells at all time periods examined.

Binding studies with ^{3}H-PDBu. Binding of ^{3}H-PDBu to HTE cells was saturable at a concentration of 150 mM. The Kd was calculated as approximately 19.73 $\pm$ 3.80 nM (Mean $\pm$ S.D. of 6 experiments) with an average of 3.35 x 10^5 $\pm$ 0.87 binding sites per cell. Studies using cells exposed to a range of concentrations of crocidolite for 1, 3 or 24 hours prior to incubation with ^{3}H-PDBu failed to cause changes in the binding activity. Thus, asbestos appears to activate PKC without competing for the phorbol ester receptor. A more plausible mechanism may be through activation of membrane phospholipases by asbestos as fibers come in contact with the plasma membrane. Recent work from this laboratory showing the rapid accumulation of diacylglycerol and inositol polyphosphates, both products of phosphoinositide hydrolysis, in HTE cells exposed to mitogenic concentrations of crocidolite supports this view (Sesko et al., manuscript in preparation).

Studies described above shed light on mechanisms of cell signalling by asbestos which are of fundamental importance to the process of tumor promotion in the respiratory tract. It is conceivable that overproduction of PKC by asbestos-exposed tracheobronchial epithelial cells gives rise to disordered growth control as has been demonstrated in fibroblasts over-expressing PKC (Housey et al., 1988). Our results are consistent with those reported by Brown and colleagues (these proceedings) and Holian and Roney (these proceedings) who show elevation of diacylglycerol in a variety of cell types after exposure to asbestos and inhibition of asbestos-induced release of $O_2^{\overline{\cdot}}$ from guinea pig alveolar macrophages treated with inhibitors of PKC, respectively.

Acknowledgements

This research was supported by grants from the National Cancer Institute, National Institute of Environmental Health Sciences, and National Heart, Lung and Blood Institute. We thank Virginia Kelleher for preparation of the manuscript.

References

Brown RC, Hoskins JA, Cole KJ, Evans CE, Sara EA (1989) Effects of asbestos fibres on some aspects of the cell second messenger system. pp. 375-381 (these proceedings).

Holian A, Roney PL (1989) Superoxide anion production induced by chrysotile asbestos in the guinea pig alveolar macrophage. pp. 223-229 (these proceedings).

Housey GM, Johnson MD, Hsiao WLW, O'Brian CA, Murphy JP, Kirschmeier P, Weinstein IB (1988) Overproduction of protein kinase C causes disordered growth control in rat fibroblasts. Cell 52:343-354.

Jaken S (1987) Phorbol receptors in intact cells and subcellular fractions. Methods in enzymology 141:275-287.

Landesman JM, Mossman BT (1982) Induction of ornithine decarboxylase in hamster tracheal epithelial cells exposed to asbestos and 12-O-tetradecanoylphorbol-13-acetate. Cancer Res 42:3669-3675.

Marsh JP, Mossman BT (1988) Mechanisms of induction of ornithine decarboxylase activity in tracheal epithelial cells by asbestiform fibers. Cancer Res 49:709-714.

Mossman BT, Cameron GS, Yotti L (1985) Cocarcinogenic and tumor promoting properties of asbestos and other minerals in tracheobronchial epithelium. In Cancer: A Comprehensive Survey: Cancer of the Respiratory Tract. Predisposing Factors. (Mass, MH et al eds) Raven Press, New York, Vol 8, pp 217-238.

Mossman BT, Craighead JE, MacPherson BV (1980a) Asbestos-induced epithelial changes in organ cultures of hamster trachea: Inhibition by the vitamin A analog, retinyl methyl ether. Science 207:311-313.

Mossman BT, Ezerman EB, Adler KB, Craighead JE (1980b) Isolation and spontaneous transformation of hamster tracheal epithelial cells. Cancer Res 40:4403-4409.

Nishizuka Y (1986) Studies and perspectives of protein kinase C. Science 233:305-312.

Perderiset M, Marsh JP, Mossman BT (1989) Asbestos activates protein kinase C in hamster tracheal epithelial cells. Submitted.

Sesko AM, Cabot M, Mossman BT (1988) Accumulation of diacylglycerol precedes asbestos-induced cellular proliferation. Supp. 12E:84.

Thomas TP, Gopalakrishna, Anderson WB (1987) Methods in enzymology 141:399-411.

QUESTIONS TO THE CELL BIOLOGISTS AND PROPOSALS FOR A DEFINITION AND CLASSIFICATION OF CARCINOGENIC FIBERS DERIVED FROM ANIMAL STUDIES

F. Pott
Medical Institute for Environmental Hygiene
at the University of Düsseldorf
Auf'm Hennekamp 50
D 4000 Düsseldorf
F.R.G.

Conclusions from Carcinogenicity Studies

Animal experiments with various mineral fibers have led to the conclusion that a fiber represents a carcinogenic agent if it is sufficiently long, thin, and durable. Moreover, there are some indications that surface properties may be important not only for cytotoxicity but also for carcinogenicity.

There is a continuous transition from the non-carcinogenicity of fibers which are too short, too thick or too soluble to the maximum possible carcinogenic potency of fibers which have the optimum length and which, additionally, are highly durable; however, the precise data are not known (see Fig. 1). For regulation purposes, it would be important to classify this continuous transition. We need to define parameters for dividing fibers into groups having, for example, up to 1 %, 5 % or 25 % of the maximum carcinogenic potency in humans.

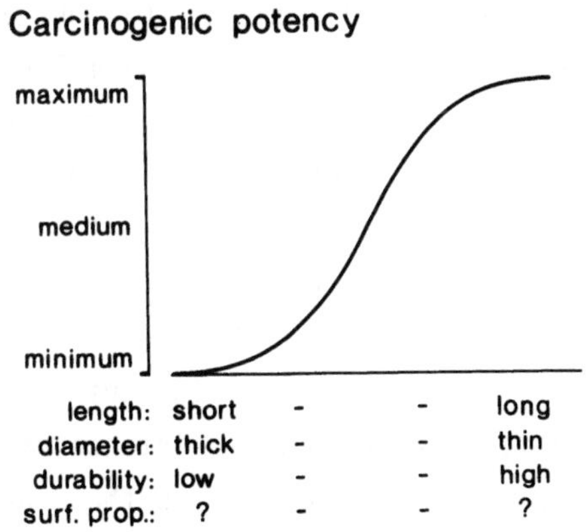

Fig. 1. Simplified illustration of the carcinogenic potency of fibers dependent on at least three parameters.

NATO ASI Series, Vol. H30
Effects of Mineral Dusts on Cells
Edited by B.T. Mossman and R.O. Bégin

Furthermore, the carcinogenicity studies in rats have shown that the mineralogic name or trade name of a fibrous material does not give a sufficient indication of its carcinogenic potency (Davis et al., 1986; Pott et al., 1987; Stanton et al., 1981; Wagner et al., 1987). The designations of the materials are names for a family which may contain very different members because of differences regarding their particle size distributions, their durabilities (Bellmann et al., 1987) and possibly other properties (Mossman, 1983; Jaurand, 1988). One member of a certain "fiber family" may be a "murderer", but others may not. Up to now, agencies have classified carcinogenic substances only according to their name because normally the names of chemicals describe homogeneous substances. However, a general classification of fibrous dusts such as attapulgite or glass fibers as "carcinogenic" is not appropriate although there are some animal experiments which clearly show a carcinogenic effect of the (special) samples tested. The contrast between regulation and facts worries a scientist who is eager to find a suitable general definition of carcinogenic fibers. He would very much appreciate some help in answering a number of the questions already put to the cell biologists at the symposium "Man-Made Mineral Fibers in the Working Environment" (Pott, 1987).

Questions to the Cell Biologists

Much effort has been devoted to the problem of determining how the carcinogenic potency of fibers varies with their length and diameter (without a definite answer so far), but there are no statements on the minimum length of time which fibers have to be in touch with cells in order to alter them so that they can develop into a tumor without the further presence of fibers. It seems reasonable to assume that a substance which does not act by its metabolites but by the elongated shape of its particles should have a stronger effect when the fibrous particles are more than less durable. But it would be of interest to exonerate long, thin fibers from the suspicion of being carcinogenic if they cannot survive the minimum persistence necessary for tumor induction. Minimum persistence can be defined as the shortest time which the fibers have to lodge in a certain area of the bronchial wall or serosa tissue (pleura or peritoneum). The shortest necessary period of persistence can be seen as a kind of threshold value. When this threshold for the incubation period is not reached, there will be no effect even from fibers which seem to have a high carcinogenic potency because of their shape. The persistence of a fiber ends by dissolution, disintegration or elimination, or even by simple translocation in the body. Persistent fibers have to be durable, but durable fibers are not necessarily persistent. In this connection, the following questions arise:

- Is persistence for just one cell division phase, in principle, sufficient for a transforming effect? Or does the (unknown) mechanism need a longer period?
- Is the minimum persistence period for a transforming effect related to the cell division period of the host species and is it possible to quantify the difference between the life-span of bronchial epithelial cells and serosa cells of humans and rats?
- Does a longer persistence time lead to a proportionately greater effect by affecting more cell generations which would result in a higher tumor incidence and/or shorter latency period?
- Are in vitro experiments with an incubation period of some days or weeks sufficient to examine the significance of the durability of fibers with regard to their carcinogenic potency?

Proposals for a General Definition of Carcinogenic Fibers and for a Classification of Fiber Types

There is a growing variety of inhalable fibers. On the one hand, adequate regulations are necessary to prevent diseases and avoid cancer risks from fibers which in special cases may be similar to those caused by asbestos. On the other hand, it is not possible to examine all the existing and new types of fibers by suitable long-term carcinogenicity tests (irrespective of the open question as to the suitable models for testing fibers). We, therefore, need a convention on general definitions of (1) unequivocally or probably carcinogenic fibers, (2) suspected or possibly carcinogenic fibers which need further clarification and (3) non-carcinogenic fibers.

To stimulate the discussion, the following proposal for a definition of carcinogenic fibers is suggested.

(1) Unequivocally or probably carcinogenic to humans

This definition is considered to be consistent with the following classifications of agencies or commissions:

IARC, Lyon:	Groups	1, 2 A
U.S. EPA:	"	A, B 1, B 2
MAK C., F.R.G.:	"	A 1, A 2
European C.:	Categories	1, 2

Inorganic fibers characterized by

length	> 5 µm
diameter	< 2 µm*
aspect ratio	> 5 : 1
durability in rats**	> 2 years

(2) Suspected or possibly carcinogenic to humans
(This preliminary classification needs further clarification.)

This definition is considered to be consistent with the following classifications of agencies or commissions:

IARC, Lyon:	Group	2 B
U.S. EPA:	"	C
MAK C., F.R.G.:	"	III B
European C.:	Category	3

Inorganic fibers characterized by

length	2 - 5 µm
diameter	2 - 3 µm*
aspect ratio	> 5 : 1
durability in rats**	6 months to 2 years

---

* including particles able to split up or separate fibers with these diameters

** length of durability with regard to the carcinogenic potency in humans

Consequently, all inorganic fibers which are shorter than 2 µm, thicker than 3 µm, or which dissolve or disintegrate in the body in less than 6 months would be exonerated from the suspicion of being carcinogenic.

It is not possible to discuss this proposal here (it has been done in German). However, because of the unsolved questions regarding standardization of the examination of fiber durability, an alternative model will be presented briefly.

An Alternative Proposal

As a first step in the rule making, all inorganic and organic materials which contain fibers (aspect ratio > 5 : 1) longer than 5 µm and thinner than 3 µm would be put in the group of suspected or possibly carcinogenic to humans.

As a second step, fiber types already established to be carcinogenic to humans or to experimental animals should be classified accordingly and those proven to be non-carcinogenic in a sensitive animal test model should be exonerated. By this procedure, some non-asbestos fibers could be classified as carcinogenic, for example, all varieties of erionite, rare varieties of attapulgite, some types of glass fiber (e.g. Manville 104/475), tested "positive" types of rock fiber, tested "positive" ceramic fibers (e.g. Fiberfrax), tested silicon carbide fibers. Other fiber types could be exonerated and classified as non-carcinogenic, for example, special types of attapulgite, sepiolite and wollastonite.

These proposed classifications are based mostly on the results of serosa tests because only few other data are available. The negative or weak positive inhalation experiments with crocidolite have shown that this test model is not sensitive enough for excluding a relatively high carcinogenic potency of fibers. But the classification of an inorganic fiber type as non-carcinogenic can be justified, when 10^9 fibers longer than 5 µm and thinner than 3 µm did not increase the incidence of sarcomas/mesotheliomas after intraperitoneal injection in rats. With this test, the effects of fiber length, diameter, durability and possibly important surface properties are examined. However, it is not always possible to administer

10^9 fibers altogether in 200 to 300 mg of a dust sample when the percentage of thick or short fibers or of non-fibrous particles is high. There also appear to be some problems in testing organic fibers.

As a third step, fibers which have not been examined sufficiently by experiments but are considered to act analogously to tested and classified samples may be classified using the same arguments applied in analogy. All ceramic and silicon carbide fibers appear to be very durable. Consequently, all types could be classified as carcinogenic if they are sufficiently long and thin.

Alongside classification, quantitative risk estimation is very important for practical purposes. While there may be great differences between fiber types which are in principle carcinogenic, no adequate procedure has been developed for risk estimation of results gained through the intraperitoneal test.

References

Bellmann B, Muhle H, Pott F, König H, Klöppel H and Spurny K (1987) Persistence of man-made mineral fibres (MMMF) and asbestos in rat lungs. Ann.occup.Hyg. 31:693-709

Davis JMG, Addison J, Bolton RE, Donaldson K and Jones AD (1986) Inhalation and injection studies in rats using dust samples from chrysotile asbestos prepared by a wet dispersion process. Br.J.exp.Path. 67: 113-129

Jaurand MC (1988) A particulate state carcinogenesis: Recent data on the mechanisms of action of fibres. In: Proceedings of the Symposium Mineral Fibres in the Non-occupational Environment. Lyon, 8-10 September 1987, World Health Organization. International Agency for Research on Cancer, Lyon (in press)

Mossman BT (1983) In vitro approaches for determining mechanisms of toxicity and carcinogenicity by asbestos in the gastrointestinal and respiratory tracts. Environ.Health Perspect. 53:155-161

Pott F (1987) Problems in defining carcinogenic fibres. Ann.occup. Hyg. 31:799-802

Pott F, Ziem U, Reiffer FJ, Huth F, Ernst H and Mohr U (1987) Carcinogenicity studies on fibres, metal compounds, and some other dusts in rats. Exp.Pathol.(Jena) 32:129-152

Stanton MF, Layard M, Tegeris A, Miller E, May M, Morgan E and Smith A (1981) Relation of particle dimension to carcinogenicity in amphibole asbestoses and other fibrous minerals. J.natl.Cancer Inst. 67:965-975

Wagner JC, Griffiths DM and Munday DE (1987) Experimental studies with palygorskite dusts. Br.J.Ind.Med. 44:749-763

RISK FACTORS OF ASBESTOS-RELATED CANCERS

A. Lange, C. Łaba, D. Garncarek-Lange, J. Tomeczko, H. Matej, B. Nowakowska
Institute of Immunology and Experimental Therapy
Czerska 12
53-114 Wrocław
Poland

Introduction

Exposure to asbestos increases significantly risk of cancer (Selikoff et al. 1979). Notably, induces also an impairment of cellular immunity and appearance of antinuclear antibodies (ANA) (Lange 1982). Our previous work showed that lack of skin reactivity (anergy) can be of value in weaving out workers being at higher risk of asbestosis (Lange et al. 1986). The latter study has been recently extended and also includes evaluation of risk factors associated with cancer appearance. In the present work a statistical analysis was performed in order to distinguish asbestos workers who are more than those who are less susceptible to asbestos-related cancers.

Asbestos Workers and Methods

388 asbestos workers were examined in 1973-1974 (Lange 1980 a,b) and since that time their fate was followed. Altogether 32 workers died between 1976-1986. This group includes 3 accidental and 29 natural deaths. Among natural deaths 14 patients died due to cancer. The cause of death was categorized after review of best available information and only one case, among cancer cases, was categorized according to death certificate. Asbestos-related cancers represent neoplasms which are known to be associated with asbestos exposure (Kagan and Jacobson 1983; Kogan et al. 1982; Selikoff et al. 1979) and may

NATO ASI Series, Vol. H30
Effects of Mineral Dusts on Cells
Edited by B. T. Mossman and R. O. Bégin

result from a direct access of the fibres to the involved organ. Therefore, lung cancers (5 cases), mesothelioma, cancer of the tonsil and uterus were considered in this paper as asbestos-related cancers. To this group cancer of the pancreas was also included, as diagnosis based on death certificate is not certain (Selikoff and Seidman 1981) and spread of the tumor from the digestive tract can not be excluded. Asbestos workers who died due to other than asbestos-related cancer causes were classified as ungrouped in the discriminant analysis. To match the age of cancer cases which at the time of the initial observation were above 40 years of age, only asbestos workers above this age were taken to statistical analysis. Untill 1973-1974 they were exposed to asbestos for at least 8 years. Altogether 195 cases were taken into consideration, but only 111, including 6 asbestos related-cancer cases, were evaluated with the use of discriminant analysis due to the missing data. Fig. 1 shows distribution of all the data. Time between initial examination and the occurence of cancer was from 3 - 13 years. ANA, anergy and elevated level of serum IgG and IgA are frequently seen in asbestos workers (Lange 1982). HLA B27 and HLA B8 are risk factors of lung fibrosis (Darke et al. 1979; Matej and Lange 1976; Merchant et al. 1975) and ANA appearance in asbestos workers, respectively (Matej et al. 1978). Smoking increases risk of lung cancer (Selikoff et al. 1979). Bronchitis is a sign of local inflammatory damage to the bronchi tree. All above data were chosen for analytical evaluation of their significance as risk factors of asbestos-related cancers. Immunological methods were given in previous papers (Lange 1980 a,b; Lange et al. 1986; Matej et al. 1978).

<u>Statistical methods and results</u>. The final goal of the present work was two-fold: 1) analysis of the significance of the chosen features as discriminating variables in cancer prediction, 2) evaluation of discriminating function as a method of classification of asbestos workers who are more and less susceptible to asbestos-related cancers.

Variable	Mean	Std Dev	Minimum	Maximum	N
SEX	1.38	.49	1	2	195
SMOKING	174.29	237.56	0	1200	195
BRON	1.56	.50	1.00	2.00	189
ANERGY	1.25	.43	1.00	2.00	166
ANA	1.08	.27	1.00	2.00	192
IGA	280.91	127.14	57	1209	179
IGG	1446.07	364.28	137	2436	179
B7B8	1.33	.47	1.00	2.00	149
B27	1.10	.30	1.00	2.00	149

Fig. 1. Distribution of variables registered in 1973-1974. Variables are described as follows: sex (female- 1, male- 2); chronic bronchitis (no productive cough or less than 3 months in two consecutive years- 1, productive cough at least for 2 years and 3 months per year- 2); anergy (skin reactivity to 2 units of PPD and/or 5 units of streptokinase- streptodornase >= 5mm- 1, reactivity to 2 units of PPD and 5 units of streptokinase-streptodornase < 5 mm- 2); ANA (absent- 1, present- 2); B7B8 (lack of HLA B7 and B8- 1, HLA B7 and/or B8 positive- 2); B27 (lack of HLA B27- 1, HLA B27 positive- 2). Index of smoking = the number of years with the habit times number of cigarettes smoked per day. Serum IgG and IgA level is given in mg%

Statistical analysis has been made with the SPSS/PC+ package of statistical programs (Nourusis 1986). At first correlation matrix for all variables has been computed (Fig. 2). These computations pointed out the following relationships: 1. Smoking habit prevailed in men and showed negative correlation with immunological abnormalities of a general characteristic namely with anergy and serum IgG level. In opposite, smoking is positively correlated with the incidence of asbestos-related cancers. Negative correlation of smoking with serum IgG level and anergy could be indirect due to the positive correlation of this habit with the male sex. The latter was negatively correlated with anergy and to a lower degree with IgG (Fig. 2). Therefore, the correlation matrix has been independently computed for males and females. Presence of negative correlation between smoking and IgG r = -0.2157 and r = -0.3440 and between smoking and anergy r = -0.1990 and r = -0.1855 have been documented in males and females, respectively.

Correlations:

	AS CANCERS	SEX	SMOKING	BRON	ANERGY
AS CANCERS	1.0000	.1445*	.1472*	.1590*	-.1279
SEX	.1445*	1.0000	.5640**	.1628*	-.1401*
SMOKING	.1472*	.5640**	1.0000	.1131	-.1434*
BRON	.1590*	.1628*	.1131	1.0000	.0039
ANERGY	-.1279	-.1401*	-.1434*	.0039	1.0000
ANA	-.0660	-.1052	-.0696	-.0090	.1614*
IGA	-.0439	.1060	.0665	.0363	-.0746
IGG	.1040	-.1192	-.2263**	.1079	.0764
B7B8	-.1623*	-.1023	-.0300	-.0654	.1696*
B27	.3433**	.1707*	.0711	.0232	-.0349

	ANA	IGA	IGG	B7B8	B27
AS CANCERS	-.0660	-.0439	.1040	-.1623*	.3433**
SEX	-.1052	.1060	-.1192	-.1023	.1707*
SMOKING	-.0696	.0665	-.2263**	-.0300	.0711
BRON	.0090	.0363	.1079	-.0654	.0232
ANERGY	.1614*	-.0746	.0764	.1696*	-.0349
ANA	1.0000	.0663	.1518*	.0118	-.1013
IGA	.0663	1.0000	.2266**	.1410*	-.0529
IGG	.1518*	.2266**	1.0000	.0820	-.0282
B7B8	.0118	.1410*	.0820	1.0000	-.2342**
B27	-.1013	-.0529	-.0282	-.2342**	1.0000

Minimum pairwise N of cases: 128
1-tailed Signif: * - .05 ** - .01

Fig. 2. Correlation matrix of variables registered at the time when asbestos workers entered the cohort (1973-1974) and incidence of asbestos exposure related cancers (as cancers) diagnosed from 1976 - 1986

2. ANA is correlated with anergy and serum IgG level (Fig. 2). Variable HLA B8 and/or B7 was correlated with anergy but not with asbestos-related cancers (Fig. 2). Notably, HLA B27 antigen was positively correlated with asbestos-related cancers ($r=0.3433$, $p<0.01$). Presence of these correlations within subgroups of features indicated that some of them may be grouped together on the basis of sharing a common factor. Therefore, cluster analysis was performed (unweighted pair-group method using arithmetic averages) which revealed the following clusters: (i) HLA B7 and/or B8 form a cluster with HLA B27 through a negative correlation, (ii) ANA and anergy, (iii) serum immunoglobulin levels IgG and IgA, (iv) smoking,

sex and to some degree bronchitis (Fig. 3).

```
    * * * * *   HIERARCHICAL CLUSTER ANALYSIS   * * * * *
Vertical Icicle Plot using Average Linkage (Between Groups)

   B  B  A  A  I  I  B  S  S
   2  7  N  N  G  G  R  M  E
   7  B  A  E  G  A  O  O  X
      8     R        N  K
            G           I
            Y           N
                        G

   9  8  5  4  7  6  3  2  1
1 +*************************
2 +****  *******************
3 +****  ****  *************
4 +****  ****  ****  *******
5 +****  ****  ****  *  ****
6 +****  *  *  ****  *  ****
7 +****  *  *  *  *  *  ****
8 +*  *  *  *  *  *  *  ****
```

Fig. 3. Cluster membership of independent variables at different stages

To identify the factors which could represent already found relationships between subsets of variables, the factor analysis was performed. Four factors have been extracted. They are interpretable on the basis of the contribution (proportinal to the coefficients shown in the Fig. 4) to each of them of different subsets of variables with an overhelming extent.

Factors from the factor analysis were used as discriminating variables in the discriminant analysis. The discriminant function has been built up to obtain a score that distinguishes asbestos workers who are more susceptible to asbestos-related cancer (group 2) from those who are not (group 1), on the basis of variables registered far before the apparent disease.

In the discriminant analysis the weights of factors computed from the factor analysis were estimated so that they could provide the best separation between the two groups. The weights expressed as the standardized discriminant function

---- FACTOR ANALYSIS ----

Rotated Factor Matrix:

	FACTOR 1	FACTOR 2	FACTOR 3	FACTOR 4
SMOKING	.85993			
SEX	.83930			
BRON	.34869			
B7B8		.77941		
B27		-.68563		
IGG			.74496	
IGA			.74171	
ANERGY				.77587
ANA				.60617

Fig. 4. Factor analysis of independent variables (Factor 1 represents a common factor for the outcome of cigarette smoking and sex related features, Factor 2 represents genetic trait associated with HLA locus B antigens, Factor 3 represents features involved in humoral immunity of a general (serum IgG level) and local in the lung (serum IgA level) character, Factor 4 represents the common factor governing the main immunological consequences of asbestos exposure)

coefficients (Fig. 5) are of a great analytical importance. When the sign is ignored, each coefficient represents the relative contribution of its associated variable to that function. The linear discriminant equation is to distinguish asbestos workers who died due to asbestos-related cancers (group 2) from those who survived (group 1). The two groups must differ in their values. Indeed, the group centroid for asbestos workers who died was -1.78849 and for those who survived it was 0.11296. The discriminant score for each case could be calculated. Using these scores, it was possible to obtain a rule for classifying cases into one out of two groups. The technique used in SPSS/PC+ is based on Bayes' rule which is of a value in the present work, as the groups are of grossly different sizes. The negative product of multiplying discriminating variable by its corresponding coefficient tends to locate the respective case closer to the centroid of the

---- D I S C R I M I N A N T   A N A L Y S I S ----

** Canonical Discriminant Function Coefficients **

	Unstandardized FUNC 1	Standardized FUNC 1
FACTOR 1	-.09958	-.09701
FACTOR 2	.98701	.95856
FACTOR 3	-.23729	-.23582
FACTOR 4	.06329	.05685
(constant)	.07711	

Canonical Discriminant Functions evaluated at Group Means

Group	FUNC 1
1	.11296
2	-1.78849

Fig. 5. Results of discriminant analysis: the weighting coefficients and averages of the group means computed for cases who developed asbestos-related cancers (group 2) and for those who did not (group 1)

asbestos-related cancer group, and the positive one to the group of those who survived. It is apparent from the data presented in Fig. 4 and Fig. 5 that variables represented by Factor 1 and Factor 3 constitute cancer risk factors, however, the weight of Factor 1 (smoking, male sex and bronchitis) and Factor 3 (serum IgG and IgA level) is rather small. Factor 2 contributes greatly to the function (high discriminant function coefficient) and constitutes cancer risk factor in HLA B27 positive cases (negative sign in factor analysis coefficient), but it is protective in HLA B8 and/or B7 positive cases. Product of Factor 4 which represents features of anergy and autoimmunity has a protective value (Fig. 4), however, of low significance having low standardized discriminant function coefficient (Fig. 5). Finally, the classification of cases on the basis of the discriminant function was performed. It was an empirical measure of the success of discrimination by observing the proportion of correct classifications. The percent of properly classified cases was 95.05 %. Four out of 6 asbestos-related cancer cases (group 2) and 92 out of 95 cases without clinically apparent cancer were correctly classified. This proves that the variables selected in this work were good discriminants.

Discussion

The present data revealed that cases which developed asbestos-related cancers were at the time preceeding for 3 to 13 years clinically apparent disease rather good responders to recall antigens and lacked ANA. Furthermore, prevalence of asbestos-related cancers in HLA B27 positive and HLA B8 and/or B7 negative population was shown. Discriminant analysis pointed out the significance of HLA B7,8 and 27 associated features in asbestos cancerogenesis. Anergy and ANA are depicting cases less prone to asbestos carcinogenesis, however, the weight of these variables is rather small. In opposite, elevation of IgG in context of elevated IgA as a result of a stimulation at the site of exposure at the mucous membrane level (Lange 1980b), points out cases which are more sussceptible to asbestos-related cancer. Analytical value of this discriminant function has been proved by classifying the cases used to derive the function in the first place and comparing predicted group membership with actual group membership. High proportion of correct classifications gave a credit to the significance of the variables which were chosen to built up the function.
The present results are interpretable on the ground of general immunology. It has been already reported that HLA B8 positive normal young adults are lower responders to T-cell mitogens (McCombs and Michalski 1982). We have already shown, that ANA appeare more frequently in HLA B8 positive asbestos workers (Matej et al. 1978). HLA B7 positivity in general is associated with some immunopathological situations. Therefore, it is understandable that these two antigens may be associated with asbestos-related immunological abnormalities, but interestingly not with asbestos-related cancers. HLA B27 was previously found

to be associated with lung fibrosis but rather in cases with mild fibrosis what probably causes that the association is rather week (Darke et al. 1979). The present data suggest that this antigen is also a marker of susceptibility to asbestos-related cancers. Both associations with fibrosis and cancer seem to be indirect. It was already shown that HLA B27 is a marker of granulocyte function (Leirisalo et al. 1980). These phagocytic cells play a role in inflammatory processes in the lung and also are suspected to contribute to cancerogenesis generating active oxygen species when taking up asbestos fibres (Mossman et al. 1986). The associations found between HLA B27 positivity and susceptibility to asbestos-related cancers, mostly lung cancer, may be specific for asbestos cancerogenesis as a similar association was not found when lung cancer cases coming from a general population were evaluated (Tiwari and Terasaki 1985).

Acknowledgements

The authors are indebted to Professor N. Victor from the Ruprecht-Karls University in Heidelberg for providing us with the SPSS package of statistical programmes and for his interest and encouragement and to Dr. H. Haas for his critical comments on the manuscript. This work was supported in part by the grant from the Polish Academy of Sciences.

References

Darke C, Wagner MMF, Grant McMillan GH (1979) HLA-A and B antigen frequencies in an asbestos exposed population with normal and abnormal chest radiographs. Tissue Antigens 13:228-232

Kagan E, Jacobson RJ (1983) Lymphoid and plasma cell malignancies: asbestos-related disorders of long latency. Am J Clin Path 80:14-20

Kogan FM, Deminov AG, Guselnikova NA, Gurvich VB (1982) Changes in occupational morbidity among workers in the asbestos mining industry in relation to the improvement of working

conditions. Arch Immunol Ther Exp 30:277-283
Lange A (1980a) An epidemiological survey of immunological abnormalities in asbestos workers. I.Non-organ and organ specific autoantibodies. Environ Res 22:162-175
Lange A (1980b) An epidemiological survey of immunological abnormalities in asbestos workers. II.Serum immunoglobulin levels. Environ Res 22:176-183
Lange A (1982) Anergy and autoimmunity in asbestos-exposed people. Arch Immunol Ther Exp 30:211-218
Lange A, Garncarek D, Tomeczko J, Ciechnowski G, Bisikiewicz R (1986) Outcome of asbestos exposure (lung fibrosis and antinuclear antibodies) with respect to skin reactivity: an 8-year longitudinal study. Environ Res 41:1-13
Leirisalo M, Repo H, Tiilikainen A, Kosunen TU, Laitinen O (1980) Chemotaxis in yersinia arthritis. HLA-B27 positive neutrophils show high stimulated motility in vitro. Arthr Rheum 23:1036-1044
Matej H, Lange A (1976) First International Symposium on HLA and Disease. Abstracts. Paris p 256
Matej H, Lange A, Garncarek D, Smolik R, Roszak E (1978) HLA and antinuclear antibody incidence in asbestos workers. Arch Immunol Ther Exp 26:201-205
McCombs CC, Michalski JP (1982) Lymphocyte abnormality associated with HLA-B8 in healthy young adults. J Exp Med 156:936-941
Merchant JA, Klouda PT, Soutar CA, Parkes WR, Lawler SD, Turner-Warwick M (1975) The HLA system in asbestos workers. Brit Med J 1:189-191
Mossman BT, Marsh JP, Shatos MA (1986) Alteration of superoxide dismutase activity in tracheal epithelial cells by asbestos and inhibition of cytotoxicity by antioxidants. Lab Invest 54:204-212
Nourusis MJ (1986) Advanced statistics SPSS/PC+ for the IBM PC/XT/AT. SPSS Inc, Chicago USA
Selikoff IJ, Hammond EC, Seidman H (1979) Mortality experience of insulation workers in the United States and Canada, 1943-1976. Ann N Y Acad Sci 330:91-116
Selikoff IJ, Seidman H (1981) Cancer of the pancreas among asbestos insulation workers. Cancer 47:1469-1473
Tiwari JL, Terasaki PI (eds) (1985) HLA and Disease Associations. Springer-Verlag Berlin

MESOTHELIOMA THRESHOLD

E.B. Ilgren, Imperial Cancer Research Fund
Cancer Studies Unit, University of Oxford, UK

This is an overview of the first comprehensive critical review of the world's literature of spontaneously occurring and experimentally induced mesotheliomas in animals. The complete work ca. 600 pp. will be published in the Spring of 1989. It has been written to examine in particular the highly controversial question of "mesothelioma threshold." This is taken to mean a dose level beneath which mesotheliomas will not occur following the administration of a so-called "mesotheliogenic" agent. Such agents include not only fibrous, naturally occurring, asbestiform and non-asbestiform substances but also man-made mineral fibers and non-fibrous materials such as hydrocarbons, hormones, metals, and viruses. The "threshold question" has been approached firstly by tabulating all of the presumptive spontaneously occurring mesotheliomas ever recorded either in sporadic reports or within control cohorts; virtually all of the long-term, whole animal experiments using an asbestiform agent ever performed, irrespective of whether a mesothelioma was produced; and any chronic tumorigenicity tests of non-asbestiform fibrous or non-fibrous agents which have ever yielded a mesothelioma. The data from these studies have been put onto an interactive relational database ("Paradox") and sorted by agent, mode of administration, species, and where appropriate, dose. Dosage levels have been arranged in ascending order firstly for each individual study and then for all studies combined. Indications of "threshold" have been taken either "directly" from the individual experiments which used either two or more dose levels or "indirectly" by assessing all of the studies <u>inter se</u> across a specified dose range. The tumor yields observed in either case were compared with those seen in the "spontaneous" control groups. The outcome of each dose response study was then crudely categorized as displaying either a negative-, a No-Observed-

NATO ASI Series, Vol. H30
Effects of Mineral Dusts on Cells
Edited by B.T. Mossman and R.O. Bégin

Effect-Level-(NOEL)-, or a plateau-response. Such analyses suggested the possible existence of a NOEL for the major types of asbestiform fiber (amosite, crocidolite, and chrysotile); for one non-asbestiform, naturally-occurring fiber (erionite); for several synthetic fibers (glass and quartz); and for at least one chemical agent, namely Dimethylnitrosourea. Previously, the failure to conclude that thresholds may exist may have been related to the failure to recognize that a variety of factors can confound the interpretation of a "low dose tumor response." Such factors include:

1. The distinct background incidence of spontaneously occurring mesotheliomas seen in both negative and positive control cohorts.
2. The misinterpretation of the term "low" dose which can actually represent a very substantial number of "significant" tumorigenic fibers.
3. The marked "accentuation" of dose when fibers are injected into the pleural and peritoneal cavities and thereby directly applied to the serosa.
4. The difficulty encountered in morphologically distinguishing a certain proportion of mesotheliomas from other lesions including hyperplastic, preneoplastic, and benign mesothelial tumors as well as non-mesothelial neoplasms such as metastatic carcinomas and primary sarcomas.
5. The extreme tumor induction period observed in several earlier "low dose" studies.

The aforementioned strongly suggest that the concept of threshold needs to be carefully reconsidered particularly in the light of today's safety concerns about low levels of mineral dust in our environment.

Spontaneously occurring mesotheliomas, ascertained from individual case reports and historical series, were found in a variety of animal species ranging from fish to rodents and higher mammals. Cases are described in the literature from 1909 to the present in at least 17 different studies. Two mesotheliomas were found in fish, whilst, amongst the rodents, mesotheliomas were most commonly noted in the rat. At least five different inbred strains (ACI, BUF, OM, WIS. F344) as well as two wild rat subspecies (Oregon, Norvegicus) were affected. The following sites were involved in the rat: the tunica of the testes (1.3-5%) and the ovaries (0.1-0.2%); the pleura (0.3-0.8%); the peritoneum (0.9-1.7%); and unspecified sites (0.1-1.7%). Spontaneous mesotheliomas were also found in a hamster, a Mastomy, a dog, a leopard, and an eland. The second most commonly affected species was the cow. In this species, 10 mesotheliomas were found in the peritoneum; 3 in the pleura; 2 in both the tunica and the peritoneum; 4 in both the pleura and the peritoneum; and 1 in an unspecified site. Most of the bovine examples occurred in old age, although several were seen in the fetus and the newborn calf.

Some workers dismiss mesotheliomas of the tunica as unimportant, claiming that these are not related to those of the peritoneum (Davis, 1988 PC). There are at least two arguments against this. Gould (1977) has said that "(such) tumors can be classified as mesotheliomas because of their morphological similarity to previously described mesotheliomas, i.e. the demonstrable continuity of tumor cells with adjacent normal mesotheliomas and a failure to show epithelial mucin." Secondly, there is evidence that these lesions can spread through the peritoneal cavity. Coene (1981) noted that "these peritoneal tumors originated on the testicular mesothelium" having developed in response to ethylene oxide. Moreover, Gould (1977) said that "pathological material from these (AAF-treated) and other Buffalo rats was studied by H. L. Stewart (pers. comm., 1974) who advises that... he had seen mesotheliomas arise spontaneously in that strain from an extension of the peritoneal mesothelium (tunica vaginalis) of the testis... (and) occasionally spread to the peritoneal cavity"... Gould (1977) continued to say, in relation to his own study, that "many of these mesotheliomas are more extensive than those in previous reports... 9 of the 384 rats had mesotheliomas. In five of the rats, the tumors were limited to the genital serosa, but in the others they involved the peritoneum as well... Rats with the most extensive lesions had similar masses covering large areas of visceral and parietal peritoneum... There was one instance of local lymphatic invasion." Goodman et al (1979) have also described extension beyond the tunica.

Mesotheliomas have also been found in negative and positive control cohorts run in parallel with experimental groups. Examples in untreated controls, have occurred principally in the rat, the tunica of the testis being involved in up to 2% of cases, and the pleura in 0.2 to 3.3%. Stanton et al. (1981) have also recorded spontaneous pleural sarcomas in 0.9 to 4.5% of untreated control cohorts. Presumably spontaneous mesotheliomas have also been found in positive controls amongst vehicle alone-, dust alone-, non-carcinogenic implant alone-, and sham alone-treated animals. Amongst the vehicle alone- treated controls, diet alone (aside from Basal Diet) has produced mesotheliomas in 0.6 to 1.0% of animals in two studies. Following saline treatment, mesotheliomas have been found following intraperitoneal injection in the rat in 3.6 to 7.8% and in the mouse in up to 6.5% of several studies. Following intrapleural saline injection in the rat, mesotheliomas were found in 2 to 2.5% of animals in three studies. In one report, gelatin produced mesotheliomas in 1.5% of intrapleurally treated rats whilst vegetable oil alone produced mesotheliomas in 3.3% of rats following intragastric administration. Following intraperitoneal injection of presumably non-mesotheliogenic dusts, e.g. sanidine, talc, and volcanic ash, mesotheliomas have been found in several rat studies in up to 12.5% of cases. Intrapleural inoculation of talc has produced pleural sarcomas in the rat in 3.3 to 3.8% of animals as recorded in four studies. So called "non-carcinogenic" intrapleural implants have produced pleural sarcomas in 0.6 to 8.0% of rats in ten studies. Finally, following intraperitoneal sham injection, peritoneal sarcomas were found in 3.3% of rats and 5.1% of mice. The comments of other investigators also support the idea that a distinct background incidence of mesotheliomas exists. Thus, Wagner et al. (1987) said that "in an exposed group of 40 rats... one mesothelioma was acceptable as being in the expected range for the particular animals used... In our experience, after intrapleural inoculation of our strain of Fischer F344 rats, an incidence of mesotheliomas of more than 10% is regarded as a cause of concern." In a similar way, a distinct background incidence has been noted by Stanton and his colleagues. Thus, Stanton and Layard (1977) said that "pleural sarcomas have occurred in rats treated only by simple pleuracotomy. Our best estimate of these non-specific background pleural sarcomas in treated controls is in the range of 2-4%." Stanton et al. (1981) continued "a comparison of the incidence of the pleural sarcomas in the 72 individual experiments has shown that the incidence of pleural sarcomas in a particular experimental group was significantly greater than that in the combined control group only if it exceeded 30%." In summary, there appears to be a distinct background incidence of spontaneously occurring tumors capable of confounding the interpretation of "low dose" studies. From the literature, it would appear that in order to obtain a significant response, a greater than 10% incidence of pleural mesotheliomas, a 30% incidence of pleural sarcomas, and a greater than 8-10% incidence of peritoneal mesotheliomas may have to be obtained after asbestos injection.

A final analysis of the dose response studies has been presented in Table 1 in which the different agents, number of studies recorded, the modes of administration, species, and possible outcomes are listed. It can been seen from these analyses that a NOEL may exist for the major forms of asbestos, i.e. amosite, chrysotile, and crocidolite for at least one species and one mode of administration.

Pott et al. (1987) have suggested, on the basis of their actinolite and chrysotile dose response studies, that there was no threshold for the carcinogenicity of asbestos. Thus, "even 0.01 mg. actinolite or chrysotile (UICC/B) had a clear carcinogenic effect after i.p. injection (experiment 15). This is the maximum amount of airborne chrysotile which is permitted in 1m^3 at the workplace in the FRG. Certainly, the dose-response relationship found with the i.p. test cannot be extrapolated to the lung of humans; however, the results confirm in principle the hypothesis that there is no threshold for the carcinogenicity of asbestos down to an extremely low dose range."

Firstly, what Pott et al. consider to be an extremely low dose may still represent millions of fibres and secondly, it is inaccurate to compare intraperitoneal and inhalation studies so that if: "an extrapolation cannot be made to the lung of humans" then the Pott et al. data do not confirm anything in particular, especially considering the fact that the fibres are not being inhaled but applied directly to the serosa. It should also be noted that the Pott et al. (1987) data are based on "preliminary results 28 months after intraperitoneal injection" (Table 3) using "rats with tumors in the abdominal cavity... based partly on macroscopic diagnoses only" (Figure legend to Table 3). It is important to further define "tumors in the abdominal cavity" since Pott et al. have said that the "spontaneous incidence of mesotheliomas and sarcomas in the abdominal cavity is low when one excludes those animals which have a malignant tumor of the uterus simultaneously in the Wistar rats. (However) a malignant uterus tumor seems to metastasise relatively often into the abdominal cavity: in the group injected with NaCl solution (102 rats)... 6 of 15 animals with a malignant tumor of the uterus showed additional mesothelioma or sarcoma of the peritoneum, but in only two rats was a mesothelioma or sarcoma found without a malignant uterus tumor." From this one can conclude that by excluding animals with coincident uterine and peritoneal mesotheliomas/sarcomas, a relatively large number (ca. 6-8%) of mesotheliomas may also have been excluded. Pott et al. (1988) continued to say that "it may be possible--in groups treated with dust--that a tumor was classified as a metastasis... even though it was induced by the substance administered and occurred accidentally alongside a malignant neoplasm of the uterus." This again suggests that the actual incidence of spontaneous mesotheliomas found in the positive control cohorts may have been underestimated as does the following: "Table 1 lists in chronological order the results of 13 experiments using the intraperitoneal test. All rats with sarcoma, mesothelioma, or carcinoma of the abdominal cavity were counted as tumor-bearing rats... The most frequent diagnoses were sarcomas as in previous experiments (Pott et al., 1976; Scheuer et al., 1973); only a few carcinomas were found. These three tumor types cannot always be differentiated with certainty histologically; sometimes two tumor types occur together." (Pott et al., 1987). Further potential confounding may also occur in cohorts repeatedly injected with dusts since positive controls did not include animals subjected to repeated injections. The need to perform such controls is inferred by Smith et al. (1974). "Depending on the nature and dose of the test materials, granulomatous responses can obliterate the pleural cavity within a few weeks. Accordingly, we have not used and do not recommend repeated injections." Further problems may arise due to technical difficulties. Thus, "after correct intraperitoneal injection of NaCl-solution no irritant or inflammatory reaction occurs. But in non-anesthetised rats, a puncture of the intestines with the injection needle cannot always be avoided and may subsequently lead to adhesive inflammation" (Pott et al., 1987). Such adhesions may, in turn, lend to additional potential confounding.

In summary, these types of problems have led to many of the currently encountered difficulties in the interpretation of the low-tumor-yield responses seen following the administration of so-called "low" fibre doses. This in turn may have contributed to the development of scientifically untenable concepts such as the "single fibre" theory.

TABLE 1
Final Analysis: Dose-Response Studies

Agent	No. of Studies Recorded	Mode	Species	Possible Outcome
Actinolite	(2)	(2) IPt	(2) Rat	(2/2) Plat. [2I]
Amosite	(14)	(3) IPL	(2/3) Hamster	(1/2) Neg. [I]
				(1/2) NOEL ? [C]
			(1/3) Rat	(1/1) NOEL ? [C]
		(4) IPt	(1/4) Mouse	(1/1) Plat [I]
			(3/4) Rat	(3/3) NOEL [3I]
		(7) IH	(6/7) Rat	(4/6) Neg. [3I]
				(2/6) NOEL ?[1I]
			(1/7) Monkey	(1/1) Neg. [I]
Anthophyllite	(2)	(1) IPt	(1/1) Rat	(1/1) NOEL ? [I]
		(1) IH	(1/1) Rat	(1/1) NOEL ? [I]
"Asbestos"	(2)	(2) IH	(1/2) Guinea Pig	(1/1) Neg. [I]
			(1/2) Hamster	(1/1) Neg. [I]
Chrysotile	(33)	(6) IPL	(2/6) Hamster	(2/2) NOEL ?[1I] [1C]
			(1/6) Monkey	(1/1) Neg. [I]
			(3/6) Rat	(3/3) NOEL ?[2I]
		(10) IPt	(8/10) Rat	(1/8) Neg. [I]
				(3/8) Plat. [3I]
				(4/8) NOEL ?[3I]
			(2/10) Mouse	(2/2) NOEL ?[1I]
		(13) IH	(12/13) Rat	(10/12) Neg. [I]
				(2/12) NOEL ? [I]
			(1/13) Monkey	(1/1) Neg. [I]
		(4) IT	(2/4) Hamster	(2/2) Neg. [I]
			(1/4) Sheep	(1/1) Neg. [I]
			(1/4) Rat	(1/1) Neg. [I]
Crocidolite	(14)	(4) IPt	(4/4) Rat	(1/4) Plat. [I]
				(3/4) NOEL ?[2I]
		(6) IPL	(2/6) Hamster	(2/2) NOEL ?[1I] [1C]
			(4/6) Rat	(4/4) NOEL ?[4I]
		(4) IH	(4/4) Rat	(2/4) Neg. [2I]
				(2/4) NOEL ?[1I] [1C]
Erionite	(9)	(5) IPt	(2/5) Mouse	(2/2) Plat. [2I]
			(3/5) Rat	(2/3) Plat. [2I]
				(1/3) NOEL ?[1I]
		(1) IPL	(1/1) Rat	(1/1) NOEL ?[1I]
		(3) IH	(3/3) Rat	(1/3) Neg. [1I] [2I]
				(2/3) NOEL ?[1C]
Glass	(10)	(1) IH	(1/1) Hamster	(1/1) NOEL ?[1I]
		(2) IPL	(1/2) Rat	(1/1) Neg. [1I]
			(1/2) Hamster	(1/1) NOEL ?[1I]
		(7) IPt	(7/7) Rat	(2/7) NOEL ?[1I]
				(2/7) Plat. [1I]
				(3/7) Neg. [3I]
Quartz	(1)	(1) IPt	(1/1) Rat	(1/1) NOEL ? [C]

TABLE 1
Final Analysis: Dose-Response Studies
(continued)

Agent	No. of Studies Recorded	Mode	Species	Possible Outcome	
Ceramic	(1)	IPt	Rat	NOEL ?	[C]
Karain Dust	(1)	IPt	Mouse	Plat.	[I]
Mordenite	(1)	IPt	Mouse	Neg.	[I]
Nemalite	(1)	IPt	Rat	Plat.	[I]
DMN	(1)	IPt	Rat	NOEL ?	[I]
Virus	(1)	IPt	Chicken	Plat.	[I]
Chrysotile and Crocidolite	(1)	IH	Rat	Neg.	[I]

Abbreviations: IPt: Intraperitoneal
IPl: Intrapleural
IH: Inhalation
Plat: Plateau
Neg: Negative
NOEL: No Observed Effect Level
[I]: Based on an analysis of an Individual Study.
[C]: Based on an analysis of the studies Combined.

REFERENCES

Coene, F. (op.cit.); Snellings et al. (1981/1981). Evidence of Carcinogenicity in Experimental Animals. NIOSH, Current Intelligence Bulletin (22/5/81).

Davis, J. (1988). Personal Communication.

Goodman, D., Ward, J., Squire, R., Chu, K. & Linhart, M. (1979). Neoplastic & Non-Neoplastic Lesions in Aging F344 Rats. Toxicol. & Appl. Pharmacology 48, 237-248.

Gould, H. (1977). Mesotheliomas of the Tunica Vaginalis Propria & Peritoneum in Fischer Rats. Vet. Pathology 14, 372-379.

Pott, F., Friedrichs, K. & Huth, F. (1976). Results of Animal Experiments Concerning the Carcinogenic Effects of Fibrous Dusts and their Interpretation with Regard to Carcinogenesis in Humans. Zbl. Babkt. Hyg. I. Abt. Orig. B162, 467-505.

Pott, F., Roller, M., Ziem, U., Reiffer, F., Bellmann, B., Rosenbruch, M. & Huth, F. (1988). Carcinogenicity Studies on Natural & Man-Made Fibres with the Intraperitoneal Test in Rats. Symposium on Mineral Fibres in the Non-Occupational Environment, Lyon, September 8, 1987., 1-4.

Pott, F., Ziem, U., Reiffer, F., Huth, F., Ernst, H., & Mohr, U. (1987). Carcinogenicity Studies on Fibres, Metal Compounds & Some Other Dusts in Rats. Exp. Pathology 32, 129-152.

Scheuer, V., Huth, F. & Pott, F. (1973). Untersuchungen zum Morphologischen Erscheinungsbild Experimenteller Tumoren

Bei Ratten nach Intraperitonealer Injektion von Asbestauben. Arch. Geschwulstforsch. 41, 120-136.

Smith, W. & Hubert, D. (1974). The Intrapleural Route as a Means for Estimating Carcinogenicity. In: Experimental Lung Cancer Carcinogenesis in Bioassays. (eds.) Karb, E. & Parke, J., Springer-Verlag, New York, 91-101.

Stanton, M. & Layard, M. (1977). The Carcinogenicity of Fibrous Minerals. Nat. Bur. Standards Special Pub. 506, 143-151.

Stanton, M., Layard, M., Tegeris, A., Miller, E., May, M., Morgan, M. & Smith, A. (1981). Relation of Particle Dimension to Carcinogenicity in Amphibole Asbestoses & Other Fibrous Minerals. JNCI, 67, 965-975.

Wagner, J., Griffiths, D. & Munday, D. (1987). Experimental Studies with Palygorskite Dusts. Br. J. Ind. Med. 44, 749-763.

SUMMARY AND CLOSING REMARKS

J. Bignon
CHU Henri Mondor
94010 Creteil Cedex, FRANCE

Historically, the series of workshops on "In Vitro Effects of Mineral Dusts" was initiated in 1979 by R.C. Brown, I.P. Gormley, M. Chamberlain and R. Davies at the MRC Pneumoconiosis Unit in Cardiff, Wales, U.K., with the main objective of exploring the possibilities of adapting to mineral dusts some in vitro short-term tests already currently in use in the field of chemical toxicology. At that time, the main preoccupation was to test asbestos fibres for their cytotoxicity. The cell systems used (red blood cells, peritoneal macrophages, various cell lines), were rather primitive. The second and third international workshops, held in the U.S.A. and F.R.G., respectively, clearly indicated that biologists were competing very successfully in order to find out the best cell systems specifically adapted to mineral dusts for studying not only cytotoxicity but also release of mediators by alveolar macrophages and their implication in fibrogenesis, and exploring genotoxicity and cell transformation. Two preoccupations were central to these research efforts: first, to understand mechanisms of pulmonary toxicology relating to inhaled mineral dusts; second, to obtain, if possible, ideal short-term tests for predicting the disease potentials of mineral dusts in animals and in humans. The fourth workshop has been entitled: "Effects of Mineral Dusts on Cells", which indicates that, besides in vitro studies, we are involved with early and late in vivo responses, either in animals or in humans.

When I agreed to act as summarizer of this workshop, I was assured that there was no expectation that I would provide a session by session critique of its proceedings. It is quite impossible to summarize the numerous data presented! I will highlight several important points which have been raised during these three days and some others which I have more or less forgotten or only suggested. Lastly, I will try to place the status of cell responses to mineral dusts research in perspective in regard to its possible clinical and toxicological implications.

From the origin, our workshops concerning health-related effects of mineral dusts attempted and succeeded in bringing together mineralogists, biologists, pathologists and medical doctors. Indeed, it is quite impos-

NATO ASI Series, Vol. H30
Effects of Mineral Dusts on Cells
Edited by B.T. Mossman and R.O. Bégin

sible to understand the cellular and sub-cellular effects of dusts if we do not know their precise mineralogic nature and if we have no clinico-pathological and epidemiological indications of their effects in humans.

The mineral world is a rather complicated one - structural and physico-chemical characteristics of mineral dusts being responsible for the types and degrees of pulmonary toxicity. The carcinogenic potential of mineral fibres is still a major problem. We have heard from Dr. Pott a very simple definition of a carcinogenic fibre- "a long, thin and durable one". So, to the initial definition of Stanton, he has added a third parameter, durability. Probably the next step will be to add a fourth one: surface properties. A growing body of data, as illustrated by our meeting and by reading the literature, demonstrates the important role of this parameter. The decreased toxicity of chrysotile and crocidolite observed after blocking some surface reactive sites is an illustration of this phenomenon. Several other presentations have documented the new concept that the surface of various mineral species is occupied by reducing sites capable of producing superoxide and other oxidant radicals. However, oxygen-derived reactive species can also be produced by phagocytic cells such as alveolar macrophages and polymorphonuclear leukocytes. So far, it is not clear what is the most noxious to epithelial cells and tissues: oxidants produced by minerals or released by cells. These important reflections on the role of oxidants in lung injury taken as a whole raises the interesting possibility of therapeutic implications for the protection of cells and tissues by using antioxidants such as superoxide dismutase, catalase, gluthathione, deferoxamine....

Inhaled dusts are able to induce in humans, as well as in animals, different types of thoracic diseases, pathologically well defined, but possibly related to different mechanisms: on the one hand, inflammatory lung conditions such as pulmonary and pleural fibrosis or degenerative disease like emphysema; on the other hand, many dusts, not only fibrous, are presently suspected to be responsible for malignant diseases, particularly lung carcinoma. Most of these dust-related diseases are progressive and clinically obvious after a long latency period: at least ten years for lung fibrosis and pleural plaques - much more (20 or 30 years) for lung cancer and even more (40 years) for mesothelioma. This must be kept in mind, because actually the animal results we have listened to corresponded to early biological events. By contrast, human data concerned mainly later states, when the disease was clinically apparent. This is probably one of the reasons for some discrepancies between animal and human data! The discrepancies are still more marked when we try to relate _in vitro_ responses to the natural course of human diseases. So far, analysis of the initial biological endo-alveolar cellular and biochemical events has not provided a clue for explaining such differences. It is possible that molecular biology, jointly with classical biochemistry, will allow us to discover what mediators are specific to one disease and in what circumstances they are synthesized and released. But presently, it is too early to answer that question.

During this workshop, eight presentations concerned clinical investigations by bronchoalveolar lavage in patients with pneumoconiosis. It is a rather new area, particularly relevant to dust-related biological alveolar responses in humans. Such studies will increase in the future as bronchoalveolar lavage easily performed in humans gives access to alveolar cells and biochemistry providing data which will help to better understand mechanisms, clinical course and prognosis of dust-related diseases. New biological techniques, including molecular biology, will allow us to understand some unexplained clinico-biological events, such as massive pulmonary fibrosis or alveolar lipoproteinosis after exposure to cytotoxic minerals such as silica. In this particular pathological condition observed in experimental and as well as in human silicosis, we do not know if the lipid coating at the mineral surfaces protects cells or if they are the consequence of some pneumocyte 2 dysfunction induced by some mineral species. For the near future, such clinical research will be important. Indeed, few patients can bring more information than a herd of sheep. It is the way we understand clinical investigation with its necessary prolongation in animals and cell systems in order to authenticate the role of etiological factors and demonstrate biological mechanisms.

Inhalation of toxic environmental particles, such as asbestos or silica, causes immune lung disorders which can be analysed by bronchoalveolar lavage. For understanding the immunotoxic events responsible for the alveolar inflammatory reaction induced by mineral dusts, we have to focus on alveolar macrophages which appear to play the key role in this complicated alveolar immune network. I will not summarize in great detail what we have heard, from the most simple assay such as counting alveolar cell populations to the most sophisticated ones, identifying different mediators by using biochemistry and molecular biology. Among the many mediators released by stimulated alveolar macrophages, we listened to new results concerning growth factors which are produced by activated macrophages under the action of mineral particles. Different contributions concerned other factors such as arachidonic acid metabolites, TNF_1 and others, but so far without specific answer. The regulation of the fibroblast responses has been rather forgotten, although it is the main cell producing collagen in lung fibrosis. Different carbohydrate receptors at the surface of fibroblasts are probably under the control of glycosidases, particularly sialidases which are able to unmask receptors under sialic acid terminal residues. This warrants further investigation.

Contrasting with the great amount of data concerning lung fibrogenesis, there was no communication nor comment on dust-related emphysema, though the high incidence of emphysema in coal workers is presently under discussion. So, in the future we will have to understand how some mineral species, types of occupational exposure, or host factors may contribute to the production of lung fibrosis while others contribute to lung destruction. The role of different proteases and inhibitors will have to be taken under consideration.

There are about 40 different cell types in the lung - some of them like fibroblasts and mesothelial cells have already been cultured _in vitro_. Different presentations have reported studies trying to culture other pulmonary cells such as pneumocytes 2, Clara cells and tracheobronchial cells. I was rather impressed by the improvement obtained in culturing tracheal and bronchial explants from rodent or human lungs. This represents a realistic model, not only for toxicological purposes, but also for pathophysiological investigations in relation to mucociliary clearance. This model seems to be useful for exploring lung cancer mechanisms as illustrated by several presentations. As experimental pathologists have not yet succeeded in obtaining a valid animal model for lung carcinoma induced by cigarette smoking, tracheal explants would probably represent a suitable system for both studying synergism between mineral dusts and tobacco smoke and exploring the role of metabolic factors (nutrition, vitamins...).

The last session concerned _in vitro_ carcinogenesis induced by mineral dusts, mainly asbestos and other mineral fibres. There is evidence that mineral fibres have a promoting effect on epithelial cell transformation. Several laboratories are exploring this point by studying activation of membrane receptors and stimulation of protein kinase C and secondary messengers under the action of mineral dusts. Intracellular hydrolysis of phospholipids, formation of diacylglycerol, as well as calcium transfer warrant further investigation in relation to this problem. However, during the last few years, it appeared that mineral fibres were also acting at the chromosome and DNA level, inducing chromosomal abnormalities and abnormal DNA repairs. The research in progress concerning the action of mineral dusts, particularly of fibrous shape, on oncogene activation, gene loss and gene transfection is very promising for understanding solid-state carcinogenesis, which already seems to belong to a complete carcinogenic mechanism. It has been the contributions of _in vitro_ studies which brought such important information which could not be deduced from animal studies. This is justification for intensifying research in this area.

TO CONCLUDE,
All of us have been actors for improving techniques which allow us to better understand the mechanisms of mineral toxicity in cells, organs and organisms. This advance of science was possible because _in vitro_ effects were compared and linked to responses in animals and humans. The application of the _ex vivo_ method to animal and human biological materials obtained by bronchoalveolar lavage represents great progress, particularly in relation to fibrogenesis. We need to expand this research in order to better understand mechanisms of dust-related emphysema and cancer.

As more and more nonfibrous minerals are presently suspected to be carcinogenic, more basic research has to be supported in the field of mineral dust lung and cell toxicology. It is the only way to protect workers and consumers! I hope that some very important persons from our supporting Administrations will understand this message, which concerns not only _in vitro_ research but also animal, clinical and epidemiological research programs. Indeed, _in vitro_ effects alone are not sufficient for assessing health hazards related to mineral dusts exposure. In _vivo_ contributions are needed to better characterize all the factors affecting human susceptibility to inhaled mineral dusts: penetration, clearance and migration of mineral dusts into the body; in vivo potentiation, neutralization or decrease of toxic potential; role of concomitant exposures to chemicals, host and behavioral factors, etc..

Otherwise, the development and validation of in vitro screening tests and animal bioassays will be increasingly important in order to test new mineral materials, particularly natural or synthetic fibers which are coming into widespread use, both as substitutes for asbestos or for technological innovations requiring materials with high tensile strength. Thus, our fourth workshop must be considered as a step in this field of science. As the powerful techniques of molecular biology are being applied to questions that were impossible to answer with the traditional approaches, we may foresee that the first one will bring new informative and comprehensive data. However, it is crucial to keep in mind that the greatest advances may be expected from laboratories where molecular biologists, experimental pathologists, epidemiologists and clinicians work together to solve all the interesting and important problems in this area of toxicology.

NATO ASI Series H

Vol. 1: **Biology and Molecular Biology of Plant-Pathogen Interactions.** Edited by J. A. Bailey. 415 pages. 1986.

Vol. 2: **Glial-Neuronal Communication in Development and Regeneration.** Edited by H. H. Althaus and W. Seifert. 865 pages. 1987.

Vol. 3: **Nicotinic Acetylcholine Receptor: Structure and Function.** Edited by A. Maelicke. 489 pages. 1986.

Vol. 4: **Recognition in Microbe-Plant Symbiotic and Pathogenic Interactions.** Edited by B. Lugtenberg. 449 pages. 1986.

Vol. 5: **Mesenchymal-Epithelial Interactions in Neural Development.** Edited by J. R. Wolff, J. Sievers, and M. Berry. 428 pages. 1987.

Vol. 6: **Molecular Mechanisms of Desensitization to Signal Molecules.** Edited by T. M. Konijn, P. J. M. Van Haastert, H. Van der Starre, H. Van der Wel, and M. D. Houslay. 336 pages. 1987.

Vol. 7: **Gangliosides and Modulation of Neuronal Functions.** Edited by H. Rahmann. 647 pages. 1987.

Vol. 8: **Molecular and Cellular Aspects of Erythropoietin and Erythropoiesis.** Edited by I. N. Rich. 460 pages. 1987.

Vol. 9: **Modification of Cell to Cell Signals During Normal and Pathological Aging.** Edited by S. Govoni and F. Battaini. 297 pages. 1987.

Vol. 10: **Plant Hormone Receptors.** Edited by D. Klämbt. 319 pages. 1987.

Vol. 11: **Host-Parasite Cellular and Molecular Interactions in Protozoal Infections.** Edited by K.-P. Chang and D. Snary. 425 pages. 1987.

Vol. 12: **The Cell Surface in Signal Transduction.** Edited by E. Wagner, H. Greppin, and B. Millet. 243 pages. 1987.

Vol. 13: **Toxicology of Pesticides: Experimental, Clinical and Regulatory Perspectives.** Edited by L. G. Costa, C. L. Galli, and S. D. Murphy. 320 pages. 1987.

Vol. 14: **Genetics of Translation. New Approaches.** Edited by M. F. Tuite, M. Picard, and M. Bolotin-Fukuhara. 524 pages. 1988.

Vol. 15: **Photosensitisation. Molecular, Cellular and Medical Aspects.** Edited by G. Moreno, R. H. Pottier, and T. G. Truscott. 521 pages. 1988.

Vol. 16: **Membrane Biogenesis.** Edited by J. A. F. Op den Kamp. 477 pages. 1988.

Vol. 17: **Cell to Cell Signals in Plant, Animal and Microbial Symbiosis.** Edited by S. Scannerini, D. Smith, P. Bonfante-Fasolo, and V. Gianinazzi-Pearson. 414 pages. 1988.

Vol. 18: **Plant Cell Biotechnology.** Edited by M. S. S. Pais, F. Mavituna, and J. M. Novais. 500 pages. 1988.

Vol. 19: **Modulation of Synaptic Transmission and Plasticity in Nervous Systems.** Edited by G. Hertting and H.-C. Spatz. 457 pages. 1988.

Vol. 20: **Amino Acid Availability and Brain Function in Health and Disease.** Edited by G. Huether. 487 pages. 1988.

NATO ASI Series H

Vol. 21: **Cellular and Molecular Basis of Synaptic Transmission.**
Edited by H. Zimmermann. 547 pages. 1988.

Vol. 22: **Neural Development and Regeneration. Cellular and Molecular Aspects.**
Edited by A. Gorio, J. R. Perez-Polo, J. de Vellis, and B. Haber. 711 pages. 1988.

Vol. 23: **The Semiotics of Cellular Communication in the Immune System.**
Edited by E. E. Sercarz, F. Celada, N. A. Mitchison, and T. Tada. 326 pages. 1988.

Vol. 24: **Bacteria, Complement and the Phagocytic Cell.**
Edited by F. C. Cabello und C. Pruzzo. 372 pages. 1988.

Vol. 25: **Nicotinic Acetylcholine Receptors in the Nervous System.**
Edited by F. Clementi, C. Gotti, and E. Sher. 424 pages. 1988.

Vol. 26: **Cell to Cell Signals in Mammalian Development.**
Edited by S. W. de Laat, J. G. Bluemink, and C. L. Mummery. 322 pages. 1989.

Vol. 27: **Phytotoxins and Plant Pathogenesis.**
Edited by A. Graniti, R. D. Durbin, and A. Ballio. 508 pages. 1989.

Vol. 28: **Vascular Wilt Diseases of Plants. Basic Studies and Control.**
Edited by E. C. Tjamos and C. H. Beckman. 590 pages. 1989.

Vol. 29: **Receptors, Membrane Transport and Signal Transduction.**
Edited by A. E. Evangelopoulos, J. P. Changeux, L. Packer, T. G. Sotiroudis, and K. W. A. Wirtz. 387 pages. 1989.

Vol. 30: **Effects of Mineral Dusts on Cells.**
Edited by B. T. Mossman and R. O. Bégin. 470 pages. 1989.

MIX
Papier aus verantwortungsvollen Quellen
Paper from responsible sources
FSC® C105338

If you have any concerns about our products,
you can contact us on
ProductSafety@springernature.com

In case Publisher is established outside the EU,
the EU authorized representative is:
Springer Nature Customer Service Center GmbH
Europaplatz 3, 69115 Heidelberg, Germany

Printed by Libri Plureos GmbH
in Hamburg, Germany